Computer Algebra Recipes for Classical Mechanics

Richard H. Enns
George C. McGuire

Computer Algebra Recipes for Classical Mechanics

CD-ROM
INCLUDED

Birkhäuser • Springer
Boston   New York

Richard H. Enns
Simon Fraser University
Department of Physics
Burnaby, BC V5A 1S6
Canada

George C. McGuire
University College of the Fraser Valley
Department of Physics
Abbotsford, BC V2S 7M9
Canada

QA
808
.E55
2003

Library of Congress Cataloging-in-Publication Data

Enns, Richard H.
  Computer algebra recipes for classical mechanics / Richard Enns and George McGuire.
    p. cm.
  Includes bibliographical references and index.
  ISBN 0-8176-4291-9 (alk. paper) — ISBN 3-7643-4291-9 (alk. paper)
  1. Mechanics, Analytic–Data processing. 2. Mathematical models. I. McGuire, George, 1940- II. Title.

QA808 .E55 2002
531–dc21   2002028181
                                                                                   CIP

AMS Subject Classifications: 26B12, 53A17, 70Bxx, 70D05, 70D10, 70H03, 70Kxx, 82Cxx

Printed on acid-free paper.
©2003 Birkhäuser Boston

All rights reserved. This work may not be translated or copied in whole or in part without the written permission of the publisher (Birkhäuser Boston, c/o Springer-Verlag New York, Inc., 175 Fifth Avenue, New York, NY 10010, USA), except for brief excerpts in connection with reviews or scholarly analysis. Use in connection with any form of information storage and retrieval, electronic adaptation, computer software, or by similar or dissimilar methodology now known or hereafter developed is forbidden.
The use of general descriptive names, trade names, trademarks, etc., in this publication, even if the former are not especially identified, is not to be taken as a sign that such names, as understood by the Trade Marks and Merchandise Marks Act, may accordingly be used freely by anyone.

ISBN 0-8176-4291-9   SPIN 10867527
ISBN 3-7643-4291-9

Typeset by the authors.
Printed in the United States of America.

9 8 7 6 5 4 3 2 1

Birkhäuser Boston • Basel • Berlin
A member of BertelsmannSpringer Science+Business Media GmbH

We dedicate the Computer Algebra Recipes series to our wives, Karen and Lynda, who provided loving support, made helpful suggestions, displayed amazing tolerance, and humored our fluctuating moods as this series evolved from a skeletal concept to full-bodied reality.

Preface

The purpose of computing is insight, not numbers.
R.W. Hamming, *Numerical Methods for Scientists and Engineers* (1973)

Computer Algebra Recipes for Classical Mechanics is a self-contained guide to problem-solving and exploration in classical mechanics using the Maple 8 computer algebra system (CAS). Its organization and underlying philosophy is similar to that of our general mathematical modeling text, *Computer Algebra Recipes: A Gourmet's Guide to the Mathematical Models of Science* [EM01]. The heart of this classical mechanics volume consists of an eclectic set (the **menu**) of more than one hundred useful and stimulating computer algebra worksheets (the **recipes**) which are systematically organized to correlate with the topics covered in standard undergraduate mechanics texts. Our book is not meant to supplant these traditional mechanics books. Rather it is designed to complement them by showing how computer algebra can not only solve standard physical models quickly, accurately, and efficiently, but can be used to develop and explore more complex physical systems.

The menu is arranged into three categories, viz., the **Appetizers**, the **Entrees**, and the **Desserts**, corresponding to the natural progression of undergraduate science and engineering students through successively higher levels of classical mechanics. In presenting this menu, no prior knowledge of Maple is assumed, the relevant command structures being systematically introduced on a need-to-know basis. The recipes are thoroughly annotated and generally presented as interesting or amusing stories or anecdotes. Each recipe takes the reader from the analytic formulation of a representative type of mechanics problem to its solution (analytical or numerical) to a graphical visualization of the solution. The graphical representations vary from static 2-dimensional pictures, to contour and vector field plots, to 3-dimensional graphs which can be rotated, to animations in time. For the reader's convenience, all the recipes are included on the CD that accompanies this book.

Our book is not just an ordinary guide to problem-solving and exploration in classical mechanics. It is a gourmet's guide! The range of mechanics problems that can be solved with the recipes that we present in this text is only limited by your imagination. We have designed each recipe so that by altering the parameter values, or the initial conditions, or the very nature of the model itself, tens of thousands of other mechanics problems can be easily generated and solved. This should prove extremely useful to instructor and student alike.

Acknowledgments

We would like to express our deep appreciation to the reviewer who so carefully read the manuscript, examined the Maple files, and made detailed suggestions for improving this text. Any remaining ambiguities or errors are, of course, the responsibility of the authors. Suggestions for new or improved mechanics recipes from interested readers are most welcome and may be sent to either one of the following e-mail addresses:

- renns@dccnet.com
- mcguire@ucfv.bc.ca.

Contents

Preface vii

INTRODUCTION 1
  A. Computer Algebra Systems 1
  B. The Spiral Approach to Learning Mechanics 2
  C. Maple Help 3
  D. Introductory Recipe 4
    D.1 Farewell Felonious Fly 5
  E. How to Use this Text 8

I THE APPETIZERS   9

1 Vectors and Kinematics   11
  1.1 Vector Addition 11
    1.1.1 Bogey 5 11
    1.1.2 The Forces of Business Meet the Forces of Nature 15
    1.1.3 Alaskan Cruise 19
    1.1.4 The Gorge Near George 23
  1.2 Vector Multiplication 29
    1.2.1 The Smoke Jumpers of Erehwon 30
    1.2.2 War Games of the Mind 33
    1.2.3 Flying by the Seat of Your Pants 37
    1.2.4 Mike's Train Ride 42
  1.3 Supplementary Recipes 45
    01-S01 Bogey 5 Revisited 45
    01-S02 A Rolling Wheel Gathers No Moss, But 46
    01-S03 Flight of Brunhilda Bumblebee 46
    01-S04 Colleen Better Beware 46
    01-S05 Justine's Clever Throw 46
    01-S06 Torpedos Away! 47
    01-S07 Mike's Vector Brain Teaser 47
    01-S08 Jennifer's Vector Identity Assignment 47
    01-S09 The Learnu Molecule 47

	01-S10 Feeling the Electric Force	48
	01-S11 Mike's Potpourri of Unit Vectors	48
	01-S12 Closest Possible Encounter	48
	01-S13 Envelope of Safety	48
	01-S14 Avoiding a Mortar Shell	48

2 Newtonian Mechanics — 49

- 2.1 Motion Involving Constant Forces 49
  - 2.1.1 Jack and Jill Go Up the Hill 50
  - 2.1.2 Will Jack Fall Down and Break His Crown? 52
  - 2.1.3 Mr. X's New Ride 56
  - 2.1.4 This Governor Is Not a Politician 61
- 2.2 Energy and Momentum 64
  - 2.2.1 Amazing But True 64
  - 2.2.2 How Arthur Won the Nobel Prize 69
  - 2.2.3 A Heck of a Wreck 73
  - 2.2.4 G Forces 75
- 2.3 Rotational Dynamics 82
  - 2.3.1 The Case of the Falling Pencil 82
  - 2.3.2 The Atwood Supreme 87
  - 2.3.3 Fast Freddie's Trick Shot 91
- 2.4 Supplementary Recipes 97
  - 02-S01 Gabrielle's Toy Car 97
  - 02-S02 Mike's Race 97
  - 02-S03 The Dirty Bird Window Washer 97
  - 02-S04 Suspension Bridge 98
  - 02-S05 Erehwonese Serving Platter 98
  - 02-S06 A Sticky Encounter 99
  - 02-S07 Rockets Away 99
  - 02-S08 Erehwon Space Probe Explosion 100
  - 02-S09 Gabrielle's Slippery Blocks 100
  - 02-S10 Another Toybox Problem 100

II THE ENTREES — 101

3 Vector Calculus — 103

- 3.1 Curvilinear Coordinates 103
  - 3.1.1 The Case of the Artistic Slug 103
  - 3.1.2 Abigail Ant Roams the Beach Ball 109
  - 3.1.3 This Doughnut Isn't for Eating 114
- 3.2 Vector Operators 119
  - 3.2.1 Enon on the Hill 119
  - 3.2.2 Are You Conservative, Mr. Vector Field? 125
- 3.3 Supplementary Recipes 131

       03-S01 Bertie Bumblebee Leaves His B and B 131
       03-S02 Alice in Cylinder Land 132
       03-S03 Car Racing in the Great White North 132
       03-S04 The Making of Another Conservative Field 132
       03-S05 More Divs and Curls 132
       03-S06 These Operators Have Many Identities 133
       03-S07 Thoughts on Flux, From Heraclitas to Gauss 133
       03-S08 Felonious Fly (Jr.) Flees 133
       03-S09 Designing Gabrielle's Toy Box 134

4 **Newtonian Dynamics I**   **135**
   4.1 Velocity-dependent forces 135
       4.1.1 Justine Takes a Dive 135
       4.1.2 A Rolling Road 138
       4.1.3 Pushing the Envelope 143
       4.1.4 Benny Boffo's Hole in One 146
   4.2 Position-dependent forces 153
       4.2.1 Mike's Mobile Modes 153
       4.2.2 A "Hard" Spring's Journey 159
       4.2.3 Wouldn't Mr. Kepler Be Pleased 165
       4.2.4 The Not-so-Simple Pendulum 172
   4.3 Supplementary Recipes 176
       04-S01 Nonlinear Drag on Trout Lake 176
       04-S02 More Golf, Anyone? 176
       04-S03 Dr. No Lets Go 177
       04-S04 George's Linear Inchworm 177
       04-S05 Period of an Anharmonic Oscillator 177
       04-S06 In Search of the Central Force Law 178
       04-S07 Orbital Precession 178
       04-S08 A Perturbing Solution 178
       04-S09 The Force of Love 179
       04-S10 Lord of the Rings? 179
       04-S11 The Growing Raindrop 179
       04-S12 The Hanging Chain 180
       04-S13 Nonlinear Drag on Trout Lake Revisited 180

5 **Newtonian Dynamics II**   **181**
   5.1 Time-dependent forces 181
       5.1.1 Mr. Q Feels the Lorentz Force 181
       5.1.2 Jane Rescues Tarzan 185
       5.1.3 The Route to Chaos 188
       5.1.4 Blowing in the Wind, Monte Carlo Style 194
   5.2 Accelerated Reference Frames 198
       5.2.1 Merry-go-round Merriment 198
       5.2.2 Falkland Fiasco 201

- 5.3 Supplementary Recipes 205
  - 05-S01 Will Ellen be Yellin'? 205
  - 05-S02 Poincare Sections and Strange Attractors 205
  - 05-S03 Tug of War: Math vs. Physics 206
  - 05-S04 Statistical Approach to Blowing in the Wind 206
  - 05-S05 Kids Will Be Kids 207
  - 05-S06 Falkland Fiasco, Revisited 207
  - 05-S07 Power Spectrum: The Idea 207
  - 05-S08 Power Spectrum: Driving Miss Duffing 208
  - 05-S09 Rocket Sled 208

III THE DESSERTS 209

6 Lagrangian & Hamiltonian Dynamics 211
- 6.1 Some Lagrangian Examples 211
  - 6.1.1 Rock in the Rim 212
  - 6.1.2 Airtrack Mechanics 215
  - 6.1.3 Can You Top This Nutation? 219
  - 6.1.4 This Double Pendulum is "Dynamite" 224
- 6.2 Calculus of Variations 227
  - 6.2.1 Of Wine Goblets and Whisky Tumblers 228
  - 6.2.2 Suzy Spider's Speedy Strand 231
- 6.3 Some Hamiltonian Examples 236
  - 6.3.1 Turning a Bar of Soap into a Conical Pendulum 236
  - 6.3.2 Lots of Lissajous Figures 240
  - 6.3.3 The KAM Torus and Hamiltonian Chaos 244
- 6.4 Supplementary Recipes 250
  - 06-S01 Parametric Excitation 250
  - 06-S02 The Oscillating Spherical Pendulum Ride 250
  - 06-S03 Coupled Pendulums 251
  - 06-S04 Eulerian Angles 251
  - 06-S05 George's Nonlinear Inchworm 251
  - 06-S06 The Fermi–Pasta–Ulam Problem 252
  - 06-S07 The Rotating Pendulum 252
  - 06-S08 The Toda Potential 253
  - 06-S09 Abby Moves in Great Circles 253
  - 06-S10 Suzy Spider's Alternate Route 254
  - 06-S11 Fermat's Principle and the Bending of Light 254

Bibliography 255

Index 257

INTRODUCTION

Science means simply the aggregate of all the recipes that are always successful. All the rest is literature.
Paul Valéry, French poet and essayist (1871–1945)

A. Computer Algebra Systems

Computer algebra systems (CASs) have started to revolutionize how we learn and teach scientific subjects which make an extensive use of mathematics. CASs not only allow one to carry out the numerical computations of standard programming languages, but to also perform complicated symbolic manipulations as well. The purpose of this text is to show how a CAS can be used to breathe new life into a venerable subject, classical mechanics, by introducing the reader to a 21st century approach to problem-solving and exploration. A CAS can perform a wide variety of mathematical operations, including

- analytic differentiation and analytic/numerical integration,

- analytic/numerical solution of ordinary/partial differential equations,

- Taylor/series expansions of functions to arbitrary order,

- analytic/numerical solution and manipulation of algebraic equations,

- production of 2- and 3-dimensional vector field and contour plots,

- animation of analytic and numerical solutions.

The computer algebra work sheets, or "recipes", in this book are based on the **Maple 8** software system, one of the most powerful CASs currently available. Any reader desiring to use an alternate CAS or a different release of Maple should have little difficulty in modifying the recipes to his or her own taste.

B. The Spiral Approach to Learning Mechanics

Traditionally, a spiral approach is used in the teaching of classical mechanics, a more sophisticated treatment of the subject being presented at each turn of the spiral as the students acquire greater mathematics skills. Our computer algebra "menu" reflects this philosophy, the recipes being organized into three progressively higher levels. In the **Appetizers**, the recipes assume a familiarity with the fundamental concepts of mechanics (e.g., kinematics, Newton's laws with constant forces, conservation of energy and momentum, etc.) and a mathematical knowledge of vectors, ordinary derivatives, integrals, and linear algebra. The **Entrees** deal with examples from intermediate mechanics (e.g., Newtonian dynamics with variable forces and accelerated frames of reference) and make use of partial derivatives, vector calculus, curvilinear coordinates, and linear and nonlinear ordinary differential equations (ODEs). In the advanced mechanics recipes which form the **Desserts**, the focus is on examples from Lagrangian and Hamiltonian mechanics.

The heart of our computer algebra menu consists of the individual computer algebra recipes. These recipes have been designed to illustrate the concepts and methods of classical mechanics and to stimulate the reader's intellect and imagination. Associated with each recipe is an intrinsically important mechanics example and often an interesting or amusing story featuring many of the same characters who appeared in *Computer Algebra Recipes: A Gourmet's Guide to the Mathematical Models of Science*. These storybook characters, who are fictitious composites of the many delightful students that we have encountered in our teaching careers, will guide you, the reader, through the steps of the recipe.

Every story or topic in the text contains the Maple code or recipe to explore that particular topic. To make life easier for the reader, we have also placed the recipes on the CD-ROM enclosed within the back cover of this text. The recipes are ordered according to the chapter number, the section number, and the story number. For example, the recipe **01-2-3** is associated with chapter 1, section 2, subsection (story) 3. Although the recipes can be directly accessed on the CD by clicking on the appropriate worksheet number, we strongly recommend that you access them through the hyperlinked menu file 00mechmenu.mws also provided on the CD. This file is keyed to the chapter headings and story titles[1] in the text. Complete instructions may be found in the menu file. The computer code in the text is accompanied by detailed explanations of the underlying mechanics concepts and/or methods and what the recipe is trying to accomplish. The corresponding code on the CD is generally the unannotated version, provided to save the reader the arduous task of typing the code.

The recommended procedure for using this text is first to read a given topic or story for overall understanding and enjoyment. If you are having any difficulty in understanding a piece of the code from the text, then you may want to execute the corresponding Maple worksheet and try variations on the code.

---

[1] e.g., **Flying by the Seat of Your Pants** for the recipe **01-2-3**.

Keep in mind that the same objective may often be achieved by a different combination of Maple commands than those used by the authors. You can use Maple's Help, explained in the following section, to learn about the commands available. After reading the topic, you should execute the worksheet (if you have not already done so) to make sure the code works as expected. At this point feel free to explore the topic. Try rotating any three-dimensional graphs or running any animations in the file. See what happens when changes in the model or Maple code are made and then try to interpret any new results. Our book is intended to be open-ended and merely serve as a guide to what is possible in classical mechanics modeling using a CAS, the possibilities being limited only by the background and desires of the reader.

At the end of each chapter, **Supplementary Recipes** are presented in the form of problems, their fully annotated solutions (recipes) being included on the CD. These recipes are also hyperlinked to the menu file with a simple numbering system. For example, **01-S02** is the second supplementary recipe in Chapter 1. Supplementary recipes can be used in two different ways. They can be regarded as problems to be solved by using the mechanics concepts and computer algebra techniques presented in the main text recipes. Your solutions can then be compared with those that we have presented. Even if you are successful, you probably will be interested in the many little computer algebra features that we introduce in our solutions. On the other hand, the **Supplementary Recipes** can be regarded as still more interesting applications of computer algebra to classical mechanics. Enjoy exploring classical mechanics with them!

C. Maple Help

In this text, the Maple commands are introduced on a need-to-know basis. If the reader wishes to learn more about these commands, or about other possible commands which might prove useful in solving a particular mechanics problem, Maple's Help should be consulted. The Help system allows the reader to explore Maple commands and features, listed by name or subject. One can search by topic or carry out a full text search. We illustrate both procedures by using the Topic Search to find the correct form of the command for taking a square root, and then using the Full Text Search to find the command for analytically solving an ODE. In either case, begin by using the mouse and clicking on Maple's Help which opens a help window.

(a) **Topic Search**
- Click on Topic Search.
- You wish to find the Maple command for taking the square root. Depending on the programming language, the command could be `sqr`, `sqrt`, `root`, ...In this case type `sq` in the Topic box. Maple will display all the commands starting with `sq`. Double click on `sqrt` or, alternately, single click on `sqrt` and then on OK. A description of the square root command will appear on the computer screen.

(b) **Full Text Search**

- Click on Full Text Search.
- Type ode in the Word(s) box and click on Search.
- Double click on dsolve. A description of the `dsolve` command for solving ODEs will appear.

In either case, Maple Help usually gives several examples of how the command is used. Assuming that you are already in the Help menu, a very useful method of remembering the relevant command structure after the Help window is closed is to proceed as follows:

- Click on the Edit button and then on the Copy Examples button.
- Close all the help windows to take you back to your Maple worksheet.
- Click the mouse on the Edit button and then on the Paste button.
- Run the file, if necessary, to see how the command works.

If on executing a Maple command, the output yields a mathematical function that is unfamiliar to you, e.g., `polylog`, you may find out what this function is by clicking on the word to highlight it, then on Help, and finally on Help on "polylog". The same Help window may also be opened by typing in a question mark followed by the word and a semicolon, e.g., `?polylog;`

Maple's Help is not perfect and occasionally you might be frustrated, but generally it is quite helpful and should be consulted whenever you get stuck with Maple syntax or are seeking just the right command to accomplish a certain mathematical task. Maple learning and programming guides are also available [Cha02] [MBM02a] [MBM02b]. Let us emphasize that in this book we merely scratch the surface of what can be done with the Maple symbolic computing package.

D. Introductory Recipe

As already mentioned in the **Preface**, in the following chapters we shall systematically present recipes which correlate with the topics developed in standard undergraduate classical mechanics texts. To give the reader a preliminary idea of what a typical recipe looks like and to introduce some basic Maple syntax, we present a somewhat sad story problem involving the demise of Felonious Fly. The locale of this incident is on the terraformed planet of Erehwon. Stories involving Erehwon are often somewhat fanciful and one should be alert to the practice on that planet of occasionally spelling names backwards. This introductory recipe is not on the accompanying CD-ROM, so after reading the story the reader should open up Maple and type the recipe in and execute it.

D.1 Farewell Felonious Fly

An unsavory insect, Felonious Fly, has been ensnared by an exotic Erehwon flytrap and is struggling to escape. Suppose that his position (in cm) is given at time t seconds after being trapped by the 1-dimensional model equation,

$$x = (t-1)^4 \cos(3t/2)\, e^{-(t^2/10)} / \sqrt{1+t^2}.$$

If, after observing Felonious's motion, you feel that our model equation is unrealistic, feel free to change the analytic form of x. This is one of the great advantages of using a CAS. Unlike the situation with pen and paper, a CAS allows you to rapidly explore changes in the model. Now the problem is stated.

Calculate Felonious's velocity (v) and acceleration (a) at arbitrary time t and plot them along with x in the same figure, using a different color and line style for each curve. Take a time interval from $t = 0$ up to a time where his motion effectively ceases. Determine the furthest distance that Felonious gets from the origin after being trapped and the time at which this occurs. First obtain a qualitative answer and then a quantitative result. Finally, animate his motion over the same time interval as in the above plot.

To solve this problem, we proceed as follows. In the following input line, we clear Maple's internal memory of any previously assigned values by entering the **restart** command after the prompt (>) symbol. All Maple command lines must be ended with either a colon (:), which (generally) suppresses any output, or a semi-colon (;), which allows the output to be viewed.

```
>    restart:
```

Felonious's position x is then entered in the next command line. We make use of the assignment (:=) operator, placing x on the left-hand side of the operator and the time-dependent form on the right-hand side. Assigned quantities can be mathematically manipulated. The symbols *, /, +, -, and ^ are used for multiplication, division, addition, subtraction, and exponentiation, respectively. The square root (**sqrt**) command has already been introduced, and the cosine (**cos**) and exponential (**exp**) commands are intuitively obvious.

```
>    x:=(t-1)^4*cos(3*t/2)*exp(-t^2/10)/sqrt(1+t^2);
```

$$x := \frac{(t-1)^4 \cos(\frac{3t}{2})\, e^{(-\frac{t^2}{10})}}{\sqrt{1+t^2}}$$

Using the derivative (**diff**) command, x is differentiated once with respect to t to yield Felonious's velocity v and twice to yield his acceleration a.

```
>    v:=diff(x,t);
```

$$v := \frac{4(t-1)^3 \cos(\frac{3t}{2})\, e^{(-\frac{t^2}{10})}}{\sqrt{1+t^2}} - \frac{3}{2}\frac{(t-1)^4 \sin(\frac{3t}{2})\, e^{(-\frac{t^2}{10})}}{\sqrt{1+t^2}} - \frac{1}{5}\frac{(t-1)^4 \cos(\frac{3t}{2})\, t\, e^{(-\frac{t^2}{10})}}{\sqrt{1+t^2}} - \frac{(t-1)^4 \cos(\frac{3t}{2})\, e^{(-\frac{t^2}{10})}\, t}{(1+t^2)^{(3/2)}}$$

```
>    a:=diff(x,t,t);
```

$$a := \frac{12(t-1)^2 \cos(\frac{3t}{2}) e^{(-\frac{t^2}{10})}}{\sqrt{1+t^2}} - \frac{12(t-1)^3 \sin(\frac{3t}{2}) e^{(-\frac{t^2}{10})}}{\sqrt{1+t^2}}$$
$$-\frac{8(t-1)^3 \cos(\frac{3t}{2}) e^{(-\frac{t^2}{10})} t}{(1+t^2)^{(3/2)}} - \frac{49}{20} \frac{(t-1)^4 \cos(\frac{3t}{2}) e^{(-\frac{t^2}{10})}}{\sqrt{1+t^2}}$$
$$-\frac{8}{5} \frac{(t-1)^3 \cos(\frac{3t}{2}) t e^{(-\frac{t^2}{10})}}{\sqrt{1+t^2}} + \frac{3}{5} \frac{(t-1)^4 \sin(\frac{3t}{2}) t e^{(-\frac{t^2}{10})}}{\sqrt{1+t^2}}$$
$$+\frac{3(t-1)^4 \sin(\frac{3t}{2}) e^{(-\frac{t^2}{10})} t}{(1+t^2)^{(3/2)}} + \frac{1}{25} \frac{(t-1)^4 \cos(\frac{3t}{2}) t^2 e^{(-\frac{t^2}{10})}}{\sqrt{1+t^2}}$$
$$+\frac{2}{5} \frac{(t-1)^4 \cos(\frac{3t}{2}) t^2 e^{(-\frac{t^2}{10})}}{(1+t^2)^{(3/2)}} + \frac{3(t-1)^4 \cos(\frac{3t}{2}) e^{(-\frac{t^2}{10})} t^2}{(1+t^2)^{(5/2)}}$$
$$-\frac{(t-1)^4 \cos(\frac{3t}{2}) e^{(-\frac{t^2}{10})}}{(1+t^2)^{(3/2)}}$$

These results, particularly the acceleration, would be tedious to derive by hand. With Maple, they are done quickly and without any errors. If the form of x is changed, the new forms of v and a are obtained almost immediately by re-executing the above `diff` command lines.

We will now plot x, v, and a together, using the following `plot` command.

```
>    plot([x,v,a],t=0..10,color=[red,blue,green],linestyle=
>    [SOLID,DASH,DASHDOT],thickness=2,labels=["time","x, v, a"]);
```

The command line has been broken over two text lines to fit into the width of the page. The time range has been taken from $t = 0$ to 10, the latter time being sufficient for all motion to effectively cease. A red solid line is used for x, a blue dashed line for v, and a green dash–dot line for a. Note that x, v, and a as well as the `color` and `linestyle` options are enclosed with square brackets ([ ]). A series of Maple objects (separated by commas) enclosed in square brackets is called a "Maple list". Maple preserves the order and repetition of elements in a list. In the present case, lists are required so that each curve has the same color and linestyle each time the recipe is executed. The `thickness` option is used to create thick curves. The default thickness is 0. Finally, axis labels are added. The double quotes denote that each enclosed item is a "Maple string". A string is a sequence of characters that has no value other than itself. It cannot be assigned to, and will always evaluate to itself. Other plot options are available (enter plot,options in Maple's Topic Search to find out more) and we shall be using many of them in ensuing recipes. On executing the plot command, we obtain the results shown in Figure 1.

INTRODUCTORY RECIPE

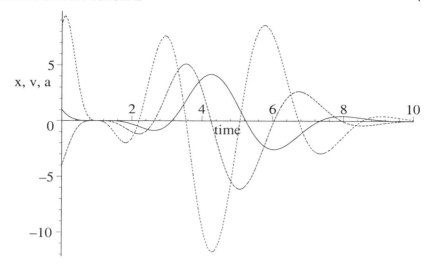

Figure 1: Solid curve: x, Dashed curve: v, Dash–dot (tallest) curve: a.

From the plot, Felonious's maximum excursion from the origin after being trapped is slightly more than 4 cm and occurs at a time of about 4 seconds. A slightly more accurate answer may be obtained by placing your cursor on the maximum of the x curve on the computer screen and clicking the left mouse button. A small window will open in the context bar in the upper left of the computer screen with the coordinates of the cursor displayed. The first value will be that of t, the second that of x.

To obtain a quantitatively accurate answer, we note that at the maximum excursion from the origin Felonious's velocity is zero. We use the floating point solve (`fsolve`) command to numerically solve the transcendental equation $v = 0$ for t in the time range $t = 3$ to 5 sec. We assign the name $tmax$ to the answer which has a default accuracy of 10 digits.

> `tmax:=fsolve(v=0,t,t=3..5);`

$$tmax := 4.257032680$$

This yields a time $tmax \simeq 4.26$ seconds. The corresponding x value is obtained by applying the evaluation (`eval`) command to x at $t = tmax$.

> `xmax:=eval(x,t=tmax);`

$$xmax := 4.180190700$$

The maximum distance that Felonious gets from the origin is about 4.18 cm.

To animate Felonious's motion, the `plots` library package must be loaded. The `with( )` command is used to load Maple library packages. Library packages are very important because they save you the effort of programming specialized plotting and mathematical operations. Approximately 90% of Maple's mathematical knowledge resides in the Maple library. Normally, we would place a colon on the following command line to suppress the output, but here we show a partial list of the large number of plot commands that are available in the `plots` package. The first command, *animate*, in the output list is the one that we shall be using. There is also a warning message that the name `changecoords` has been redefined. This warning appears even if a colon is used. If desired, warnings can be removed by using a colon and inserting the command `interface(warnlevel=0)` prior to loading the library package. We shall not use this command or reproduce warnings in the rest of the text recipes.

```
> with(plots);
```
Warning, the name changecoords has been redefined

[*animate, animate3d, animatecurve, arrow, changecoords, complexplot,* ...
... *sparsematrixplot, sphereplot, surfdata, textplot, textplot3d, tubeplot*]

In the `animate` command, Felonious's position at time t is entered as a two-component list with the first entry being the x coordinate, the second the (0) y-coordinate. The time interval is taken to be $t = 0$ to 10 as in the earlier plot. The animation will consist of 100 frames. Felonious is represented as size 18 blue circle and the axes are labeled.

```
> animate([x,0],t=0..10,frames=100,style=point,symbol=circle,
> symbolsize=18,color=blue,labels=["x","y"]);
```

On executing the `animate` command line, the opening frame of the animation will appear on the screen. The animation can be initiated by clicking on the plot and then on the start arrow in the Maple tool bar. You will observe Felonious's futile attempt to escape the clutches of the voracious Erehwon flytrap. The animation may be made to repeat by clicking on the looped arrow and stopped by clicking on the solid square. Other options are available in the tool bar.

E. How to Use this Text

Begin at the beginning ... and go on till you come to the end.
Lewis Carroll, *Alice's Adventures in Wonderland* (1865)

Although we have provided some of Maple's basic syntax in our introductory recipe, we recommend that the computer algebra novice start at the beginning of the **Appetizers**, even if your mechanics background is above that of the recipes presented there. It is in these early chapters that more of the basic features of the Maple system are introduced. Further, you might be surprised at how even initially simple mechanics problems can be made more interesting and often much more challenging because of the fact that a computer algebra system is being used. Whatever approach you adopt to using this book, we hope that you savor the wide variety of classical mechanics recipes that we have created. Bon Appetit!

Your computer algebra chefs, Richard and George.

Part I

THE APPETIZERS

*I do not know what I may appear to the world;
but to myself I seem to have been only like a boy playing on
the seashore, and diverting myself in now and then finding a
smoother pebble or a prettier shell than ordinary, whilst the great
ocean of truth lay all undiscovered before me.*
Isaac Newton, English physicist (1642–1727)

*Nature is an endless combination and repetition of a very few laws.
She hums the old well-known air through innumerable variations.*
Ralph Waldo Emerson, U.S. essayist, poet, philosopher (1803–1882)

*The time has come, the Walrus said,
To talk of many things:
Of shoes -- and ships -- and sealing wax --
Of cabbages -- and kings --
And why the sea is boiling hot --
And whether pigs have wings.*
Lewis Carroll, *The Walrus and the Carpenter* (1872)

Chapter 1

Vectors and Kinematics

In this introductory chapter, our recipes illustrate how Maple may be used to add and multiply vectors and how they may be manipulated in solving two- and three-dimensional kinematic problems. Although the `LinearAlgebra` library package could also be used to deal with vectors, in this chapter we shall exclusively employ the `VectorCalculus` package for this purpose. Using this latter package allows us to view the output vectors in terms of the unit vectors relevant to the chosen coordinate system as well as to perform such mathematical operations as the dot and cross products. Only Cartesian coordinates will be considered, other coordinate systems being explored in Chapter 3 along with the vector operators gradient, divergence, and curl.

1.1 Vector Addition

Our first vector worksheet or recipe involves a 2-dimensional example from the world of golf and illustrates the addition of displacement vectors by (a) analytically making use of unit vectors (b) graphically placing the tail of one vector at the tip of another. The second recipe in this section involves a 3-dimensional force problem, while the final two recipes on vector addition involve non-trivial kinematic examples involving relative velocity.

1.1.1 Bogey 5

Golf is a good walk spoiled.
Attributed to Mark Twain, American author and humorist (1835–1910)

One of the fairways on the golf course at the Metropolis Country Club runs due east and is perfectly flat from the tee to the hole. When playing this fairway recently Colleen, who is left-handed, "sliced" her tee shot 22° north of east, landing in thick rough 120 yards from the tee. Compounding her woes, a huge maple tree (120 yards east and 42 yards north of the tee) blocked a direct shot to the hole. So, Colleen was forced to blast her next "recovery" shot 15° east of south, the ball landing 75 yards from her. Her third "chip" shot to the green carried 64 yards, the ball landing 10 yards from the hole on a direct line from hole to tee. She then took two putts along this line to hole out, thus scoring a "bogey" 5 on the "par 4" hole.[1] What is the distance from tee to hole for this fairway? On attempting to record her score, Colleen realized that she had inadvertently left her scorecard and good pen lying on the ground near the maple tree where her tee shot landed. What was the (approximate) shortest distance that she had to walk back to retrieve her pen and scorecard?

Although this problem is sufficiently simple that it can be done with pen and paper and a calculator, we have asked Vectoria, a recent physics graduate from the Metropolis Institute of Technology (MIT) and a friend of Colleen's, to illustrate how the Maple CAS may be used to answer these questions.

In the following input line, Vectoria begins her recipe with the `restart` command as well as loading the `plots` and `VectorCalculus` library packages. Like the authors of this text, Vectoria likes to place all the library packages on the same line as the `restart` command. Whenever graphs are being drawn, the `plots` package is very useful, because it contains so many plotting commands, amongst them the `arrow` and `display` commands. For the present recipe, Vectoria wants to graphically represent each golf shot displacement vector with an arrow (thus the `arrow` command for plotting an arrow) and display (thus the `display` command) the arrow plots in a single figure. In Cartesian coordinates, the `VectorCalculus` package will produce an output[2] in terms of the unit vectors e_x, e_y, and e_z along the x, y, and z axes, respectively.

> `restart: with(plots): with(VectorCalculus):`

Vectoria enters the distances achieved on each of Colleen's three fairway shots and on her two putts and assigns the names $s1$, $s2$, $s3$, and $s4$, respectively. Since the output here is the same as the input, she suppresses the output for each entry by using a colon.

> `s1:=120: s2:=75: s3:=64: s4:=10:`

Arguments for all trigonometric functions must be given in radians (1 radian = $180/\pi$ degrees), so the angles for Colleen's first two shots are entered and converted to radians. Vectoria uses the floating point evaluation command, `evalf`, to express $\theta 1$ and $\theta 2$ in decimal form (default is 10 digits). Maple is case sensitive, the command Pi for π being capitalized.

> `theta1:=evalf(22*Pi/180); theta2:=evalf(15*Pi/180);`

$$\theta 1 := 0.3839724354$$
$$\theta 2 := 0.2617993878$$

---

[1] In reality, the distances and angles would be only approximately known, but may be regarded as exact for the purposes of this recipe. We are also using the term "bogey" in the American sense as one stroke more than par on a hole.

[2] Vector symbols are not displayed in the output. If the `LinearAlgebra` package is employed instead, the output is in terms of column or row vectors rather than unit vectors.

1.1. VECTOR ADDITION

Vectoria enters the displacement vector for Colleen's first shot, using the shorthand[3] syntax < , > for entering the vector. The first argument is the component in the x (easterly) direction, the second argument the component in the y (northerly) direction. She assigns the name a to this vector $\vec{a}$.

> `a:=<s1*cos(theta1),s1*sin(theta1)>;`
$$a := 111.2620626\, e_x + 44.95279121\, e_y$$

The vector $\vec{a}$ is expressed in terms of the unit vectors in the x and y directions. The relevant distance $s1$ and angle $\theta 1$ have been automatically substituted and the products evaluated in the output. On Colleen's tee shot, the easterly and northerly components were about 111 and 45 yards, respectively. The displacement for the second shot is similarly entered and assigned the name b.

> `b:=<s2*sin(theta2),-s2*cos(theta2)>;`
$$b := 19.41142838\, e_x - 72.44443697\, e_y$$

The vectors $\vec{a}$ and $\vec{b}$ are added, the resultant $\vec{r}$ being assigned the name r.

> `r:=a+b;`
$$r := 130.6734910\, e_x - 27.49164576\, e_y$$

Of course, in the output the x component of $\vec{r}$ is just the sum of the x components of $\vec{a}$ and $\vec{b}$, and similarly for the y components. On Colleen's third (chip) shot, the ball traveled 64 yards and ended up on the direct line from tee to hole. So the angle of this shot must be given by arcsin(27.49/64). To perform this calculation, Vectoria extracts the second component of r with `r[2]`. She removes the minus sign with the absolute value command, `abs`. The result is then divided by $s3$ and the arcsine performed,

> `theta3:=arcsin(abs(r[2])/s3);`
$$\theta 3 := 0.4440021156$$

yielding $\theta 3 \approx 0.44$ rads. The displacement vector for the third shot is then

> `c:=<s3*cos(theta3),s3*sin(theta3)>;`
$$c := 57.79454484\, e_x + 27.49164576\, e_y$$

Finally, the displacement vector $\vec{d}$ for the two putts is entered and the total displacement vector $\vec{r}2 = \vec{r} + \vec{c} + \vec{d}$ calculated.

> `d:=<s4,0>;`
$$d := 10\, e_x$$

> `r2:=r+c+d;`
$$r2 := 198.4680358\, e_x + 0.\, e_y$$

The distance from tee to hole is given by the easterly (x) component of $\vec{r}2$, so Vectoria concludes that this distance must have been about 198 yards. She notes that the northerly (y) component was, of course, zero.

To summarize Colleen's shots and illustrate the graphical addition of vectors, Vectoria now creates a vector addition diagram. The `arrow` command is used

---

[3] A longer form for entering the vector is `Vector([,])`.

to create a graph of each displacement vector. She enters the abbreviations w, hl, hw, and s for the width, head length, head width, and shape of the arrow.

> w:=width: hl:=head_length: hw:=head_width: s:=shape:

For the tee shot, the base of the vector is taken at the tee which is entered as the zero vector $<0,0>$ in aa. The second argument is the vector $\vec{a}$. Vectoria chooses to color this arrow red. By trial and error, Vectoria chooses specific numerical values for the arrow width, head length, and head width. The default width is 1/20, the head length 1/5, and the head width twice the length of the arrow. She takes the arrow shape to be double_arrow, which is actually the default shape in 2-dimensions. The reader can experiment with other shapes.

> aa:=arrow(<0,0>,a,color=red,w=2,hl=5,hw=5,s=double_arrow):

For the second shot, Vectoria follows the standard procedure for graphically adding vectors. She takes the first argument in bb to be $\vec{a}$, which corresponds to placing the second displacement vector with its tail at the tip of the first vector. The second entry is the vector $\vec{b}$ itself. This arrow is colored green and the same arrow parameter values are used as in aa. Similarly, blue and brown arrows are entered in cc and dd for the third shot and the putts.

> bb:=arrow(a,b,color=green,w=2,hl=5,hw=5,s=double_arrow):

> cc:=arrow(a+b,c,color=blue,w=2,hl=5,hw=5,s=double_arrow):

> dd:=arrow(a+b+c,d,color=brown,w=2,hl=5,hw=5,s=double_arrow):

The hole at $\vec{r}2$ is plotted as a red circle with the pointplot command, its shape and size being controlled with the symbol and symbolsize options. The maple tree is plotted as a green circle, its location being entered as a Maple list.

> hole:=pointplot(r2,style=point,symbol=circle,
> symbolsize=16,color=red):
> tree:=pointplot([120,42],style=point,symbol=circle,
> symbolsize=30,color=green):

Using the textplot command, she annotates the diagram, entering each piece of text as a Maple string. The coordinates of each entry are determined by trial and error. The seven list entries are enclosed in curly brackets, such brackets denoting a Maple "set". Maple does not preserve order or repetition in a set.

> comment:=textplot({[10,-5,"tee"],[35,25,"slice"],
> [135,45,"tree"],[140,20,"recovery"],[175,-20,"chip shot"],
> [180,10,"putts"],[205,10,"hole"]},color=blue):

Finally, Vectoria superimposes the set of plots in a single figure with the display command. To keep the scale in the x and y directions the same, she uses the option scaling=constrained. Otherwise, Maple will choose its own scaling. The minimum number of tickmarks along the x and y axes is also controlled as well as the view ($x = 0..210$ and $y = -30..50$ yards).

> display({aa,bb,cc,dd,hole,tree,comment},scaling=constrained,
> tickmarks=[4,3],view=[0..210,-30..50]);

1.1. VECTOR ADDITION

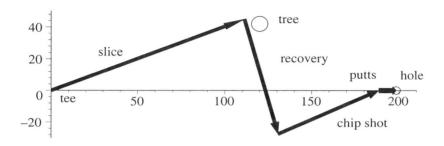

Figure 1.1: Displacement vector addition for Colleen's golf shots.

Figure 1.1 shows the graphical displacement vector addition.

To retrieve her pen and score card, Colleen had to return to the vicinity of where her tee shot landed. Vectoria calculates the vector difference $\vec{f} = \vec{r}2 - \vec{a}$. The shortest distance will be along this line except, perhaps, for a slight detour around the trunk of the maple tree. This distance $(dist)$ is obtained by using the theorem of Pythagorus, taking the square root of the sums of the squares of the first and second components of $\vec{f}$.

> f:=r2-a;

$$f := 87.2059732\,e_x - 44.95279121\,e_y$$

> dist:=sqrt(f[1]^2+f[2]^2);

$$dist := 98.11032158$$

So, Colleen had to walk back about 98 yards to retrieve her pen and scorecard. This evidently caused some minor annoyance to the group of business people who were impatiently waiting to tee off behind them. This group included Syd and Benny Boffo, prominent realtors and entrepreneurs, who were trying to land a building contract from the TV executives whom they were entertaining.

1.1.2 The Forces of Business Meet the Forces of Nature

An empty head is not really empty; it is stuffed with rubbish.
Hence the difficulty of forcing anything into an empty head.
Eric Hoffer, U.S. philosopher, *Reflections on the Human Condition* (1973)

Despite the annoying delay of their golf match, Syd and Benny Boffo have sufficiently impressed their clients that they have won the contract to build a 40 meter vertical TV transmission tower and ancillary facilities. The tower is to be stabilized against high winds by connecting its top to the level ground below with three taut cables. Taking the z-axis to be along the tower and the coordinate origin at the top, cable 1 exerts a force of magnitude $f1 = 3500$ N on the tower top and is anchored to the ground at $(x = -30, y = 0, z = -40)$ meters. Cable 2 exerts a force $f2 = 2500$ N on the tower top and is anchored to the ground at (20, -30, -40) meters. If the third cable is to exert a force $f3 = 3000$ N on the tower top and its anchor position is at $(x, y, -40)$ meters, what values should x and y have to produce a resultant force directed along the tower axis? What is the resultant force magnitude f due to the three cables?

Syd and Benny have asked Jennifer, a faculty member in the Institute of Applied Mathematics at MIT, to provide the answers to these questions. Like her former student Vectoria, Jennifer begins by loading the plots and VectorCalculus library packages.

```
> restart: with(plots): with(VectorCalculus):
```

The magnitudes of the forces exerted by the three cables are known, but the force directions remain to be determined. Jennifer needs to find the three unit vectors pointing from the origin along each cable. Then, each force vector will be equal to the product of its magnitude and the corresponding unit vector.

Because such operators are so useful, Jennifer decides to use functional, or "arrow", operators to carry out these two steps as well as some later calculations in the recipe. The next command line, assigned the name u, is for forming unit vectors. The operation on the right-hand side (rhs) of the arrow[4] will be applied to an input vector $\vec{v}$ when the command u(v) is subsequently entered. On the rhs of the arrow, the vector is divided by its magnitude to form the unit vector.

```
> u:=v->v/sqrt(v[1]^2+v[2]^2+v[3]^2):
```

Similarly, if f is a force magnitude and $\hat{u}$ is a unit vector, the following arrow operator will generate the force vector on subsequently calculating $F(f, u)$.

```
> F:=(f,u)->f*u:
```

To keep the intermediate numbers in the calculation a convenient size, Jennifer enters the magnitudes of the four forces in units of 100 Newtons. She adds a comment (prefixed by the sharp symbol #) to this effect at the end of the command line.

```
> f1:=35: f2:=25: f3:=30: f4:=f:    #units of 100 N
```

Again for convenience, she enters the anchor position vectors in units of 10 m. Thus, cable 1 is anchored at $[-3, 0, -4]$, so it position vector relative to the origin at the top of the tower is entered as v1=<-3,0,-4>. The corresponding unit vector u1 and force vector F1 are calculated, the output being expressed in terms of the unit vectors along the x and z axes.

```
> v1:=<-3,0,-4>: u1:=u(v1); F1:=F(f1,u1);
```

$$u1 := \left(\frac{-3}{5}\right) e_x - \frac{4}{5} e_z$$

$$F1 := (-21) e_x - 28 e_z$$

---

[4]Created on the keyboard with the hyphen (-) followed by the greater than symbol (>).

1.1. VECTOR ADDITION

Similarly, the unit and force vectors for cables 2 and 3, as well as for the resultant, are evaluated,

> v2:=<2,-3,-4>: u2:=u(v2): F2:=F(f2,u2);

$$F2 := \frac{50\sqrt{29}}{29} e_x - \frac{75\sqrt{29}}{29} e_y - \frac{100\sqrt{29}}{29} e_z$$

> v3:=<x,y,-4>: u3:=u(v3): F3:=F(f3,u3);

$$F3 := \frac{30x}{\sqrt{x^2+y^2+16}} e_x + \frac{30y}{\sqrt{x^2+y^2+16}} e_y - \frac{120}{\sqrt{x^2+y^2+16}} e_z$$

> v4:=<0,0,-4>: u4:=u(v4): F4:=F(f4,u4);

$$F4 := -f\, e_z$$

The sum of the force components associated with the three cables must add up to the force component of the resultant force in each of the x, y, and z directions. Jennifer creates a function eq to carry out this operation. The selection operation [i] picks out the ith component of each force. When eq(i) is subsequently entered, the ith component equation will be generated.

> eq:=i->F1[i]+F2[i]+F3[i]=F4[i]:

The three component equations are generated with the sequence (seq) command and enclosed in curly brackets to form a Maple set. The system of equations is assigned the name eqs.

> eqs:={seq(eq(i),i=1..3)};

$$eqs := \left\{ \frac{50\sqrt{29}}{29} - 21 + \frac{30x}{\sqrt{x^2+y^2+16}} = 0,\ -\frac{75\sqrt{29}}{29} + \frac{30y}{\sqrt{x^2+y^2+16}} = 0,\right.$$
$$\left. -\frac{100\sqrt{29}}{29} - 28 - \frac{120}{\sqrt{x^2+y^2+16}} = -f \right\}$$

Jennifer numerically solves the system of three equations for the (Maple) set of unknowns, x, y, and f,

> sol:=fsolve(eqs,{x,y,f});

$$sol := \{f = 70.41880627,\ x = 1.964878906,\ y = 2.335861675\}$$

and assigns the values in sol. This is necessary because otherwise, e.g., entering x; would produce the symbol x, not its numerical value.

> assign(sol):

Remembering that the length and force units were scaled, she calculates the x and y coordinates of the anchor position of cable 3 and the magnitude of the resultant force in real units.

> xcoord=10*x*meters; ycoord:=10*y*meters; fmag:=100*f*newtons;

$$xcoord = 19.64878906\ meters$$
$$ycoord := 23.35861675\ meters$$
$$fmag := 7041.880627\ newtons$$

Thus $x \approx 19.6$ meters, $y \approx 23.4$ meters, and the magnitude of the resultant force is about 7042 newtons. Jennifer wants to graphically show the force vectors along the three cables as well as the resultant force pointing down the tower axis. To this end, she creates a functional operator, assigned the name a, for generating arrow graphs. Since the plot will be 3-dimensional, she decides to use cylindrical arrows. When a(F,c) is entered, where F is the force vector and c the color, an arrow plot will be generated, although not displayed.

>   a:=(F,c)->arrow(F,width=2,shape=cylindrical_arrow,color=c):

Using the display command, the four force arrows are now produced and displayed together. The scaling is constrained and a particular angular orientation chosen for the 3-dimensional viewing box which has framed axes.

>   display({a(F1,red),a(F2,blue),a(F3,green),a(F4,black)},
>   scaling=constrained,orientation=[-36,60],axes=framed,
>   tickmarks=[3,3,3],labels=["Fx","Fy","Fz"]);

The 3-dimensional force vector diagram is shown in Figure 1.2, the resultant

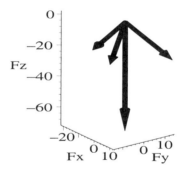

Figure 1.2: Forces along three cable directions and the resultant force.

force vector pointing vertically downwards. The forces in this diagram are in scaled units (i.e., units of 100 Newtons). The axis labels Fx, Fy, and Fz refer to the x, y, and z force components.

On completing her recipe, Jennifer has asked us to point out that if the reader clicks his or her mouse on the computer plot, the angular coordinates, θ and ϕ given in the orientation option, will appear in a small window at the top left of the computer screen. The plot may be rotated on the screen by dragging on the picture with the mouse or by clicking on the small up and down arrows in the angular coordinates window. The cylindrical nature of the force arrows will be more apparent on rotation of the plot. If this option were not included, some of the force arrows would tend to virtually disappear at certain orientations of the figure.

1.1.3 Alaskan Cruise

They say travel broadens the mind; but you must have the mind.
G. K. Chesterton, English essayist, novelist, and poet (1874–1936)

When George went on a holiday boat cruise to Alaska during the writing of this book, he was strongly tempted to take his laptop computer with him so he could continue creating interesting computer algebra recipes. Ultimately, he was persuaded by his wife not to take it, so one can only speculate on what might have happened if he had succumbed to the temptation. Perhaps, the following scenario might have occurred.

George and his wife are seated at Captain Jules Verne's table for the evening meal. After some desultory small talk, the captain turns his attention to George.

"Your cabin steward has informed me that you are a physicist and that you have been observed working on your laptop computer", the captain says to George. "So, you might be interested in using it to solve the following brain teaser. On a previous trip, our vessel was traveling steadily north at a speed of 18 knots.[5] At some instant in time, a coast guard vessel was observed six nautical miles due west of us, traveling on a steady course at a speed of 26 knots. Some time later the coast guard vessel was observed to pass somewhere behind our ship, the distance of closest approach being two nautical miles. Given only this information, can you answer the following questions? If you can, I will treat you to a bottle of superb Australian Shiraz at tomorrow night's supper."

(a) What was the course of the coast guard vessel?

(b) How much time elapsed between first sighting and closest approach?

(c) What was the distance between ships six minutes after closest approach?

So, the next morning when his wife and her friends have gone shopping in the port of Juneau, George turns on his laptop and tackles Captain Verne's brain teaser. He first loads the `plots` and `VectorCalculus` library packages.

> `restart: with(plots): with(VectorCalculus):`

At time $t = 0$ (time of first sighting), Verne's ship is taken to be at the origin (0,0) and the coast guard vessel at (-6,0). For $t > 0$, Verne's ship will move along the positive y-axis (pointing north), while the coast guard vessel's velocity vector will make a (as yet unknown) positive angle θ with the y direction. George enters the position vector $\vec{r}1$ of the cruise ship and $\vec{r}2$ of the coast guard vessel at time t.

> `r1:=<0,18*t>; r2:=<-6+26*sin(theta)*t,26*cos(theta)*t>;`

$$r1 := 18\,t\,\mathrm{e_y}$$

$$r2 := (26\sin(\theta)\,t - 6)\,\mathrm{e_x} + 26\cos(\theta)\,t\,\mathrm{e_y}$$

The separation of the two ships is given by the vector difference $\vec{r}1 - \vec{r}2$.

---

[5] For the landlubber, knot is the abbreviation for nautical mile (6076.1 ft) per hour.

```
> sep:=r1-r2;
```
$$sep := (-26\sin(\theta)\,t + 6)\,e_x + (18\,t - 26\cos(\theta)\,t)\,e_y$$

At the minimum distance (*mindist*) between ships, the magnitude of the separation vector is equal to two nautical miles. This expression is then simplified.
```
> mindist:=sqrt(sep[1]^2+sep[2]^2)=2;
```
$$mindist := \sqrt{(-26\sin(\theta)\,t + 6)^2 + (18\,t - 26\cos(\theta)\,t)^2} = 2$$
```
> mindist:=simplify(mindist);
```
$$mindist := 2\sqrt{250\,t^2 - 78\sin(\theta)\,t + 9 - 234\cos(\theta)\,t^2} = 2$$

The time T at which the minimum distance occured is obtained by differentiating *mindist* with respect to time,
```
> diff(mindist,t);
```
$$\frac{500\,t - 78\sin(\theta) - 468\cos(\theta)\,t}{\sqrt{250\,t^2 - 78\sin(\theta)\,t + 9 - 234\cos(\theta)\,t^2}} = 0$$

and solving the result of the last command line (indicated by %) for t.
```
> T:=solve(%,t);
```
$$T := -\frac{39}{2}\,\frac{\sin(\theta)}{-125 + 117\cos(\theta)}$$

The time T of closest approach is expressed in terms of the unknown angle θ. George substitutes this time into *mindist* and simplifies the result, yielding an equation for θ alone.
```
> simplify(subs(t=T,mindist));
```
$$3\sqrt{2}\,\sqrt{-\frac{(-9 + 13\cos(\theta))^2}{-125 + 117\cos(\theta)}} = 2$$

The previous output is analytically solved for θ, yielding two answers in Θ.
```
> Theta:=solve(%,theta);
```
$$\Theta := \arccos\left(\frac{8}{13} - \frac{\sqrt{97}}{39}\right),\ \arccos\left(\frac{8}{13} + \frac{\sqrt{97}}{39}\right)$$

Two possible angles Θ for the route of the coast guard vessel are produced. George numerically evaluates the angles (in radians) labeling them $\Theta 1$, $\Theta 2$.
```
> Theta1:=evalf(Theta[1]); Theta2:=evalf(Theta[2]); #radians
```
$$\Theta 1 := 1.199472022$$
$$\Theta 2 := 0.5197982033$$

Using the convert(units) command, a functional operator is created to convert the angles from radians into degrees. This operator is applied to $\Theta 1$, $\Theta 2$.
```
> Angle:=Theta->convert(Theta,units,radian,degree):
> Angle1:=Angle(Theta1); Angle2:=Angle(Theta2); #degrees
```
$$Angle1 := 68.72468450$$
$$Angle2 := 29.78224324$$

1.1. VECTOR ADDITION

So the coast guard vessel either traveled along a path making an angle of 68.7° or 29.8° with the northerly direction. The corresponding times (in hours) of closest approach are obtained by evaluating T at the two angles.

> `T1:=eval(T,theta=Theta1); T2:=eval(T,theta=Theta2);`

$$T1 := 0.2201306271$$
$$T2 := 0.4129779309$$

George suspects that the shorter of the two times (thus the larger of the two angles) corresponds to the coast guard vessel passing behind the cruise ship. The larger of the two times (the smaller of the two angles) probably corresponds to the coast guard vessel passing in front of the cruise ship. To confirm this conjecture, George works with both answers.

The two possible position vectors ($\vec{r}2a$ and $\vec{r}2b$) at time t for the coast guard vessel are obtained by evaluating $\vec{r}2$ at $\Theta 1$ and $\Theta 2$.

> `r2a:=eval(r2,theta=Theta1); r2b:=eval(r2,theta=Theta2);`

$$r2a := (24.22803862\,t - 6)\,e_x + 9.434094803\,t\,e_y$$
$$r2b := (12.91433012\,t - 6)\,e_x + 22.56590520\,t\,e_y$$

George now uses the `pointplot` command in `gr1` to plot the location of the cruise ship as size 12 blue circles at times $t = T1$ and $t = T2$. The coordinates of each location are evaluated at each time and entered as a list ([,]). The two lists are then combined into a "list of lists" (of the structure [[,],[,]]).

> `gr1:=pointplot([[eval(r1[1],t=T1),eval(r1[2],t=T1)],`
> `[eval(r1[1],t=T2),eval(r1[2],t=T2)]],style=point,`
> `symbol=circle,symbolsize=12,color=blue):`

To plot the two possible paths of the coast guard vessel, George uses a parametric plot structure in `gr2`. The components of the position vector $r2a$ are entered in the form `[r2a[1],r2a[2],t=0..T1]`. The presence of the time range in the third position of the list indicates a parametric plot, time being the parameter. Similarly, George forms `[r2b[1],r2b[2],t=0..T2]` and then creates a list of lists. The possible paths of the coast guard vessel will be plotted as straight lines up until their respective times of closest approach to the cruise ship. Because no color is specified, Maple colors the two lines differently.

> `gr2:=plot([[r2a[1],r2a[2],t=0..T1],[r2b[1],r2b[2],t=0..T2]]):`

The two graphs are superimposed in a single figure with the `display` command. The horizontal and vertical scaling are constrained, the axes are boxed, and direction labels added to the axes. To make a nicer plot than that obtained by default, George takes the horizontal view to be from $x = -6.5$ to $x = 1$ miles and the vertical view from $y = -1$ to $y = 10$ miles.

> `display({gr1,gr2},scaling=constrained,axes=boxed,`
> `labels=["East-->","North"],view=[-6.5..1,-1..10]);`

The resulting picture is shown in Figure 1.3. The two circles indicate the cruise ship's location at the two possible times of closest approach. The two straight lines are the possible paths of the coast guard vessel from the time of first

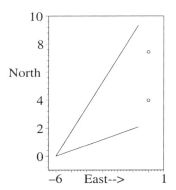

Figure 1.3: Circles: cruise ship's location at possible times of closest approach; Angled paths: possible routes of coast guard vessel.

sighting at $x = -6$, $y = 0$ to the respective times of closest approach. Clearly, the upper line with the smaller angle of 29.8° with the northerly direction corresponds to a situation where the coast guard vessel would pass in front of the cruise ship. The distance of closest approach in this case would be the distance from the upper end of this line to the upper circle. Since Captain Verne said that the coast guard vessel passed behind the cruise ship, George must reject this solution. The coast guard vessel must have traveled along the lower straight line path in the figure, making an angle of 68.7° with North. The time of closest approach in this case occured about 0.22 hours, or thirteen minutes, after first sighting. Qualitatively, the distance of closest approach is the distance from the upper end of the lower line to the lower circle.

To find the distance between ships six minutes after closest approach, George proceeds as follows. He obtains a general expression for the distance between the coast guard vessel and the cruise ship at time t by using the maximum (max) command to select the larger of the two angles and substituting this angle into the left-hand side (lhs) of *mindist*. George calculates the distance in this manner because the two answers in Θ can appear in a different order from one execution of the recipe to the next.

```
> dist:=subs(theta=max(Theta1,Theta2),lhs(mindist));
```

$$dist := 2\sqrt{250\,t^2 - 78\sin(1.199472022)\,t + 9 - 234\cos(1.199472022)\,t^2}$$

The distance expression is now plotted over the time interval $t = 0$ to twice the minimum of the two times $T1$ and $T2$. Again the min command is used because the values of $T1$ and $T2$ can be interchanged from one execution of the recipe to the next.

```
> plot(dist,t=0..2*min(T1,T2));
```

The resulting plot is shown in Figure 1.4. The time of closest approach (0.22

1.1. VECTOR ADDITION

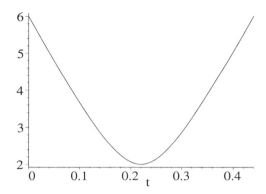

Figure 1.4: Separation distance as a function of time.

hours) for the coast guard vessel passing behind the cruise ship corresponds to the minimum in the curve. The separation distance six minutes, or 0.1 hours, later can be obtained either approximately or exactly. Placing his cursor on the curve at $t = 0.22 + 0.1 = 0.32$ and clicking the mouse, George reads the X, Y coordinates, 0.32, 3.26, of the cursor in a small box at the top left of his computer screen. So the separation distance at $t = 0.32$ hours was about 3.26 nautical miles. Not content with this approximate result, he obtains a more accurate answer by evaluating the distance at the minimum of $(T1, T2)$ plus 0.1 hours,

> `separation:=eval(dist,t=min(T1,T2)+0.1);`

$$separation := 3.256336266$$

yielding a separation distance of 3.256... nautical miles.

Having solved Captain Verne's brain teaser, George can hardly wait to sample that bottle of Australian Shiraz at tonight's supper.

1.1.4 The Gorge Near George

No man ever steps in the same river twice, for it's not the same river and he's not the same man.
Heraclitas, Greek philosopher (535–475 BC)

Having just speculated on what George might have done if he had taken a laptop computer on his Alaskan cruise, let's consider a more complex boat velocity problem tangentially involving another George. While attending a wedding reception in a lush vineyard overlooking the Columbia river gorge near George, Washington, Richard was struck by the panoramic view of the seemingly placid river far below. The Columbia is now laced with power generating dams, but one can speculate on what the current must have been like through the gorge prior to the era of the dams. A classic problem involving relative velocity is to determine the angular heading that a boat must maintain to move directly across a flowing river.

In the following recipe, a power boat is to cross the Columbia river as it flows west (negative x direction) towards the ocean. The river has a width $w = 1000$ meters and an assumed speed profile given by

$$Vr = Vr0[(w/2)^2 - y^2]/[w^2(a^2y^2+1)],$$

where $Vr0 = 50$ m/s, $a = 1/100$ m^{-1}, and $y = 0$ corresponds to the middle of the river. The boat is able to maintain a fixed angular heading and a speed which is n times the river's average speed. The boat leaves the south bank at the point $(x = 0, y = -w/2)$ and intends to land on the north bank at $(0, w/2)$. The questions to be answered are as follows:

(a) What does the velocity profile of the river look like?

(b) What are the maximum and average speeds of the river?

(c) How does the angular heading depend on n?

(d) How does the time it takes the boat to cross the river depend on n?

(e) How does the boat path vary as n is changed?

(f) Animate the motion of the boat for $n = 3$.

After loading the familiar library packages, let's enter the river's velocity profile.

```
>   restart: with(plots): with(VectorCalculus):
>   Vr:=Vr0*((w/2)^2-y^2)/(w^2*((a*y)^2+1));
```

To determine the maximum speed, we differentiate Vr with respect to y,

```
>   diff(Vr,y);
```

$$-\frac{2\,Vr0\,y}{w^2\,(a^2\,y^2+1)} - \frac{2\,Vr0\,(\frac{w^2}{4}-y^2)\,a^2\,y}{w^2\,(a^2\,y^2+1)^2}$$

and set the output of the previous command line equal to zero and solve for y.

```
>   ymax:=solve(%=0,y);
```

$$ymax := 0$$

As expected, the maximum river speed occurs at $ymax = 0$, i.e., at the river's center. The maximum speed $Vmax$ follows on evaluating Vr at $y = ymax$.

```
>   Vmax:=eval(Vr,y=ymax);
```

$$Vmax := \frac{Vr0}{4}$$

The average river speed is obtained by integrating (using the int command) the velocity profile over the width of the river and dividing by the width.

```
>   Vave:=int(Vr,y=-w/2..w/2)/w;
```

$$Vave := \frac{1}{2}\,\frac{Vr0\,(\arctan(\frac{a\,w}{2})\,w^2\,a^2 - 2\,a\,w + 4\arctan(\frac{a\,w}{2}))}{w^3\,a^3}$$

1.1. VECTOR ADDITION

The parameter values are entered and Vr is plotted in Figure 1.5. A title and axis labels having been added to the figure, the words entered as Maple strings.

> `w:=1000: a:=1/100: Vr0:=50:`
> `plot(Vr,y=-w/2..w/2,tickmarks=[3,3],title="velocity`
> `profile of the river",labels=["distance","vel"]);`

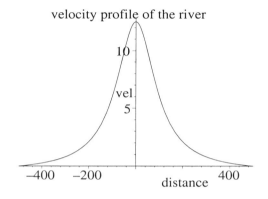

Figure 1.5: Plot of Vr with a title added to the figure.

Numerically evaluating $Vmax$ and $Vave$ to four digits accuracy,

> `Vmax:=evalf(Vmax,4); Vave:=Vave; Vave2:=evalf(Vave,4);`

$$Vmax := 12.50$$
$$Vave := \frac{13}{5}\arctan(5) - \frac{1}{2}$$
$$Vave2 := 3.070$$

the maximum river speed is 12.5 m/s and the average river speed is about 3.07 m/s. We have also expressed $Vave$ in terms of whole numbers, as using this form will make many of the subsequent output lines simpler in appearance. The velocity vector $\vec{vr}$ for the river is formed.

> `vr:=<-Vr,0>;`

$$vr := -\frac{250000 - y^2}{20000\left(\frac{y^2}{10000} + 1\right)} e_x$$

The power boat's speed is n times $Vave$.

> `Vb:=n*Vave;`

$$Vb := n\left(\frac{13}{5}\arctan(5) - \frac{1}{2}\right)$$

If θ is the angular heading, the boat's velocity vector $\vec{vb}$ is as follows.

> `vb:=<Vb*cos(theta),Vb*sin(theta)>;`

$$vb := n\left(\frac{13}{5}\arctan(5) - \frac{1}{2}\right)\cos(\theta)\, e_x + n\left(\frac{13}{5}\arctan(5) - \frac{1}{2}\right)\sin(\theta)\, e_y$$

From an observer's viewpoint on shore, the boat's velocity $\vec{v}$ is the sum of its velocity $\vec{v}b$ relative to the river and the river's velocity $\vec{v}r$ relative to the shore.

```
>    v:=vb+vr;
```

Now one cannot simply integrate the velocity vector $\vec{v}$ with respect to time to obtain the boat's position vector relative to the shore, because the x velocity component depends on y and y in turn is a function of time. The y coordinate of the boat at time t is calculated by integrating the second component of $\vec{v}$ from 0 to t and subtracting $w/2$.

```
>    y:=int(v[2],tt=0..t)-w/2;
```

$$y := n\left(\frac{13}{5}\arctan(5) - \frac{1}{2}\right)\sin(\theta)\,t - 500$$

Setting $y = 0$ and solving for t yields the time *thalf* it takes to cross half-way across the river.

```
>    thalf:=solve(y=0,t);
```

$$thalf := \frac{5000}{n\sin(\theta)\,(26\arctan(5) - 5)}$$

The symmetry of the river's velocity profile implies that $x = 0$ at $t = thalf$. Thus, we integrate the first component of the velocity $\vec{v}$ over the time interval $t = 0..thalf$, set the result equal to zero, and simplify the lengthy output (not shown here) with the radical simplification (`radsimp`) command.

```
>    eq:=0=int(v[1],t=0..thalf);
>    eq:=radsimp(eq);
```

$$eq := 0 = \frac{500\,(n\cos(\theta) - 1)}{n\sin(\theta)}$$

The equation *eq* is solved for θ as a function of n, yielding two solutions in Θ.

```
>    Theta:=solve(eq,theta);
```

$$\Theta := \arctan\left(\frac{\sqrt{-1+n^2}}{n}, \frac{1}{n}\right),\ \arctan\left(-\frac{\sqrt{-1+n^2}}{n}, \frac{1}{n}\right)$$

The first and second arguments in the arctangent correspond to the numerator and denominator of this function. For $n > 1$, the first solution in Θ corresponds to a positive angle, the second to a negative angle. The positive branch of the solution can be obtained by applying the `radsimp` command to Θ.

```
>    Theta:=radsimp(Theta);
```

$$\Theta := \arctan\left(\frac{\sqrt{-1+n^2}}{n}, \frac{1}{n}\right)$$

Clearly, the ratio n of boat speed to average river speed plays an important role in the angular "heading" Θ. For $n > 1$, the heading is a real positive angle, while for $n = 1$ the heading is zero. For $n < 1$, the argument becomes imaginary and there is no heading which will get the boat across the river. To confirm this interpretation, let's convert Θ into degrees and plot it as a function of n over

1.1. VECTOR ADDITION

the range $n = 0$ to $n = 10$, selecting a sufficient number of points to produce a smooth curve. The default number is 50. Choosing `thickness=2` results in a thick curve.

```
> plot(Theta*180/Pi,n=0..10,thickness=2,numpoints=1000,
> labels=["n","degrees"]):
```

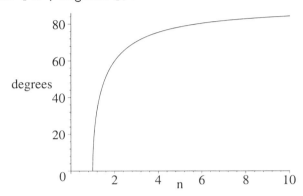

Figure 1.6: Heading in degrees as a function of n.

The resulting plot shown in Figure 1.6 supports the above remarks. Note that as $n \to \infty$, the heading asymptotically approaches 90 degrees. This can also be deduced from the expression $180\,\Theta/\pi$. For $n \to \infty$, one obtains $180\arctan(1/0)/\pi = 180\arctan(\infty)/\pi = 180(\pi/2)/\pi = 90$.

Substituting $\theta = \Theta$ into 2 *thalf* and simplifying yields the time *duration* it takes to cross the river.

```
> duration:=simplify(subs(theta=Theta,2*thalf));
```

$$duration := \frac{10000}{\sqrt{-1+n^2}\,(26\arctan(5) - 5)}$$

For $n > 1$ the time duration is real and finite. For $n = 1$, the duration is infinite, while for $n < 1$ the duration becomes imaginary. The duration is now plotted as a function of n over the range $n = 1$ to 10.

```
> gr1:=plot(duration,n=1..10,thickness=2,
> labels=["n","seconds"]):
```

A second graph is formed, connecting the points $[1, 0]$ and $[1, 1000]$ with a thick green line. The resulting vertical line at $n = 1$ is dashed with the command option `linestyle=3`.

```
> gr2:=plot([[1,0],[1,1000]],style=line,color=green,
> thickness=2,linestyle=3):
```

The two graphs are superimposed with the `display` command, producing Figure 1.7. The plot supports the analysis of the time duration formula.

```
> display(gr1,gr2,view=[0..10,0..1000]);
```

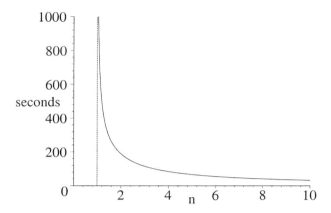

Figure 1.7: Time to cross the river as a function of n.

The x-coordinate of the boat's trajectory is obtained by substituting Θ into $v[1]$, introducing a temporary time variable tt, and integrating vx over the time interval $tt = 0...t$ and simplifying (lengthy result not shown here).

```
>   vx:=subs({theta=Theta,t=tt},v[1]);
>   x:=simplify(int(vx,tt=0..t));
```

The y coordinate of the boat's path is also calculated. The boat's trajectory depends on the ratio n as well as time t.

```
>   y:=simplify(subs(theta=Theta,y));
```

$$y := \frac{13}{5} t \arctan(5) \sqrt{-1+n^2} - \frac{\sqrt{-1+n^2}\, t}{2} - 500$$

A graph function `pl` is introduced to plot the boat's path across the river for different n values. We use a color option that produces different colors for each trajectory, which will vary randomly from one execution of the recipe to the next. Here RGB stands for red, green and blue, and rand() produces a random 12 digit number. Dividing by 10^{12} generates a fraction between 0 and 1. The three fraction entries determine the fractions of red, green, and blue to be used in producing the resulting line color.

```
>   pl:=n->plot([x,y,t=0..duration],color=
>   COLOR(RGB,rand()/10^12,rand()/10^12,rand()/10^12)):
```

Using the sequence (seq) command, plots are generated for $n = 2$ to 10 and superimposed in the same figure with the display command. The resulting picture is shown in Figure 1.8.

```
>   display(seq(pl(n),n=2..10),labels=["x","y"]);
```

The trajectory with the smallest deviation from the vertical corresponds to $n = 10$, while the one with the largest deviation corresponds to $n = 2$. Notice

1.2. VECTOR MULTIPLICATION

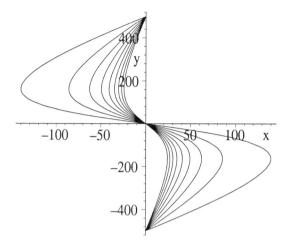

Figure 1.8: Paths traveled by boat for different n values.

that in order to better view these deviations we have not constrained the plot. To see the actual colors of the paths, the recipe must be executed.

Choosing a specific value for n, for example $n = 3$, the boat's motion across the Columbia river can be animated. In the following animate command, 100 frames are used and the boat is represented by a size 16 blue circle.

```
>   n:=3:
>   animate([x,y],t=0..duration,frames=100,style=point,
>   symbol=circle,symbolsize=16,color=blue);
```

On executing the command line, the initial frame is displayed on the computer screen. To activate the animation, click on the plot and then on the start arrow in Maple's tool bar at the top of the screen.

1.2 Vector Multiplication

In this section, the four recipes make use of the dot (or scalar) and cross (or vector) products. If $\vec{A}$ and $\vec{B}$ are two vectors with magnitudes A and B and an angle θ between them, then these products are defined as follows:

- Dot product: $\vec{A} \cdot \vec{B} = A B \cos(\theta)$,
- Cross product: $\vec{A} \times \vec{B} = A B \sin(\theta) \hat{n}$.

Here $\hat{n}$ is a unit vector obtained by putting the fingers of your right hand along $\vec{A}$ and curling them toward $\vec{B}$ in the direction of the smaller angle from $\vec{A}$ to $\vec{B}$. Your thumb will point in the direction of $\hat{n}$.

1.2.1 The Smoke Jumpers of Erehwon

What is it that breathes fire into the equations and makes a universe for them to describe...?
Stephen Hawking, English theoretical physicist (1942–)

On the terraformed planet Erehwon, the summer has been particularly hot and dry and numerous forest fires have been ignited by lightning strikes. Some of these fires are in remote mountainous areas where the only means of attacking the fire on the ground is to call in the "smoke jumpers". Each two-member fire fighting crew is flown by specially equipped ornithopter into the air space above the fire. As the ornithopter hovers in a stationary position, each smoke jumper is launched horizontally, one after another, in roughly opposite directions. The smoke jumpers then fall freely until at some instant their attached hang gliders unfurl and they can glide to the ground in the vicinity of the fire.

We have asked Greg Arious Nerd, one of Erehwon's most prominent mathematicians, to simulate the free fall portion for a particular smoke jumping crew, Eoj Ig and his girlfriend Enaj. We have further instructed Greg to incorporate the dot product into the calculation by telling him that the free fall ceases when the angle between Eoj's and Enaj's velocity vectors reaches a preset value, e.g., 90°. In this case, the dot product is zero. Air resistance is to be neglected.

Being a mathematician, Greg prefers to carry out the calculation symbolically and substitute parameter values at the end of the calculation for plotting purposes, etc. In his recipe, he labels Eoj as "particle" number 1 and Enaj as number 2, and sets up the following initial situation. Working in 3-dimensional Cartesian coordinates and using the VectorCalculus package, he gives Eoj the initial position $\vec{r}10=0$ (entering r10=<0,0,0>), initial velocity $\vec{v}10 = -Vo\,\hat{e}_x$ (v10=<-Vo,0,0>), and acceleration $\vec{a}1 = -g\,\hat{e}_z$ (a1=<0,0,-g>), where Vo is assumed to be positive and g is the acceleration due to gravity. Enaj's initial position is $\vec{r}20 = d\,\hat{e}_x$ (r20=<d,0,0>), her initial velocity is $\vec{v}20 = b\,\hat{e}_x + c\,\hat{e}_y$ (v20=<b*Vo,c*Vo,0>), and she experiences the acceleration $\vec{a}2 = -g\,\hat{e}_z$ (a2=<0,0,-g>). The initial separation distance d between Enaj and Eoj is taken to be positive, as are the velocity ratio parameters b and c. Eoj is ejected from the ornithopter at time $t=0$ and Enaj is ejected T seconds later.

> restart: with(plots): with(VectorCalculus):
> r10:=<0,0,0>: v10:=<-Vo,0,0>: a1:=<0,0,-g>:   #Eoj
> r20:=<d,0,0>: v20:=<b*Vo,c*Vo,0>: a2:=<0,0,-g>: #Enaj

At time $t > 0$, Eoj's position vector $\vec{r}1$ is obtained by the applying the well-known constant acceleration relation, $\vec{r} = \vec{r}_0 + \vec{v}_0 t + (1/2)\vec{a}\,t^2$.

> r1:=r10+v10*t+1/2*a1*t^2;

$$r1 := -t\,Vo\,\mathrm{e_x} - \frac{t^2 g}{2}\,\mathrm{e_z}$$

Ejected T seconds after Eoj, Enaj's position $\vec{r}2$ at time t is as follows.

> r2:=r20+v20*(t-T)+1/2*a2*(t-T)^2;

1.2. VECTOR MULTIPLICATION

$$r2 := (d + (t-T)\, b\, Vo)\, e_x + (t-T)\, c\, Vo\, e_y - \frac{(t-T)^2\, y}{2}\, e_z$$

To determine the separation distance between Eoj and Enaj at time t, Greg first calculates the separation vector $\vec{r}2 - \vec{r}1$.

```
> sep:=r2-r1;
```

$$sep := (d + (t-T)\, b\, Vo + t\, Vo)\, e_x + (t-T)\, c\, Vo\, e_y + (-\frac{(t-T)^2\, g}{2} + \frac{t^2\, g}{2})\, e_z$$

The distance between the two smoke jumpers at time t is calculated by taking the dot product of the separation (sep) vector with itself and then taking the square root. In the VectorCalculus package[6] the long form of the command for taking the dot product between two vectors $\vec{V}1$ and $\vec{V}2$ is DotProduct(V1,V2), while the short form uses the dot notation, viz. V1 . V2. Inserting spaces before and after the dot help to distinguish the dot product from a decimal point and make for easier readability. Greg will work with the short form as illustrated in the following command line.

```
> distance:=sqrt(sep . sep);   #dot product
```

$$distance := (4\, d^2 + 8\, d b\, Vo\, t - 8\, d b\, Vo\, T + 8\, d t\, Vo + 4\, b^2\, Vo^2\, t^2 - 8\, b^2\, Vo^2\, t T$$
$$+ 8\, b\, Vo^2\, t^2 + 4\, b^2\, Vo^2\, T^2 - 8\, b\, Vo^2\, T t + 4\, t^2\, Vo^2 + 4\, c^2\, Vo^2\, t^2 - 8\, c^2\, Vo^2\, t T$$
$$+ 4\, c^2\, Vo^2\, T^2 + 4\, g^2\, t^2\, T^2 - 4\, g^2\, t T^3 + g^2\, T^4)^{(1/2)}/2$$

Eoj's and Enaj's velocities, $\vec{v}1$ and $\vec{v}2$, are calculated at time t

```
> v1:=diff(r1,t); v2:=diff(r2,t);
```

$$v1 := -Vo\, e_x - t\, g\, e_z$$

$$v2 := b\, Vo\, e_x + c\, Vo\, e_y - (t-T)\, g\, e_z$$

The smoke jumpers fall under the influence of Erehwon's gravity until the angle between the velocity vectors $\vec{v}2$ and $\vec{v}1$ is $90\,°$. So Greg calculates the dot product between these vectors and sets the product equal to zero.

```
> eq:=v2 . v1=0;   #dot product
```

$$eq := -b\, Vo^2 + (t-T)\, g^2\, t = 0$$

The time ts at which the free fall stops and the hang gliders unfurl is obtained by solving eq for t.

```
> ts:=solve(eq,t);
```

$$ts := \frac{\frac{T g}{2} + \frac{\sqrt{T^2 g^2 + 4 b\, Vo^2}}{2}}{g}, \frac{\frac{T g}{2} - \frac{\sqrt{T^2 g^2 + 4 b\, Vo^2}}{2}}{g}$$

Two solutions are generated, the positive square root result being selected and simplified by then applying the radsimp command to ts.

```
> ts:=radsimp(ts);
```

$$ts := \frac{T g + \sqrt{T^2 g^2 + 4 b\, Vo^2}}{2 g}$$

---

[6]Dot products may also be performed by loading the LinearAlgebra package.

The distance between the smoke jumpers at time ts is calculated,

```
> distance:=simplify(subs(t=ts,distance));
```

but the lengthy expression is not shown here. Now Greg decides to choose a representative set of parameter values which will produce a nice 3-dimensional plot. He knows that on Erehwon, $g=10.0$ m/s². He assumes that Eoj and Enaj are initially separated by $d=1$ m. He gives Eoj a speed of $Vo=10$ m/s and Enaj a greater speed (taking $b=3$, $c=3$) since she is ejected $T=2$ seconds later.

```
> parameters:={g=10.0,d=1,Vo=10,b=3,c=3,T=2}:
```

Using the parameter values and evaluating the *distance* and time ts,

```
> distance:=eval(distance,parameters);
```

$$distance := 78.87331615$$

```
> ts:=eval(ts,parameters);
```

$$ts := 3.000000000$$

Eoj and Enaj are separated by nearly 79 meters when free fall ends, which occurs $ts = 3$ seconds after Eoj's ejection. Enaj is only in free fall for one second when the velocity vectors make an angle of 90° with each other. To plot their trajectories using the spacecurve command, Greg converts the position vectors $\vec{r}1$ and $\vec{r}2$ to Maple lists and evaluates them with the parameter values.

```
> r1:=convert(r1,list); r2:=convert(r2,list);
```

$$r1 := [-t\,Vo,\ 0,\ -\frac{t^2\,g}{2}]$$

$$r2 := [d + (t-T)\,b\,Vo,\ (t-T)\,c\,Vo,\ -\frac{(t-T)^2\,g}{2}]$$

```
> r1:=eval(r1,parameters);
```

$$r1 := [-10\,t,\ 0,\ -5.000000000\,t^2]$$

```
> r2:=eval(r2,parameters);
```

$$r2 := [-59 + 30\,t,\ 30\,t - 60,\ -5.000000000\,(t-2)^2]$$

As a partial check, Greg mentally notes that at time $t = 2$, Enaj is at $\vec{r}2 = \hat{e}_x$ as she should be. Greg creates a functional operator for the spacecurve plotting command. A plot of a 3-dimensional trajectory will be generated (but not shown) when the position vector $\vec{r}$ (entered as r), the color c, and the lower limit t1 of the time range are supplied as arguments to gr.

```
> gr:=(r,c,t1)->spacecurve(r,t=t1..ts,color=c,thickness=2):
```

Eoj's trajectory is colored red and begins at t1=0, while Enaj's trajectory is colored blue and begins at t1=2, hence the graph entries gr(r1,red,0) and gr(r2,blue,2) in the following display command line. Various plot options are chosen, the resulting figure being shown in Figure 1.9.

```
> display({gr(r1,red,0),gr(r2,blue,2)},axes=framed,
> orientation=[-40,70],scaling=constrained,tickmarks=[3,3,3],
> labels=["x","y","z"]);
```

1.2. VECTOR MULTIPLICATION

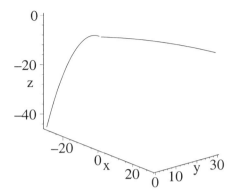

Figure 1.9: Trajectories of the two smoke jumpers during free fall.

The trajectories of the two smoke jumpers prior to the unfurling of their hang gliders are as shown. Eoj, of course, has fallen the furthest in the vertical z-direction, since he was ejected before Enaj. The plot may be rotated on the computer screen by clicking on the figure and dragging it with the mouse.

At our instructions, Greg took the angle between the velocity vectors to be 90° when the hang gliders unfurl, but you can play around with this angle as well as choosing other parameter values than those used by Greg.

1.2.2 War Games of the Mind

War is at best barbarism...War is hell.
William T. Sherman, Graduation address at Michigan Military Academy (1879)

Erehwon, being more civilized than many other worlds in the known universe, now only engages in war games of the mind. Greg Arious Nerd, whom we met in the last recipe, has set the following warm-up computer algebra exercise for candidates aspiring to represent Northern Erehwon in the next Olympic War Games. A gun on the shore of the Slimy Sea fires a shell with a muzzle speed of 2000 km/hour at an incoming hostile aeroplane which is heading towards the gun. At the instant of firing, the plane is 15 km away at an altitude of 1 km above the gun emplacement, and has a horizontal velocity of 500 km/hour and a downward velocity of 50 km/hour. On Erehwon, the acceleration due to gravity is $g = 10$ m/s². Neglect air resistance in this warm-up exercise.

(a) What is the required elevation angle for the gun?
(b) What is the time interval between firing and impact?
(c) Animate the trajectories of the shell and plane over this time interval.
(d) What is the horizontal range of the shell when impact takes place?

(e) Plot the time-dependent angle between the shell's velocity and the sea.

(f) At what time is the shell at its maximum height?

This 2-dimensional problem is tackled by Enaj, who has finished her summer job as smoke jumper, and has returned to her studies at the Erehwon Institute of Technology. Let's see how she handles this exercise. Enaj loads the plots and VectorCalculus packages and decides to include $N=100$ plots in the animation of the trajectories. This should be sufficient to produce a smooth animation.

>   restart: with(plots): with(VectorCalculus): N:=100:

Choosing to work in kilometers and hours, Enaj uses the units option in the convert command to convert the acceleration of gravity from m/s² to km/h². The acceleration due to gravity is then given by the vector $\vec{a}=-g\,\hat{e}_y$.

>   g:=convert(10,units,m/s^2,km/h^2): a:=<0,-g>;

$$a := (-129600)\,e_y$$

The plane's initial position $\vec{R}p0 = 15\,\hat{e}_x + \hat{e}_y$ and velocity $\vec{V}p = -500\,\hat{e}_x - 50\,\hat{e}_y$ are entered, and its position $\vec{R}p$ at time t after the firing of the shell is calculated.

>   Rp0:=<15,1>: Vp:=<-500,-50>:
>   Rp:=Rp0+Vp*t;

$$Rp := (15 - 500\,t)\,e_x + (1 - 50\,t)\,e_y$$

Enaj takes the shell's initial position $\vec{R}s0$ to be at the origin and enters the shell's initial speed $vs0$. Assuming that the shell's initial velocity makes an angle θ with the sea, its initial velocity $\vec{V}s0 = vs0\,\cos\theta\,\hat{e}_x + vs0\,\sin\theta\,\hat{e}_y$.

>   Rs0:=<0,0>: vs0:=2000: Vs0:=<vs0*cos(theta),vs0*sin(theta)>;

$$Vs0 := 2000\cos(\theta)\,e_x + 2000\sin(\theta)\,e_y$$

On leaving the gun, the shell moves under the influence of gravity. At time t, its position $\vec{R}s$ is given by the output of the following kinematic formula.

>   Rs:=Rs0+Vs0*t+a*t^2/2;

$$Rs := 2000\,t\cos(\theta)\,e_x + (2000\,t\sin(\theta) - 64800\,t^2)\,e_y$$

The separation vector $\vec{R}p - \vec{R}s$ is formed.

>   sep:=Rp-Rs;

$$sep := (15 - 500\,t - 2000\,t\cos(\theta))\,e_x + (1 - 50\,t - 2000\,t\sin(\theta) + 64800\,t^2)\,e_y$$

By setting the first component of sep equal to zero and solving for the time, an expression for the time T of impact in terms of the unknown angle θ is obtained.

>   T:=solve(sep[1]=0,t);

$$T := \frac{3}{100}\frac{1}{1+4\cos(\theta)}$$

Enaj substitutes $t=T$ into sep[2] and sets the result equal to zero.

>   eq:=subs(t=T,sep[2]=0);

$$eq := 1 - \frac{3}{2}\frac{1}{1+4\cos(\theta)} - \frac{60\sin(\theta)}{1+4\cos(\theta)} + \frac{1458}{25}\frac{1}{(1+4\cos(\theta))^2} = 0$$

1.2. VECTOR MULTIPLICATION

The formidable transcendental equation eq is numerically solved for θ.
> `Theta:=fsolve(eq,theta); #radians`
$$\Theta := 0.2587080519$$
Enaj finds that the gun must be elevated at an angle of about $\Theta = 0.26$ radians with the sea, or on converting to degrees, at $\Theta 2 = 14.8$ degrees.
> `Theta2:=convert(Theta,units,radian,degree); #degrees`
$$\Theta 2 := 14.82287950$$
Evaluating T at $\theta = \Theta$, the time TT of impact is about 0.006 hrs, or using the `convert(units)` command about 22 s.
> `TT:=eval(T,theta=Theta); #hours`
$$TT := 0.006164106735$$
> `TTsec:=convert(TT,units,h,sec); #seconds`
$$TTsec := 22.19078425$$
To animate the shell's trajectory, Enaj evaluates $\vec{R}s$ at $\theta = \Theta$.
> `Rs:=eval(Rs,theta=Theta);`
$$Rs := 1933.442613\, t\, \mathrm{e_x} + (511.6636240\, t - 64800\, t^2)\, \mathrm{e_y}$$
She introduces a functional operator `gr` which will plot a thick red trajectory for the shell over the time range $t = 0..i\,(TT)/N$ when `gr(Rs,red)` is subsequently entered. The time parameter i will take on the values 1, 2, ..., N in a subsequent sequence command. Similarly, a blue trajectory will be drawn for the plane when `gr(Rp,blue)` is entered. Axis labels have been included, the words "horizontal range" and "height" being entered as Maple strings.
> `gr:=(R,c)->plot([R[1],R[2],t=0..i*TT/N],color=c,`
> `thickness=2,labels=["horizontal range","height"])):`

A second functional operator `pl` is introduced to display the plots `gr(Rs,red)` and `gr(Rp,blue)` together for each value of i.
> `pl:=i->display({gr(Rs,red),gr(Rp,blue)}):`

The sequence of plots `pl(i)` from $i=1$ to N are now displayed, the illusion of animation being created by including the option `insequence=true`. The view is taken to be a horizontal range from the origin out to 15 km and a vertical range from 0 up to a height of 1.1 km. After executing the command line, Enaj initiates the animation by clicking on the computer picture and then on the start arrow in Maple's tool bar.
> `display(seq(pl(i),i=1..N),insequence=true,`
> `view=[0..15,0..1.1]);`

When the animation ceases, the completed trajectories up to the time of impact are as shown in Figure 1.10. The plane follows the straight path from the upper right, the shell the parabolic path from the origin.

Evaluating the first component of $\vec{R}s$ at $t=TT$ (the time to impact),
> `distance:=eval(Rs[1],t=TT); #kilometers`
$$distance := 11.91794663$$

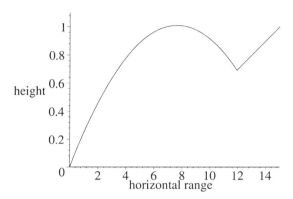

Figure 1.10: Trajectories of the shell and the plane.

Enaj finds that the horizontal distance traveled by the shell when impact takes place is about 11.9 km. Next, she determines the shell's velocity $\vec{Vs}$ at time t.

> `Vs:=eval(Vs0+a*t,theta=Theta);`

$$Vs := 1933.442613\, e_x + (511.6636240 - 129600\, t)\, e_y$$

When the shell is at maximum height, the vertical component of its velocity is zero. Setting `Vs[2]` equal to zero and solving for the time yields the time $tmax$ at which the shell is at its maximum height.

> `tmax:=solve(Vs[2]=0,t); #hours`

$$tmax := 0.003948021790$$

The shell is at maximum height at about 0.0039 hours after being fired or, on converting to seconds in the next command line, after 14.2 seconds.

> `tmax:=convert(tmax,units,h,sec); #seconds`

$$tmax := 14.21287844$$

The angle that the shell makes with the sea at arbitrary time t prior to impact with the plane can be calculated in two different ways. Enaj enters a unit vector $\hat{u} = \hat{e}_x$ parallel to the sea's surface.

> `u:=<1,0>:`

The angle $\Phi 1$ between $\vec{Vs}$ and $\hat{u}$ can be determined by forming the dot product of $\vec{Vs}$ and $\hat{u}$, dividing by the product of their magnitudes, and taking the arccosine of the result.

> `Phi1:=arccos((Vs . u)/sqrt((Vs . Vs)*(u . u)));`

$$\Phi 1 := \arccos(\frac{1933.442613}{\sqrt{0.3738200338\, 10^7 + (511.6636240 - 129600\, t)^2}})$$

Alternatively, the angle can be calculated by taking the arctangent of the ratio `Vs[2]/Vs[1]`, since the first component of $\vec{Vs}$ is parallel to the sea.

> `Phi2:=arctan(Vs[2],Vs[1]);`

1.2. VECTOR MULTIPLICATION

$$\Phi 2 := \arctan(511.6636240 - 129600\, t,\, 1933.442613)$$

To check if the results $\Phi 1$ and $\Phi 2$ are the same, Enaj converts both angles to degrees, and plots them together up to the time of impact. She chooses different colors and line styles for each curve.

```
> plot([180*Phi1/Pi,180*Phi2/Pi],t=0..TT,tickmarks=[3,4],
> labels=["hours","degrees"],color=[blue,red],linestyle=[1,3]);
```

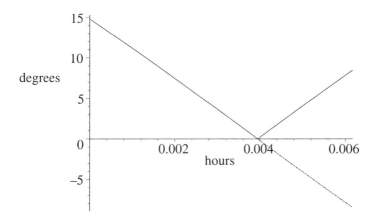

Figure 1.11: Angle of the shell's velocity with the sea's surface.

The result is shown in Figure 1.11. The solid curve is $\Phi 1$, the dashed curve (which overlaps the solid one up to about 0.004 hours) is Φ_2. The bounce in angle in $\Phi 1$ is an unphysical artifact of branch choice in its calculation. Physically, the angle changes sign because the slope of the velocity vector changes sign from positive to negative. This behavior is seen $\Phi 2$. From the plot, Enaj sees that in either case the angle that the shell makes with the sea is zero at slightly less than 0.004 hours in agreement with `tmxax` calculated earlier.

Having solved this warm-up exercise, Enaj wonders what the "real" problems are like, particularly those that she might face at the Olympic War Games.

1.2.3 Flying by the Seat of Your Pants

In flying the probability of survival is inversely proportional to the angle of arrival.
Neil Armstrong, American astronaut (1930–)

Rory, the chief test pilot for the Antipodean Air Force, was commissioned to test a new experimental rocket plane, code named Pegasus.[7] As a safety feature, the pilot's seat has been designed to be ejected from Pegasus with a

---
[7]In Greek mythology, a winged horse which sprang from Medusa's body upon her death.

small booster rocket if Pegasus's acceleration exceeds 8 g. To avoid a possible subsequent collision between Rory and Pegasus, the seat was to be ejected in a direction given by the cross product of Pegasus's acceleration and velocity vectors at the time of ejection. Unfortunately, in her maiden flight, Pegasus went out of control but Rory was fortunate enough to be ejected before the aircraft exploded. From Pegasus's flight recorder, we have learned that its position vector (in meters) at a time t seconds after take-off was

$$\vec{r} = (0.2\,t^2 + 10\,t + 50)\,\hat{e}_x + (0.1\,t^2 + 5\,t + 100)\,\hat{e}_y + 10\,e^{(0.1\,t)}\,\hat{e}_z$$

and Pegasus did not explode until 10 seconds after Rory's ejection. Similarly, from a flight recorder in the pilot's seat, we also know that the booster rocket attached to the seat fired for 20 seconds, giving Rory an acceleration of magnitude $40\,e^{-2t}$ m/s^2 over this time interval. After the booster rocket quit firing, Rory went into free fall. Neglecting air resistance and taking $g = 9.8$ m/s^2,

(a) at what time was Rory ejected from Pegasus?

(b) in what direction was he ejected?

(c) plot Rory's trajectory from the time of take-off through free fall.

The plots library package is loaded because we shall be making use of the `display`, `spacecurve`, `pointplot3d`, and `textplot3d` commands contained in this package. Loading the `VectorCalculus` package will allow us to calculate the cross product of Pegasus's acceleration and velocity vectors.

> `restart: with(plots): with(VectorCalculus):`

The acceleration g due to gravity is entered, along with Pegasus's position $\vec{r}$ at time t.

> `g:=9.8:`
> `r:=<0.2*t^2+10*t+50,0.1*t^2+5*t+100,10*exp(0.1*t)>;`

$$r := (0.2\,t^2 + 10\,t + 50)\,e_x + (0.1\,t^2 + 5\,t + 100)\,e_y + 10\,e^{(0.1\,t)}\,e_z$$

Pegasus's velocity $\vec{v}$ and acceleration $\vec{a}$ are readily determined by first differentiating $\vec{r}$ and then $\vec{v}$ with respect to time.

> `v:=diff(r,t);`

$$v := (0.4\,t + 10)\,e_x + (0.2\,t + 5)\,e_y + 1.0\,e^{(0.1\,t)}\,e_z$$

> `a:=diff(v,t);`

$$a := 0.4\,e_x + 0.2\,e_y + 0.10\,e^{(0.1\,t)}\,e_z$$

The magnitude of Pegasus's acceleration at time t is calculated.

> `a_mag:=sqrt(a . a);`

$$a\_mag := \sqrt{0.20 + 0.0100\,(e^{(0.1\,t)})^2}$$

Rory's time of ejection, $t0$, is obtained by setting this acceleration magnitude equal to $8\,g$, and numerically solving for the time.

> `t0:=fsolve(a_mag=8*g,t);`

1.2. VECTOR MULTIPLICATION

$$t0 := 66.64392751$$

So, Rory was ejected 66.6 seconds after the beginning of the flight. To determine the direction in which he was ejected, the cross product, $\vec{c}$, of Pegasus's acceleration $\vec{a}$ and velocity $\vec{v}$ is calculated using the short form &x of the cross product command. The long form would be CrossProduct(a,v).

>   c:=a &x v;

$$c := (0.20\, e^{(0.1\,t)} - 0.10\, e^{(0.1\,t)}\,(0.2\,t + 5))\, \text{e}_x$$
$$+ (0.10\, e^{(0.1\,t)}\,(0.4\,t + 10) - 0.40\, e^{(0.1\,t)})\, \text{e}_y + 0.\, \text{e}_z$$

The unit vector $\hat{n}o$ in the direction of $\vec{c}$ is calculated by dividing $\vec{c}$ by its magnitude. To simplify the output, a simplify command is applied with the command assuming t>0 included. The assuming command applies here to the single operation of simplification. If we wanted to make the assumption that $t > 0$ applied globally, we would make use of the assume command.

>   no:=simplify(c/sqrt(c.c)) assuming t>0;

$$no := (-0.4472135954)\, \text{e}_x + 0.8944271908\, \text{e}_y + 0.\, \text{e}_z$$

To plot Rory's trajectory, we need to first determine his position at the time of ejection. Evaluating $\vec{r}$ at time $t = t0$ yields this position $\vec{r}o$.

>   ro:=eval(r,t=t0);

$$ro := 1604.721890\, \text{e}_x + 877.3609450\, \text{e}_y + 7839.872449\, \text{e}_z$$

At this time he had a velocity $\vec{v}o$.

>   vo:=eval(v,t=t0);

$$vo := 36.65757100\, \text{e}_x + 18.32878550\, \text{e}_y + 783.9872449\, \text{e}_z$$

Next, we have to look at the time interval when the booster rocket was firing. The booster rocket's acceleration vector $\vec{a}o$ was equal to the given magnitude times the unit vector $\hat{n}o$.

>   ao:=40*exp(-2*tt)*no;

$$ao := -17.88854382\, e^{(-2\,tt)}\, \text{e}_x + 35.77708763\, e^{(-2\,tt)}\, \text{e}_y + 0.\, \text{e}_z$$

During this time interval, Rory's ejection velocity $\vec{v}e$ relative to the earth was equal to the vector sum of the velocity $\vec{v}o$ and the time integral of the above booster acceleration over the interval from $t0$ to some time T.

>   ve:=vo+int(ao,tt=t0..T);

$$ve := (36.65757100 + 8.944271910\, e^{(-2.\,T)})\, \text{e}_x$$
$$+ (18.32878550 - 17.88854382\, e^{(-2.\,T)})\, \text{e}_y + 783.9872449\, \text{e}_z$$

Rory's position relative to his starting point at the beginning of the flight was equal to the vector sum of the position vector $\vec{r}o$ and the time integral of $\vec{v}e$ over the interval from $t0$ to some time t.

>   re:=ro+int(ve,T=t0..t);

$$re := (-838.282614 + 36.65757100\,t - 4.472135955\,e^{(-2 \cdot t)})\,e_x$$
$$+(-344.1413070 + 18.32878550\,t + 8.944271910\,e^{(-2 \cdot t)})\,e_y$$
$$+(-44408.11667 + 783.9872449\,t)\,e_z$$

Given that the booster rocket fired for 20 seconds, Rory's final velocity ($\vec{v}ef$) and position ($\vec{r}ef$) vectors at the time $t1 = t0 + 20$ when the booster quit firing are easily evaluated.

> `t1:=t0+20:`

> `vef:=eval(ve,T=t1);`

$$vef := 36.65757100\,e_x + 18.32878550\,e_y + 783.9872449\,e_z$$

> `ref:=eval(re,t=t1);`

$$ref := 2337.873310\,e_x + 1243.936655\,e_y + 23519.61735\,e_z$$

For $t > t1$, our pilot, Rory, was moving only under the influence of gravity. In this free fall phase, his position vector was $\vec{r}p$.

> `rp:=ref+vef*(t-t1)+<0,0,-g>*(t-t1)^2/2;`

$$rp := (-838.282614 + 36.65757100\,t)\,e_x + (-344.141307 + 18.32878550\,t)\,e_y$$
$$+(-44408.11667 + 783.9872449\,t - 4.900000000\,(t - 86.64392751)^2)\,e_z$$

To plot Rory's entire trajectory, we first calculate the time tb at which Pegasus exploded and introduce a final time tf to which Rory's trajectory will be plotted. We have arbitrarily taken the latter time to be 120 seconds after the time $t1$ when the booster rocket quit firing.

> `tb:=t0+10; tf:=t1+120;`

$$tb := 76.64392751$$
$$tf := 206.6439275$$

Pegasus's position $\vec{r}b$ at the time tb it exploded is evaluated as well as Rory's position $\vec{r}pf$ at time tf.

> `rb:=eval(r,t=tb); rpf:=eval(rp,t=tf);`

$$rb := 1991.297600\,e_x + 1070.648800\,e_y + 21310.98282\,e_z$$

$$rpf := 6736.781830\,e_x + 3443.390915\,e_y + 47038.08670\,e_z$$

To plot the different portions of Rory's trajectory, a functional operator SC is formed involving the 3-dimensional `spacecurve` plotting command.

> `SC:=(r,T1,T2,c)->spacecurve(evalm(r),t=T1..T2,color=c):`

The position vector r, the ends T1 and T2 of the time range, and the color c must be provided as arguments to SC to create the plot. Note that the matrix evaluation command (`evalm`) is applied to r to convert the position $\vec{r}$ into a Maple list for plotting purposes. The subsequent command `SC(r,0,tb,green)` will plot Pegasus's trajectory as a green curve up to the time tb of explosion. Similarly `SC(re,t0,t1,red)` will plot Rory's trajectory as a red curve during the booster rocket stage, and `SC(rp,t1,tf,blue)` will plot the free fall portion of Rory's trajectory as a blue curve.

1.2. VECTOR MULTIPLICATION

To place size 20 black circles on the final plot at critical stages of the trajectory, a functional operator PP is formed to apply the pointplot3d command.

> PP:=r->pointplot3d(r,symbol=circle,symbolsize=20,color=black):

The 3-dimensional textplot command (textplot3d) is used in TP to add the words "boom!", "eject", "booster rocket quits", and "wow!" to the plot in the vicinity of the circles generated by applying PP. The location of each word is controlled by entering the three coordinates of each of the vectors $\vec{r}b$, $\vec{r}o$, $\vec{r}ef$, and $\vec{r}pf$ as a list with each word entered as a string in the fourth element. A list of lists is then formed. So that the words don't end up right on top of the circles, the coordinates are shifted slightly and we align the words to the right with the align=RIGHT option.

> TP:=textplot3d([[rb[1]-100,rb[2]-100,rb[3]-100,"boom!"],
> [ro[1]-100,ro[2]-100,ro[3],"eject"],
> [ref[1]-100,ref[2]-100,ref[3]+100,"booster rocket quits"],
> [rpf[1]-100,rpf[2]-100,rpf[3],"wow!"]],color=red,align=RIGHT):

On executing the following display command line, the completed plot is produced, a black and white version being shown in Figure 1.12. Note that we have not constrained the scaling in this figure.

> display({SC(r,0,tb,green),SC(re,t0,t1,red),SC(rp,t1,tf,blue),
> PP(ro),PP(rb),PP(ref),PP(rpf),TP},orientation=[150,75],
> tickmarks=[3,3,3],axes=framed,labels=["x","y","z"]);

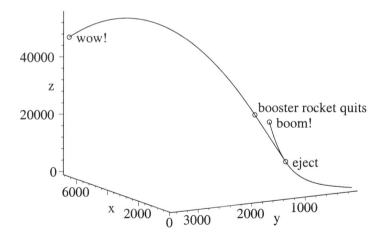

Figure 1.12: Rory's path during the ill-fated flight of Pegasus.

Notice how Rory continued to gain altitude for some time after ejection because he had already acquired an upward velocity while still seated in Pegasus.

According to our reconstruction of his trajectory, he reached an altitude of about 50000 m or 50 km, before beginning to fall. Oh, by the way, has any one actually seen Rory lately?

1.2.4 Mike's Train Ride

You see things: and say "Why?"
But I dream things that never were; and I say "Why not?"
George Bernard Shaw, Anglo–Irish playwright and critic (1856–1950)

Vectoria's husband, Mike, is a Ph.D candidate in the Institute of Applied Mathematics at MIT. As part of his tutorial on vectors for the introductory mechanics course that he is "TA-ing", he has created a computer algebra recipe to solve and explore the following problem on circular motion. According to Mike, this recipe was inspired by his reminiscences of the train ride that he took in his bachelor days from Cuzco to Puno on Lake Titicaca in the Peruvian highlands after his summer stint on an archaeological dig near the Inca ruins at Machu Pichu. Of course, the mathematical details of the following "story" are a figment of Mike's active imagination.

A short, heavy, Peruvian train starts from rest and accelerates along a circular portion of track of radius $R = 500$ m. Its angular coordinate at t seconds is $\theta = (1/2)\alpha\, t^2$, with the angular acceleration $\alpha = 1/10000$ radians/s^2. When the angle between the train's velocity ($\vec{v}$) and acceleration ($\vec{a}$) vectors reaches 0.8 radians ($\approx 45.8\,°$), a mischievous boy throws his sister's ball out the window with a speed $V = 10$ m/s in a direction given by the double cross product $\vec{a} \times (\vec{v} \times \vec{a})$. At what time is the ball thrown out the window? What is the speed and direction of the ball, and the train's velocity and acceleration, at this time? Make a suitable plot indicating the moving train and the above vectors.

Mike begins his recipe by loading the `plots` and `VectorCalculus` packages, and enters the angular acceleration α, the radius R, the speed V, and the train's angular coordinate θ at time t.

```
> restart: with(plots): with(VectorCalculus):
> alpha:=1/10000: R:=500: V:=10: theta:=alpha*t^2/2;
```

In Cartesian coordinates, the position of the train at time t is given $\vec{r} = R\cos\theta\,\hat{e}_x + R\sin\theta\,\hat{e}_y + 0\,\hat{e}_z$. A 3-dimensional vector is created here, because three dimensions are required for performing the double cross product.

```
> r:=<R*cos(theta),R*sin(theta),0>;
```

$$r := 500\cos\left(\frac{t^2}{20000}\right)\mathrm{e_x} + 500\sin\left(\frac{t^2}{20000}\right)\mathrm{e_y}$$

Differentiating $\vec{r}$ with respect to t, the train's velocity $\vec{v}$ and acceleration $\vec{a}$ are calculated.

```
> v:=diff(r,t); a:=diff(v,t);
```

$$v := -\frac{1}{20}\sin(\frac{t^2}{20000})\,t\,\mathrm{e_x} + \frac{1}{20}\cos(\frac{t^2}{20000})\,t\,\mathrm{e_y}$$

1.2. VECTOR MULTIPLICATION

$$a := (-\frac{1}{200000}\cos(\frac{t^2}{20000})t^2 - \frac{1}{20}\sin(\frac{t^2}{20000}))\,e_x$$
$$+(-\frac{1}{200000}\sin(\frac{t^2}{20000})t^2 + \frac{1}{20}\cos(\frac{t^2}{20000}))\,e_y$$

The angle Θ between the velocity and acceleration at time t is determined by taking the dot product of $\vec{a}$ and $\vec{v}$, dividing by the product of the magnitudes, and taking the arccosine of the result.

> `Theta:=arccos((a . v)/(sqrt(a . a)*sqrt(v . v)));`

The lengthy output, which is not shown, is simplified with the `symbolic` option. This option has the same effect here as using the command `assuming t>0`.

> `Theta:=simplify(Theta,symbolic);`

$$\Theta := \arccos\left(\frac{10000}{\sqrt{t^4 + 100000000}}\right)$$

Since the ball is thrown out the window when $\Theta = 0.8$ rads, Mike numerically solves for the time T at which this sad (for the sister) event takes place.

> `T:=fsolve(Theta=0.8,t) ;`

$$T := 101.4711071$$

The boy throws his sister's ball out the window at about 101 seconds after the slowly accelerating train starts to move. Mike creates a functional operator `f` to evaluate a vector v at time $t=T$ when `f(v)` is subsequently entered.

> `f:=v->eval(v,t=T):`

Using `f`, Mike calculates the train's position $\vec{r}t$ at time T, as well as its velocity $\vec{v}t$, and its acceleration $\vec{a}t$.

> `rt:=f(r); vt:=f(v); at:=f(a);`

$$rt := 435.1908598\,e_x + 246.1887804\,e_y$$
$$vt := (-2.498104810)\,e_x + 4.415929834\,e_y$$
$$at := (-0.06942780697)\,e_x + 0.01817053990\,e_y$$

The double cross product $\vec{a} \times (\vec{v} \times \vec{a})$ is evaluated in `dcp`, and the unit vector $\hat{u}$ pointing in the direction in which the ball is thrown is determined.

> `dcp:=at &x (vt &x at);`

$$dcp := 0.004746079808\,e_x + 0.01813429400\,e_y - 0.\,e_z$$

> `u:=dcp/sqrt(dcp . dcp);`

$$u := 0.2531907412\,e_x + 0.9674163783\,e_y - 0.\,e_z$$

As seen from a stationary observer outside the train, the ball's speed is the sum of the component of the train's velocity $\vec{v}t$ in the direction of $\hat{u}$ (thus, the dot product $\vec{v}t \cdot \hat{u}$) plus the speed V with which the boy threw the ball.

> `speed:=(vt . u)+V;`

$$speed := 13.63954584$$

The ball's speed is about 14 m/s. Mike now summarizes the main features by making a plot. He first creates a functional operator A for producing arrows to graphically represent the various vectors in the problem.

> A:=(sc,b,c)->arrow([rt[1],rt[2]],[sc*b[1],sc*b[2]],color=c):

Here sc is a scale factor which will scale the size of the arrow representing the vector $\vec{b}$, and c is the color. Mike will choose the scale factors by trial and error so that final plot looks nice. The base of the plotted arrow is placed at $\vec{r}t$, the components of which are entered as a list as the first argument of the arrow command. The scaled components of $\vec{b}$ are entered as a list in the second argument, and the color is the third argument. The entry A(sc1,u,blue) will produce a scaled (scale factor sc1) blue arrow to represent the unit vector $\hat{u}$ pointing in the direction in which the ball is thrown. Similarly, A(sc2,at,green) and A(sc3,vt,red) will generate differently scaled green and red arrows to represent the train's acceleration $\vec{a}t$ and velocity $\vec{v}t$ at the time the ball is thrown.

The scale factors sc1, sc2, and sc3 that Mike has chosen are now entered. Mike decides to plot the location of the train at N=10 equal time steps.

> sc1:=100: sc2:=1000: sc3:=20: N:=10:

Although he could plot the train's location by using a functional operator, Mike decides that it would be instructive to his students to show them the use of the concatenation operator || and the "do loop" for carrying out a repetitive calculation. The general syntax for this common programming structure is:

for *name* from *expression* by *expression* to *expression* while *expression* do
statement sequence;
end do:

Here the *statement sequence* is the main body of the do loop. In the following command line, *name* is the index i, the first *expression* is 0, the second *expression* is not specified, the third *expression* is N, and there is no conditional while *expression* present. Since the by *expression* is missing, the default is to increment i by 1 each time the *statement sequence* is executed.

> for i from 0 to N do

The concatenation operator || is a binary operator which requires a name (or a string) as its left operand. The left operand below is the time t. Its right operand (the index i) is evaluated and then concatenated (joined) to name (or string). For $i = 1, 2, ...$, the entry t||i generates $t1 = T/N$, $t2 = 2T/N$, etc.

> t||i:=i*T/N;

The position $\vec{r}$ of the train is evaluated at time t||i, the components being placed in a list and assigned the name pt||i. The point pt||i is then plotted in PL||i as a size 14 blue circle with pointplot, and the do loop ended.

> pt||i:=[eval(r[1],t=t||i),eval(r[2],t=t||i)];
> PL||i:=pointplot(pt||i,symbol=circle,symbolsize=14,
> color=blue);
> end do: #end of do loop

1.3 SUPPLEMENTARY RECIPES

The circular portion of track of radius R is plotted in PP with the `polarplot` command and colored brown. Mike adds appropriate labels in TP to the three vectors with the `textplot` command.

```
> PP:=polarplot(R,color=brown):
> TP:=textplot({[490,350,"u"],[360,250,"at"],[410,350,"vt"]}):
```

All the plots are superimposed with the `display` command, the resulting picture being shown in Figure 1.13. The circles indicate the train's position at equal times. The arrow u indicates the direction in which the ball is thrown, while vt and at indicate the directions of the trains velocity and acceleration at time T.

```
> display({A(sc1,u,blue),A(sc2,at,green),A(sc3,vt,red),
> seq(PL||i,i=0..N),PP,TP},scaling=constrained,
> tickmarks=[3,3],view=[0..600,0..600]);
```

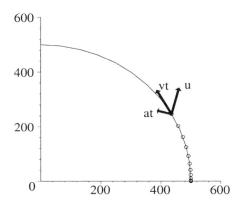

Figure 1.13: Graphical summary of Mike's calculation.

The train is accelerating so the distance between circles increases. The velocity vector $\vec{vt}$ points tangent to the train's circular path as expected. The acceleration vector $\vec{at}$ does not point towards the origin, because it is the vector sum of a (non-zero) tangential component and a radial component of acceleration. The direction of $\hat{u}$ is easily understood. By the right-hand rule, the vector product $\vec{vt} \times \vec{at}$ points out of the page and $\vec{at} \times (\vec{vt} \times \vec{at})$ points in the direction $\hat{u}$.

1.3 Supplementary Recipes

01-S01: Bogey 5 Revisited

Redo the golfing recipe by using row vectors and the `LinearAlgebra` package. A row vector is formed with the syntax < | >. Use the `Norm` command to determine the distance that Colleen had to walk back to retrieve her pen and scorecard.

01-S02: A Rolling Wheel Gathers No Moss, But ....

While rolling her toy tractor across the kitchen floor and munching on a jam-laden cracker, Gabrielle accidentally deposits a small spot of jam on the rim of the rear tractor wheel. The position of the jam spot is given at t seconds by $\vec{r} = (10t - 5\sin(2t))\,\hat{e}_x + (5 - 5\cos(2t))\,\hat{e}_y$ cm.

(a) Plot $\vec{r}$ over the time interval $t = 0$ to 10 s. Animate the motion of the jam spot, taking 200 frames and representing the spot as a size 14 red circle. Superimpose the animation on the first plot.

(b) Calculate the spot's velocity and acceleration vectors as well as its instantaneous and average speeds. Plot the two speeds in the same figure.

(c) Determine the times and positions where the spot has maximum speed. Use this information to confirm that the speed of the tractor wheel's axle is 10 cm/s, a result that could have been anticipated from the form of $\vec{r}$.

(d) Animate the motion of the tractor wheel's rim (draw with the `circle` command in the `plottools` library) with the jam spot superimposed.

01-S03: Flight of Brunhilda Bumblebee

A fast Erehwonese bumblebee by the name of Brunhilda leaves her hive and flies along a path given by $\vec{r}(t) = \hat{e}_x\, A\sin(\omega t) + \hat{e}_y\, A\cos(\omega t) + \hat{e}_z\, B t^2$ meters at t seconds, where A, B, and ω are real, positive, parameters.

(a) Show that the magnitude of Brunhilda's acceleration is constant.

(b) Taking $A = 2$, $B = 1/20$, $\omega = 3$, $T = 10$ seconds, animate Brunhilda's motion over the time interval $t = 0..T$ and superimpose her motion on the entire trajectory that she travels up to time T.

(c) Determine the total distance that Brunhilda flies up to time T and her average speed over this interval. What is Brunhilda's displacement vector from the hive at time T and what is the magnitude of this vector?

01-S04: Colleen Better Beware

After her golf match, Colleen is driving at 120 km/h in a 100 km/h zone in her sporty red Mustang. She is traveling in a direction 20° east of north and descending a short steep hill with a 10% grade. At the bottom of the hill, the road is level and heads 50° east of north. A southbound police car, with its radar unit switched on, is traveling at 90 km/h along the level stretch at the base of the hill and is approaching the sports car. What is Colleen's velocity vector with respect to the police car? What is her speed relative to the police car? Make a suitable 3-dimensional plot showing all of the velocity vectors.

01-S05: Justine's Clever Throw

Jennifer's niece, Justine, stands 4 m from the vertical wall of her school and throws a red ball at the wall. The ball leaves Justine's hand at 2 m above the level ground with an initial velocity, $\vec{v} = 10\sqrt{2}\cos(\phi)\,\hat{e}_x + 10\sqrt{2}\sin(\phi)\,\hat{e}_y$ m/s. When the ball strikes the wall, its horizontal component of velocity is reversed, but its vertical component remains unchanged. Taking $g = 9.8$ m/s^2, neglecting air resistance, and treating the ball as a point particle, what angle ϕ must Justine throw the ball so that it just clears a $3\frac{1}{2}$ m high fence located 14 m from the school wall? (Derive and solve a transcendental equation for ϕ.) Animate the ball's motion from the time it leaves Justine's hand to the instant it hits the ground on the other side of the fence. Superimpose the animation on a plot of the entire trajectory. Include the fence and school wall.

01-S06: Torpedos Away!
Here is another exercise that Enaj has been given to solve in preparation for the Olympic War Games of the Mind. A merchant ship is moving with a speed of 10 m/s at an angle 0.5 rads north of east. A submarine located 4000 m east and 3000 m south of the ship fires a torpedo at constant velocity at it. In order to place as much explosive as possible in the torpedo, the weight of its fuel is reduced to a minimum. Accordingly, the torpedo designers want to fire the torpedo with the slowest speed possible and still be able to hit the ship. What is the slowest speed and at what angle should the torpedo be fired so that a hit occurs? Animate the ship and torpedo for this situation.

01-S07: Mike's Vector Brain Teaser
In one of his tutorials, Mike has given his students the following vector problem. Two vectors $\vec{A}$ and $\vec{B}$, with $B = 2A$ and making some angle θ with $\vec{A}$, are first added to produce $\vec{A}+\vec{B}$ and then subtracted to produce $\vec{A}-\vec{B}$. The magnitude of the addition is $\sqrt{3}$ times the magnitude of the subtraction. What is the angle θ, in degrees, between $\vec{A}$ and $\vec{B}$? Taking $\vec{A}$ to be a unit vector along the horizontal axis, plot $\vec{A}$ and $\vec{B}$ in the same figure.

01-S08: Jennifer's Vector Identity Assignment
Jennifer has asked her students to use Maple to prove the following two vector identities assuming that $\vec{A}$, $\vec{B}$, and $\vec{C}$ are general three-dimensional vectors:

$$\vec{A}\times(\vec{B}\times\vec{C}) = (\vec{A}\cdot\vec{C})\,\vec{B} - (\vec{A}\cdot\vec{B})\,\vec{C}, \quad \vec{A}\times(\vec{B}\times\vec{C}) + \vec{C}\times(\vec{A}\times\vec{B}) + \vec{B}\times(\vec{C}\times\vec{A}) = 0.$$

01-S09: The Learnu Molecule
The Learnu molecule is a triatomic molecule which has been synthesized on Erehwon. Atoms 1, 2, and 3 of this molecule are located at the points $P_1(3, 1, -2)$, $P_2(-1, 2, 4)$, and $P_3(2, -1, 1)$, respectively.

(a) Using the fact that the volume of a parallelepiped defined by three distinct vectors $\vec{A}$, $\vec{B}$, $\vec{C}$ whose tails have a common vertex is given by $|\vec{A}\cdot(\vec{B}\times\vec{C})|$, find the equation for the plane passing through the atoms.

(b) Making use of the `polygon` plot command in the `plottools` package, create a 3-d plot of the triangular region enclosed by P_1, P_2, P_3. Color the triangle red and enclose its perimeter with a thick black solid line. Represent the atoms by blue circles in the same picture and label them.

(c) Determine the three bond lengths and the three bond angles (in degrees) for the Learnu molecule. Verify that the bond angles add up to 180°.

01-S10: Feeling the Electric Force
The force $\vec{F}$ exerted on a point charge q_0 (in C) in free space located at the position $\vec{r}_0$ meters due to N other charges q_i located at $\vec{r}_i$, $i = 1, 2, ..., N$, is given by Coulomb's law, $\vec{F} = k\, q_0 \sum_{i=1}^{N} q_i(\vec{r}-\vec{r}_i)/|\vec{r}-\vec{r}_i|^3$, where $k = 8.99 \times 10^9$ N·m²/C². Point charges $q_0 = 10$ nC, $q_1 = 1$ mC and $q_2 = -2$ mC are located at $\vec{r}_0 = 3\hat{e}_y + \hat{e}_z$, $\vec{r}_1 = 3\hat{e}_x + 2\hat{e}_y - \hat{e}_z$, and $\vec{r}_2 = -\hat{e}_x - \hat{e}_y + 4\hat{e}_z$. What is the angle in radians and degrees between $\vec{r}_1$ and $\vec{r}_2$? Calculate the resultant electric force in mN on q_0. Make a 3-d plot showing the labeled charge locations, the forces exerted on q_0 by q_1 and q_2, and the resultant force on q_0.

01-S11: Mike's Potpourri of Unit Vectors
Given the two position vectors $\vec{a} = 4\hat{e}_x + 3\hat{e}_y$ and $\vec{b} = -\hat{e}_y - 3\hat{e}_y + 2\hat{e}_z$, Mike has requested that you find a unit vector $\hat{u}$ perpendicular to both $\vec{a}$ and $\vec{b}$ and a second unit vector $\hat{v}$ which points towards a position halfway between $\vec{a}$ and $\vec{b}$. Determine the area of the triangle which has $\vec{a}$ and $\vec{b}$ along its edges. Using colored cylindrical arrows, constrained scaling, and suitable labels, plot $\hat{u}, \hat{v}, \vec{a}$ and $\vec{b}$ in the same figure. Determine the angles that $\hat{v}$ makes with $\vec{a}$ and $\vec{b}$.

01-S12: Closest Possible Encounter
Metropolis air traffic control has established one straight flight path which passes through the points $a(-16, 6, 12)$ and $b(-1, 2, 5)$ and a second straight flight path which passes through the points $c(1, -1, 11)$ and $d(4, 9, 15)$. The distances are in km. Derive a parametric equation for each flight path. Determine the shortest distance between the two flight paths and at which point along each path this minimum possible separation occurs. Create a 3-dimensional plot of the two flight paths and the minimum separation line.

01-S13: Envelope of Safety
A previously dormant volcanic mountain in the Pacific Northwest has erupted and is throwing rocks into the atmosphere at speeds of up to 700 m/s. Colleen's sister, Sheelo, is a part-time National Geographic photographer who has hired an aircraft to film the spectacle. Neglecting the height of the mountain and air resistance, and assuming that the rocks are thrown in all directions, analytically determine the envelope of safety outside of which the aircraft is safe from flying rocks. Taking $g \approx 10$ m/s², create a planar plot showing the envelope of safety and the animated trajectories of rocks flying with the maximum speed at different angles with the horizontal.

01-S14: Avoiding a Mortar Shell
An enemy mortar emplacement is set 8230 m horizontally from the edge of a vertical cliff which drops 107 m down to a flat plain. Help Enaj determine how close to the bottom edge of the cliff should invading troops on the plain remain in order to guarantee that they will not be directly hit by a mortar shell? The muzzle speed of the shells is 305 m/s and $g = 9.8$ m/s². Neglect air resistance. Plot the trajectory of the mortar shell for this situation.

Chapter 2

Newtonian Mechanics

The study of Newtonian mechanics involves the application of three well-known laws of motion to the movement of a body experiencing a net, or resultant, force. Newton's first, second, and third laws are as follows [MT95]:

(1) *A body remains at rest or in uniform motion unless acted upon by a force.*
(2) *A body acted upon by a force moves in such a way that the time rate of change of momentum equals the force.*
(3) *Whenever a body exerts a force on another body, the latter exerts a force of equal magnitude and opposite direction on the former.*

Since the net force on a body is the vector sum of the forces acting on it, we could use the `VectorCalculus` package to express the forces in terms of unit vectors and then add them. An equivalent approach, which does not require loading this package, is to resolve the forces into their components.

2.1 Motion Involving Constant Forces

The simplest dynamical problems are those involving translational motion of rigid bodies under the influence of constant forces. The bodies are treated as if they are "particles" with all their mass concentrated at a single point, the center of mass. Our first three recipes in this section involve forces that are constant in magnitude and direction. These include frictional, gravitational, normal, and applied forces. The first recipe is an application of Newton's first law while the second is a simple example of the second law when the mass is taken to be constant. In this case, Newton's second law takes the simpler form $\vec{F} = \frac{d}{dt}(m\vec{v}) = m\frac{d\vec{v}}{dt} = m\vec{a}$, where $\vec{F}$ is the net force on the mass m and $\vec{v}$ and $\vec{a}$ are the mass's velocity and acceleration, respectively. The third example is much more complex, involving both the second and third laws. The final recipe in this section deals with the circular motion of an object moving at constant speed. In this case, the centripetal force has a constant magnitude and always points to the center of the circle.

2.1.1 Jack and Jill Go Up the Hill

Jack and Jill went up the hill, To fetch a pail of water;
Mother Goose, children's nursery rhyme

Unlike the nursery rhyme characters, our modern-day Jack and Jill, Kevin and his sister Ruth, are trudging up a snowy slope which makes an angle θ with the horizontal with the goal of tobogganing down the other side. Kevin is pulling a toboggan (weight w) at constant speed, exerting a steady pull P at an angle ϕ with the slope. If the coefficient μ_k of kinetic friction between the toboggan and the snow is about 0.10, at what angle ϕ should Kevin pull in order to minimize the pull P? Can you guess what this angle might be? If the toboggan weighs 25 N, the slope angle is 25°, and the slope length is 50 m, how much work does Kevin do pulling the toboggan up the slope with the minimum force?

To solve this problem, the reader might wish to sketch a "free body diagram" indicating all the forces acting on the toboggan. There are four forces involved, the weight w of the toboggan acting vertically downwards, the normal force N exerted perpendicular to and by the slope on the toboggan, the frictional force F_{fr} between the toboggan and snow acting down the slope, and the uphill pull P at an angle ϕ to the slope. It is convenient to resolve the forces into components normal and parallel to the slope. For P, these components are $P \sin\phi$ and $P \cos\phi$, respectively, and for the weight $w \cos\theta$ and $w \sin\theta$.

> `restart:`

Let's consider the situation perpendicular to the slope. Since there is no motion in this direction, the force components normal to the surface must balance. This condition is entered in *eq1*.

> `eq1:=P*sin(phi)+N=w*cos(theta); #normal to surface`

$$eq1 := P\sin(\phi) + N = w\cos(\theta)$$

The frictional force is related to the normal force by $F_{fr} = \mu_k N$. In Chapter 1, we repeatedly used the selection operation `[i]` to select the ith component from a multi-component object. The selection operation may also be used to index as well as subscript names. In the following command line, entering `F[fr]` and `mu[k]` produces the subscripted quantities F_{fr} and μ_k in the output.

> `F[fr]:=mu[k]*N; #frictional force`

$$F_{fr} := \mu_k N$$

Since the toboggan is being pulled at constant speed, by Newton's first law the pull component up the slope must balance the sum of the weight component and frictional force down the slope. This condition is entered in *eq2*.

> `eq2:=P*cos(phi)=w*sin(theta)+F[fr]; #parallel to surface`

$$eq2 := P\cos(\phi) = w\sin(\theta) + \mu_k N$$

The set of simultaneous equations, *eq1*, *eq2*, are analytically solved for the set of unknowns, P and N, the solution being labeled *sol* and the result assigned.

> `sol:=solve({eq1,eq2},{P,N});`

2.1. MOTION INVOLVING CONSTANT FORCES

$$sol := \{P = \frac{w\,(\sin(\theta) + \mu_k \cos(\theta))}{\cos(\phi) + \mu_k \sin(\phi)},\ N = -\frac{w\,(\sin(\phi)\sin(\theta) - \cos(\theta)\cos(\phi))}{\cos(\phi) + \mu_k \sin(\phi)}\}$$

```
> assign(sol):
```

The pull is then given by the output of the following command line.

```
> P:=P;
```

$$P := \frac{w\,(\sin(\theta) + \mu_k \cos(\theta))}{\cos(\phi) + \mu_k \sin(\phi)}$$

To determine the angle ϕ which minimizes the pull, we differentiate P with respect to ϕ and set the result equal to zero in *eq3*.

```
> eq3:=diff(P,phi)=0;
```

$$eq3 := -\frac{w\,(\sin(\theta) + \mu_k \cos(\theta))\,(-\sin(\phi) + \mu_k \cos(\phi))}{(\cos(\phi) + \mu_k \sin(\phi))^2} = 0$$

Then *eq3* is solved for ϕ,

```
> phi:=solve(eq3,phi);
```

$$\phi := \arctan(\mu_k)$$

yielding the relation $\phi = \arctan(\mu_k)$. If Kevin pulls at this angle, he will minimize his pull. For snow, $\mu_k \approx 0.1$, so entering this value, the angle ϕ can be evaluated.

```
> mu[k]:=0.1: phi:=phi; #angle in radians
```

$$\phi := .09966865249$$

Kevin should pull at an angle of about 0.1 radians, or on converting to degrees,

```
> Phi:=convert(phi,units,radian,degree); #degrees
```

$$\Phi := 5.710593137$$

at an angle of 5.71° with respect to the slope. Some readers might have initially thought that Kevin should pull parallel to the slope, but in fact $\phi = 0$ does not yield the minimum pull. To calculate how much work Kevin does pulling with minimum force, the values of the slope length L, the weight w, and the slope angle θ are entered. The slope angle is changed from degrees into radians using the `convert(units)` command.

```
> L:=50: w:=25: Theta:=25:
> theta:=convert(Theta,units,degree,radian); #radians
```

$$\theta := \frac{5\pi}{36}$$

Since the pull force is constant, the work is equal to the component of P in the direction of the displacement times the total displacement L of the toboggan. This work is now calculated to four significant figures using `evalf( ,4)`, and the energy units are included in the result. Kevin does 635.5 Joules of work.

```
> work:=evalf(P*cos(phi)*L,4); #Joules
```

$$work := 635.5$$

Although not requested, it is instructive to plot the ratio of the pull P at an arbitrary angle ϕ to the pull when the angle $\phi = 0$ (i.e., a pull parallel to the slope). First, ϕ must be freed from the numerical value obtained earlier. This can be accomplished with the `unassign` command.

```
> unassign('phi'):
```

Using `denom(P)` to extract the denominator of P, the force ratio is now calculated and then plotted in Figure 2.1 over the range $\phi = 0$ to 0.25 radians.

```
> force_ratio:=eval(denom(P),phi=0)/denom(P);
```

$$force\_ratio := \frac{1.}{\cos(\phi) + 0.1\sin(\phi)}$$

```
> plot(force_ratio,phi=0..0.25,tickmarks=[3,3],thickness=2,
> labels=["phi","P"]);
```

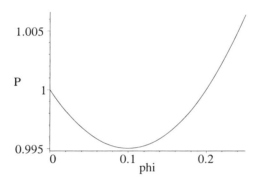

Figure 2.1: Ratio of the pull P at an arbitrary angle ϕ to that at $\phi = 0$.

From the above figure, we see that the minimum value of P is about 0.995 that exerted at $\phi = 0$. So, the reader who guessed that the minimum pull would occur at $\phi = 0$ should not feel too badly. Further note that the force ratio remains very close to 1 over a fairly wide angular range.

2.1.2 Will Jack Fall Down and Break His Crown?

Jack fell down and broke his crown, And Jill came tumbling after.
Mother Goose, children's nursery rhyme

Having reached the top of the hill, Kevin and Ruth intend to toboggan down the other side, which is characterized by two different segments of different slopes and slightly different snow conditions. Our goal in this recipe is to animate their motion down the hill. For simplicity, air resistance will be neglected and the acceleration g due to gravity taken to be 9.8 m/s². The loaded toboggan

2.1. MOTION INVOLVING CONSTANT FORCES

is assumed to start essentially from rest. We shall use parameter values characteristic of the hill that Kevin and Ruth are using, but the reader can adjust these numbers to his or her own hill. Remember, however, that if the speed gets too large, air resistance should not be neglected. We shall consider a wide variety of velocity-dependent air resistance examples in the Entrees.

We begin by loading the plots library package, because we shall be using the pointplot and display commands. The value of g is also entered.

> `restart: with(plots): g:=9.8:`

The first segment of the hill is $L1=100$ m long, makes an angle of 10° or $\theta_1 = 10\pi/180$ radians with the horizontal, and has a coefficient of kinetic friction $\mu 1 = 0.1$. The second segment is $L2=150$ m long, makes an angle of 5° or $\theta_2 = 5\pi/180$ radians with the horizontal, and because it is in the shade, has a slightly lower frictional coefficient, $\mu 2 = 0.08$.

> `L1:=100: theta1:=evalf(10*Pi/180): mu1:=0.1:`
> `L2:=150: theta2:=evalf(5*Pi/180): mu2:=0.08:`

Letting m be the mass in kg of the toboggan plus occupants, the normal force N exerted by the hill on the toboggan is given by $N = mg\cos\theta$. The net force F parallel to the slope on the loaded toboggan is equal to the difference between the weight component $mg\sin\theta$ down the hill and the frictional force μN up the hill. By Newton's second law, the acceleration of the toboggan down the hill then is $a = F/m$. The following functional operator a will calculate the acceleration when the values of μ and θ are specified. Similarly functional operators are created for calculating the velocity v and the distance s traveled when the initial velocity $v0$, acceleration a and the time t are specified.

> `a:=(mu,theta)->g*sin(theta)- mu*g*cos(theta); #acceleration`

$$a := (\mu, \theta) \to g\sin(\theta) - \mu g \cos(\theta)$$

> `v:=(v0,a,t)->v0+a*t:  #velocity`
> `s:=(v0,a)->v0*t+1/2*a*t^2:  #distance`

For the first leg of the run, we now determine the acceleration $a1$ of the toboggan and (taking $v0=0$) the distance $s1$ traveled in t seconds.

> `a1:=a(mu1,theta1); s1:=s(0,a1); #1st leg`

$$a1 := 0.7366405431$$
$$s1 := 0.3683202716\, t^2$$

Setting the distance $s1$ equal to $L1$ and numerically solving for the time,

> `T1:=fsolve(s1=L1,t=0..100); #time to complete 1st leg`

$$T1 := 16.47734325$$

it takes about 16.5 seconds to complete the first leg. The speed at the end of this leg is now calculated. This will be the input speed for the next segment.

> `v20:=v(0,a1,T1); #speed at end of 1st leg`

$$v20 := 12.13787908$$

54   CHAPTER 2. NEWTONIAN MECHANICS

The toboggan is traveling at about 12 m/s or, on using the `convert(units)` command to convert from m/s to km/h, about 44 km/h.

> `V[20]:=convert(v20,units,m/s,km/h);`

$$V_{20} := 43.69636469$$

In the following command line, we calculate the acceleration and the distance traveled in t seconds for the second leg. The initial velocity for this leg is $v20$.

> `a2:=a(mu2,theta2); s2:=s(v20,a2); #second leg`

$$a2 := 0.0731096358$$
$$s2 := 12.13787908\,t + 0.03655481790\,t^2$$

Setting $s2=L2$ and solving for the time, it takes about $T2=12$ s to cover the second leg. The total time $T1 + T2$ for the toboggan run is 28.4 s.

> `T2:=fsolve(s2=L2,t=0..100); #time to complete 2nd leg`

$$T2 := 11.92941905$$

> `total_time:=T1+T2;`

$$total\_time := 28.40676230$$

The final velocity on competing the second leg is about 13 m/s, or 47 km/hour.

> `velocity_final:=v(v20,a2,T2);`

$$velocity\_final := 13.01003456$$

> `V[f]:=convert(velocity_final,units,m/s,km/h);`

$$V_f := 46.83612442$$

The toboggan's motion can now be animated. Taking `spacing=4` will produce a graph for every fourth time step. The `step` size is taken to be 0.05 seconds. The toboggan will be represented by a circle sliding down the two straight-line hill segments. To create the illusion of sliding on a surface, a height `h=0.45` is added to the vertical coordinate of the toboggan.

> `spacing:=4: step:=0.05: h:=0.45:`

A `pointplot` command is used to generate a graph `gr` of the two connected segments of hill as well as in the functional operator `PP` to plot the toboggan's position. In `gr`, the option `style=line` connects the plotting points, which are entered as a list of lists, with straight lines. If you are a purist and object to our artistic whimsy in coloring the snow green, feel free to change the snow's color. In PP, the position of the toboggan is represented by a size 16 circle. The x and y coordinates of the toboggan must be given and the color specified.

> `gr:=pointplot([[0,0],[L1*cos(theta1),-L1*sin(theta1)],`
> `[L1*cos(theta1)+L2*cos(theta2),-L1*sin(theta1)`
> `-L2*sin(theta2)]],style=line,color=green,thickness=3):`
> `PP:=(x,y,c)->pointplot([x,y],symbol=circle,symbolsize=16,`
> `color=c):`

The following do loop generates graphs of the toboggan's position at successive times for the first leg. The do loop is allowed to run from $i=0$ to 1000. The time

2.1. MOTION INVOLVING CONSTANT FORCES

t on the ith step is calculated. A conditional if...then...end if statement is included to break the loop if $t > T1$, the time for the first leg.

```
>    for i from 0 to 1000 do
>    t:=step*i;
>    if t > T1 then break end if;
```

The coordinates $x[i]$, $y[i]$ of the toboggan on the ith step are calculated, h being added to the y coordinate. These coordinates are entered, along with the color red, in PP to plot the toboggan's position as a red circle. In d[i], the graphs PP and gr are displayed together.

```
>    x[i]:=s1*cos(theta1); y[i]:=-s1*sin(theta1)+h:
>    d[i]:=display(PP(x[i],y[i],red),gr);
```

The values of i, $x[i]$, and $y[i]$ are recorded, the final values being assigned the names N, $s1x$, and $s1y$, respectively.

```
>    N:=i: s1x:=x[i]; s1y:=y[i];
>    end do:    # animation of 1st leg
```

Incorporating the values N, $s1x$, $s1y$, produced by the first loop, and coloring the toboggan blue, the structure of the second do loop is similar to the first.

```
>    for j from 0 to 1000 do
>    t:=step*j;
>    if t > T2 then break end if;
>    x[N+j]:=s1x+s2*cos(theta2); y[N+j]:=s1y-s2*sin(theta2):
>    d[N+j]:=display(PP(x[N+j],y[N+j],blue),gr);
>    N2:=N+j;
>    end do:    #animation of second leg
```

With the necessary graphs created, executing the following display line

```
>    display(seq({d[spacing*i]},i=1..N2/spacing),insequence=true,
>    tickmarks=[3,3]);#unconstrained plot
```

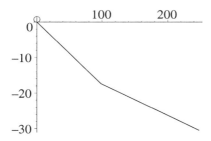

Figure 2.2: Kevin and Ruth's toboggan run.

will produce the toboggan run shown in Figure 2.2 with the toboggan represented as a red circle at the top of the run. Because the angles of the slopes have been taken to be small, the plot is not constrained. This results in the vertical scale being exaggerated compared to the horizontal scale. Because of the `insequence=true` option, clicking on the computer plot, and on the start arrow in the tool bar, causes the circle (toboggan) to slide down the hill. The circle color changes to blue as the toboggan traverses the second leg of the hill. The simulation stops when the circle reaches the end of this leg because of the `break` command in the second do loop.

Having already reached 47 km/h at this point, one can only hope that the hill flattens out so that Kevin and Ruth do not tumble off and break their crowns like the nursery rhyme characters, Jack and Jill.

2.1.3 Mr. X's New Ride

It takes a good deal of physical courage to ride a horse. This...
I have. I get it at about forty cents a flask, and take it as required.
Stephen Leacock, Canadian humorist, economist, *Reflections on Riding* (1910)

Mr. X, who owns the Metropolis Amusement Park and has been previously

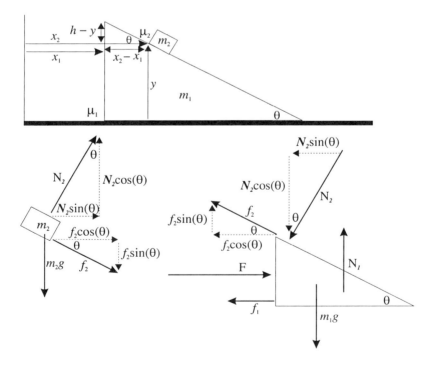

Figure 2.3: Top: Geometry for the new ride. Bottom: Free body diagrams.

2.1. MOTION INVOLVING CONSTANT FORCES

involved in creating some hair-raising rides, has an idea for a possible new ride. It is still in the early stages of formulation, but the essence of the first part of the ride can be summarized in the physicist's normal dry style by the geometry shown in Figure 2.3.

The rider's cage is represented as a small block of mass m_2 initially at the bottom of an inclined plane of mass m_1, slope angle θ, and maximum height h. By applying a horizontal force F to the left-hand edge of the inclined plane, the plane will be accelerated to the right and the cage (m_2) will accelerate up the inclined plane. The coefficient of friction between the inclined plane and the horizontal surface is μ_1 and between the cage and the inclined plane is μ_2. As consultants to Mr. X, our goal is to determine the accelerations of the inclined plane and the cage and to animate their motions. What happens to the cage when it reaches the top of the inclined plane is still to be figured out, but Mr. X muses that maybe it will be grabbed by a hook and swung in some complicated manoeuvre. Let's leave this second phase of the ride for Mr. X to mull over and address ourselves to solving the first part.

Because this example is considerably more complicated than in the previous two recipes, free body diagrams have been given in Figure 2.3 for the mass m_2 (bottom left) and the inclined plane (bottom right). Note that by Newton's third law, the frictional force f_2 exerted by m_2 on m_1 is equal and opposite to the frictional force exerted by m_1 on m_2. Similarly, the normal force N_2 exerted on m_1 by m_2 is equal and opposite to the normal force exerted on m_2 by m_1. To solve this problem, we choose a non-accelerating (inertial) frame to measure distances from. At some instant in time, the leading edge of the block m_2 is located at a vertical height y above the horizontal surface and a horizontal distance x_2 from the left-hand edge of the geometrical diagram. The trailing edge of the inclined plane is located a distance x_1 from the same reference line. The various forces in each free body diagram are then resolved into components parallel and perpendicular to the horizontal surface.

```
> restart: with(plots):
```

Acting on m_2 are the normal force N_2 exerted by the top surface of the inclined plane, the pull of gravity $m_2 g$, and the frictional force f_2 exerted on m_2 by the accelerating inclined plane. The relation of the frictional force to the normal force is entered.

```
> f[2]:=mu[2]*N[2];
```

$$f_2 := \mu_2 N_2$$

Referring to the free body diagram and using Newton's second law, the vertical and horizontal accelerations, a_{2y} and a_{2x}, of m_2 are related to the net vertical and horizontal force components by the following two command lines.

```
> vert2:=N[2]*cos(theta)-f[2]*sin(theta)-m[2]*g=m[2]*a[2*y];
```

$$vert2 := N_2 \cos(\theta) - \mu_2 N_2 \sin(\theta) - m_2 g = m_2 a_{2y}$$

```
> hor2:=N[2]*sin(theta)+f[2]*cos(theta)=m[2]*a[2*x];
```

$$hor2 := N_2 \sin(\theta) + \mu_2 N_2 \cos(\theta) = m_2 a_{2x}$$

Now, let's look at the inclined plane. It is not accelerating in the vertical direction, so the vertical force components on m_1 must balance. Therefore, from the right-hand free body diagram, the normal force N_1 is given by the following expression.

```
> N[1]:=m[1]*g+N[2]*cos(theta)-f[2]*sin(theta); #vert1
```
$$N_1 := m_1 g + N_2 \cos(\theta) - \mu_2 N_2 \sin(\theta)$$

The frictional force f_1 exerted by the horizontal surface on the bottom of the inclined plane is entered, the normal force N_1 being automatically substituted.

```
> f[1]:=mu[1]*N[1];
```
$$f_1 := \mu_1 \left(m_1 g + N_2 \cos(\theta) - \mu_2 N_2 \sin(\theta) \right)$$

Finally, Newton's second law is entered for the horizontal acceleration a_x of the inclined plane.

```
> hor1:=F-N[2]*sin(theta)-f[1]-f[2]*cos(theta)=m[1]*a[x];
```
$$hor1 := F - N_2 \sin(\theta) - \mu_1 \left(m_1 g + N_2 \cos(\theta) - \mu_2 N_2 \sin(\theta) \right) - \mu_2 N_2 \cos(\theta)$$
$$= m_1 a_x$$

From Figure 2.3, one has the geometrical relation $h-y = (x_2-x_1) \tan \theta$. Noting that h and θ are fixed, mentally differentiating this relation twice with respect to time yields the following constraint equation, relating the three acceleration components.

```
> accel_constraint:=-a[2*y]=(a[2*x]-a[x])*tan(theta);
```
$$accel\_constraint := -a_{2y} = (a_{2x} - a_x) \tan(\theta)$$

In the next three command lines, the equations $vert2$, $hor2$, and $hor1$ are solved for a_{2y}, a_{2x}, and a_x, respectively.

```
> a[2*y]:=solve(vert2,a[2*y]);
```
$$a_{2y} := \frac{N_2 \cos(\theta) - \mu_2 N_2 \sin(\theta) - m_2 g}{m_2}$$

```
> a[2*x]:=solve(hor2,a[2*x]);
```
$$a_{2x} := \frac{N_2 \left(\sin(\theta) + \mu_2 \cos(\theta) \right)}{m_2}$$

```
> a[x]:=solve(hor1,a[x]);
```
$$a_x := \frac{F - N_2 \sin(\theta) - \mu_1 m_1 g - \mu_1 N_2 \cos(\theta) + \mu_1 \mu_2 N_2 \sin(\theta) - \mu_2 N_2 \cos(\theta)}{m_1}$$

If the acceleration constraint equation were entered, the three acceleration expressions would be automatically substituted yielding a lengthy expression involving the normal force N_2. This constraint equation can be solved for N_2.

```
> N[2]:=solve(accel_constraint,N[2]);
```
$$N_2 := m_2 \left(m_1 g + \tan(\theta) F - \tan(\theta) \mu_1 m_1 g \right) \Big/ \big(m_1 \cos(\theta) - m_1 \mu_2 \sin(\theta)$$
$$+ \tan(\theta) m_1 \sin(\theta) + \tan(\theta) m_1 \mu_2 \cos(\theta) + \tan(\theta) m_2 \sin(\theta)$$
$$+ \tan(\theta) m_2 \mu_1 \cos(\theta) - \tan(\theta) m_2 \mu_1 \mu_2 \sin(\theta) + \tan(\theta) m_2 \mu_2 \cos(\theta) \big)$$

2.1. MOTION INVOLVING CONSTANT FORCES

The N_2 expression is automatically substituted into the accelerations a_{2y}, a_{2x}, and a_x. These acceleration expressions can be simplified by subsequently entering `a[2*y]`, `a[2*x]`, and `a[x]` as arguments in the arrow operator `A`.

```
> A:=a->simplify(a,symbolic):
> a[2*y]:=A(a[2*y]); a[2*x]:=A(a[2*x]); a[x]:=A(a[x]);
```

On executing the last command line, the reader will observe that the lengthy expressions for the accelerations are all completely expressed in terms of known quantities. To obtain these analytic expressions with pen and paper would have been quite an arduous task.

To animate the inclined plane and the cage, we shall now choose some nominal values for the various parameters. Here in this scaled model simulation we will take $\mu_2 = 0.5$, $\mu_1 = 0.1$, $m_1 = 10$ kg, $m_2 = 2$ kg, $g = 10$ m/s², and an angle θ of 30° or $\pi/6$ radians. The applied force $F = 300$ Newtons, the initial vertical height of m_2 is $y02=0$, the height $h=2$ m, $x01=0$, and $x02=h/\tan\theta$. The initial velocities of both the cage and the inclined plane are taken to be zero. The quantity b is a length parameter used in drawing the edges of the block m_2 in the animation.

```
> mu[2]:=0.5: mu[1]:=0.1: m[1]:=10: m[2]:=2: g:=10:
> theta:=evalf(Pi/6): F:=300: y02:=0: x01:=0: h:=2:
> x02:=h/tan(theta): V02:=0: V01:=0: b:=.1*h:
```

With all the parameter values given, the acceleration components are automatically calculated in the following command line. The horizontal acceleration a_x of m_1 (inclined plane) is 24.8 m/s², while the vertical and horizontal accelerations (a_{2y}, a_{2x}) of m_2 (cage) are about 3 m/s² and 19.7 m/s², respectively.

```
> a[x]:=a[x]; a[2*y]:=a[2*y]; a[2*x]:=a[2*x];
```

$$a_x := 24.80966844$$
$$a_{2y} := 2.976629499$$
$$a_{2x} := 19.65399491$$

Using the following functional operator S for calculating the distance traveled in time t, given the initial position $S0$, initial velocity $V0$, and acceleration a, the horizontal positions $x2$ of m_2, and $x1$ of m_1, at time t are determined as well as the vertical position y of m_2.

```
> S:=(S0,V0,a)->S0+V0*t+a*t^2/2:
> x2:=S(x02,V02,a[2*x]); x1:=S(x01,V01,a[x]); y:=S(y02,V02,a[2*y]);
```

$$x2 := 3.464101613 + 9.826997455\, t^2$$
$$x1 := 12.40483422\, t^2$$
$$y := 1.488314750\, t^2$$

The time tf for the cage to reach the top of the inclined plane is calculated by solving $y=h$ for t and applying `radsimp` to obtain a positive answer.

```
> tf:=radsimp(solve(y=h,t));
```

$$tf := 1.159224643$$

The time is about 1.16 seconds. To generate a reasonably smooth animation, this time is now divided in step into $N=100$ equal time steps.

```
> N:=100: step:=tf/N:
```

The following do loop creates $N + 1$ plots of the block m_2 and the inclined plane at each of the i time steps. In the plot command p[i], the coordinates of the four corners of the rectangular block (whose size is controlled by the parameter b) are chosen so that the block's bottom edge (created by choosing style=line) lies flush on the incline and the four edges form right angles at the corners. Similarly, in p2[i], the coordinates of the three corners of the inclined plane are entered and, using a line style and the filled=true option, the interior of the inclined plane is given a uniform aquamarine hue. The plots are superimposed on each time step in d[i], and the do loop ended.

```
> for i from 0 to N do:
> t:=step*i:
> p[i]:=plot([[x2,y],[x2-b*cos(theta),y+b*sin(theta)],[x2-b
> *cos(theta)+b*sin(theta),y+b*sin(theta)+b*cos(theta)],[x2+b
> *sin(theta),y+b*cos(theta)],[x2,y]],style=line,thickness=3):
> p2[i]:=plot([[x1,0],[x1,h],[x1+x02,0],[x1,0]],style=line,
> color=aquamarine,filled=true,axes=normal):
> d[i]:=display([p2[i],p[i]]);
> end do:
```

The following interface command is used to create a picture which fills the entire viewing window on the computer screen.

```
> interface(plotdevice=window):
```

On executing the following display command, a colored rendition of Figure 2.4 will appear on the computer screen. On clicking on the play arrow in the tool bar, the inclined plane will accelerate to the right and the block (cage) will accelerate up the inclined plane.

```
> display(seq(d[i],i=0..N-3),insequence=true,
> scaling=constrained,view=[0..x2-4+h/tan(theta),0..h+b]);
```

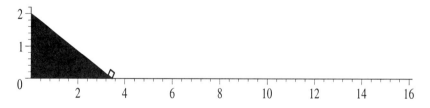

Figure 2.4: Beginning of animation of Mr. X's new ride.

2.1. MOTION INVOLVING CONSTANT FORCES

On finishing the animation, one should close the picture window and execute the following `interface` command line to return to the normal sort of picture viewing window.

> `interface(plotdevice=default):`

Whether Mr. X's proposed new ride is a good one will depend on what he has up his sleeve for the next part of the ride. However, there is no doubt that solving the first part of the ride has involved a non-trivial application of Newton's second and third laws.

2.1.4 This Governor Is Not a Politician

He knows nothing and he thinks he knows everything.
That points clearly to a political career.
George Bernard Shaw, Anglo-Irish playwright, critic (1856–1950)

In this recipe, the governor with which we will be dealing is fortunately not of

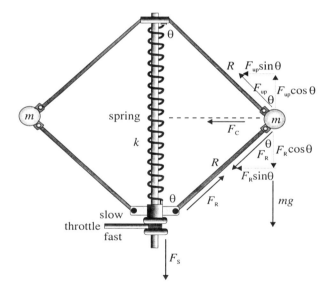

Figure 2.5: A governor for controlling the speed of a motor.

the political variety, but rather is a mechanical device for automatically controlling the speed of an engine or motor by regulating the intake of fuel. Figure 2.5 shows one type of governor which can be used to control the speed of a motor. The motor turns the central vertical shaft, causing each heavy metal ball of mass m to move in a circle in the horizontal plane with speed v. The four connecting rods, each of length R and making an angle θ with the shaft, are assumed to be light compared to the balls. The bottom two rods are connected to a flange which can move up and down and control the motor's throttle and therefore the fuel intake. The spring on the shaft has an unstretched length of $2R$, so that when θ is greater than zero, the spring is compressed. The spring has a spring constant k and is assumed to obey Hooke's linear force law.

So how does this governor control the speed? If the motor speeds up, the heavy metal balls move outwards and the connecting rods lift the movable flange upwards. This flange moves the throttle in such a way that the fuel intake is reduced and the motor is slowed down. On the other hand, if the motor slows down, the balls move inwards and the compressed spring expands, thus pushing the flange downwards and increasing the speed of the motor.

For this governor, we would like to derive an analytic expression for the speed v as a function of the angle θ and then plot this expression for $m = 5$ kg, $k = 200$ N/m, $R = 1$ m, and a gravitational acceleration $g = 10$ m/s^2. If the throttle is set to run with a spring compression of 20% of its full length, what is the speed v of the governor and the angle θ?

When the angle is θ, the two balls move in a circle of radius $r = R\sin(\theta)$.

```
> restart:
> r:=R*sin(theta);
```

$$r := R\sin(\theta)$$

For fixed θ, the vertical force components acting on a ball (e.g., the right one in Figure 2.5) must balance. Referring to the force diagram in the figure, this yields the following equation, labeled F_{vert}.

```
> F[vert]:=F[up]*cos(theta)=F[R]*cos(theta)+m*g;
```

$$F_{vert} := F_{up}\cos(\theta) = F_R\cos(\theta) + mg$$

Since each ball is moving in a circle, the centripetal force F_C acting on a ball must be the sum of the horizontal contributions from F_{up} and F_R. This produces the equation F_{hor}.

```
> F[hor]:=F[C] =F[up]*sin(theta)+F[R]*sin(theta);
```

$$F_{hor} := F_C = F_{up}\sin(\theta) + F_R\sin(\theta)$$

The equations F_{vert}, F_{hor} are solved for F_{up} and F_R and the solution assigned.

```
> sol:=solve({F[vert],F[hor]},{F[up],F[R]});
```

$$sol := \left\{ F_R = \frac{1}{2}\frac{F_C\sin(\theta)^2 - F_C + \cos(\theta)\,m\,g\sin(\theta)}{(\sin(\theta)^2 - 1)\sin(\theta)}, \right.$$
$$\left. F_{up} = \frac{1}{2}\frac{F_C\sin(\theta)^2 - F_C - \cos(\theta)\,m\,g\sin(\theta)}{(\sin(\theta)^2 - 1)\sin(\theta)} \right\}$$

```
> assign(sol):
```

In equilibrium the spring force F_S is equal to twice the vertical component $F_R\cos\theta$ of F_R since there are two connecting rods to the bottom flange. This expression is entered in F_S and simplified.

```
> F[S]:=simplify(2*F[R]*cos(theta));
```

2.1. MOTION INVOLVING CONSTANT FORCES

$$F_S := -\frac{-\cos(\theta)\,F_C + m g \sin(\theta)}{\sin(\theta)}$$

The centripetal force F_C is related to the ball speed v by the well-known relation $F_C = mv^2/r$. The radius r is automatically substituted in F_C.

```
> F[C]:=m*v^2/r; #centripetal force
```

$$F_C := \frac{m v^2}{R \sin(\theta)}$$

When the angle is θ, the spring is compressed by a length $\epsilon = 2R - 2R\cos\theta$.

```
> epsilon:=(2*R-2*R*cos(theta));
```

$$\varepsilon := 2R - 2R\cos(\theta)$$

By Hooke's law, the spring force $F_S = k\epsilon$. The expression for F_S is automatically subsituted in the following equation, labeled Fs

```
> Fs:=F[S]=k*epsilon;
```

$$Fs := -\frac{-\dfrac{\cos(\theta)\,m v^2}{R\sin(\theta)} + m g \sin(\theta)}{\sin(\theta)} = k\,(2R - 2R\cos(\theta))$$

The equation Fs is solved for the speed v in terms of the angle θ.

```
> v:=sqrt(solve(Fs,v^2));
```

$$v := \sqrt{\frac{R\sin(\theta)^2\,(m g + 2 k R - 2 k R \cos(\theta))}{\cos(\theta)\,m}}$$

If the throttle is set to run with a spring compression of 20% of its full length, then the spring force $F_S = 0.2\,k\,(2R)$. Simplifying F_S symbolically, this equation is entered in eq, which is then solved for the angle θ.

```
> eq:=simplify(F[S],symbolic)=0.2*k*(2*R);
```

$$eq := -2\,k\,R\,(-1 + \cos(\theta)) = 0.4\,k\,R$$

```
> Theta:=solve(eq,theta);
```

$$\Theta := 0.6435011088$$

In this case the angle is about 0.64 radians, or on converting to degrees,

```
> Theta2:=convert(Theta,units,radian,degree);
```

$$\Theta 2 := 36.86989764$$

about 37°. The given parameter values are now entered.

```
> m:=5: k:=200: R:=1: g:=10:
```

Taking the angle $\theta = \Theta$, the "running" speed of the balls is calculated

```
> V[running]:=eval(v,theta=Theta);
```

$$V_{running} := 3.420526275$$

and found to be 3.42 m/s. To get a feeling for how this speed varies with angle, let's convert the angle θ in radians to an angle in degrees.

```
> degrees:=theta*Pi/180;
```

The speed of the balls is now expressed in terms of θ, given in degrees.

```
> speed:=subs(theta=degrees,v);
```

$$speed := \sqrt{-\frac{1}{5} \frac{\sin(\frac{\theta \pi}{180})^2 (-450 + 400\cos(\frac{\theta \pi}{180}))}{\cos(\frac{\theta \pi}{180})}}$$

The speed is now plotted over the angular range $\theta = 0$ to $90°$, the viewing range being adjusted to produce a good picture. If the view is not specified, the default vertical range turns out to be very large.

```
> plot(speed,theta=0..90,labels=["angle","speed"],
> view=[0..90,0..100]);
```

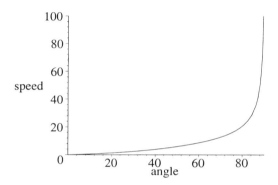

Figure 2.6: Speed v as a function of the angle in degrees.

2.2 Energy and Momentum

When the resultant force acting on a body is zero, then (from Newton's second law) the time rate of change of (linear) momentum is zero and therefore the momentum of the body remains constant. If the external forces acting on a body are conservative, then the total mechanical energy remains constant. In the next four recipes, we make use of these two important conservation laws.

2.2.1 Amazing But True

Life is amazing: and the teacher had better prepare [her]self to be a medium for that amazement.
Edward Blishen, *Donkey Work* (1983)

For the introductory mechanics course that she is currently teaching, Jennifer has created an interesting computer algebra recipe on elastic collisions, an elastic collision being one in which the kinetic energy is conserved. In particular, in this recipe she will confirm and explore two "amazing but true" results.

2.2. ENERGY AND MOMENTUM

First, Jennifer will consider a moving particle a (mass m_a) making a perfectly elastic collision with a second stationary particle b (mass m_b) along the line of their centers. She will show graphically and analytically that the ratio of masses, $x = m_b/m_a$, which maximizes the fraction of kinetic energy transferred to b is $x = 1$. Next, if the mass ratio does not have this value, Jennifer will graphically and analytically show that the amount of energy transferred to b can be increased by inserting a third stationary particle c (mass m_c) between a and b. For maximum energy transfer, one must have $m_c = \sqrt{m_a m_b}$.

Jennifer enters the relation $m_b = x m_a$, using square brackets to attach subscripts to the masses. Her first goal is to show that $x = 1$ maximizes the kinetic energy transfer from particle a to particle b.

```
> restart:
> m[b]:=x*m[a];
```
$$m_b := x\, m_a$$

The energy conservation relation is entered. Here E_{ai} is the initial total energy of particle a before the collision, while E_{af} is its final energy after the collision. Particle b initially has no energy (it's at rest) and has a final energy E_{bf}.

```
> energy:=E[ai]=E[af]+E[bf];
```
$$energy := E_{ai} = E_{af} + E_{bf}$$

Letting v_{ai}, v_{af} be the initial and final velocities of particle a, and v_{bf} the final velocity of particle b, the 1-dimensional momentum conservation (since the resultant force on the two-particle system is zero) statement is entered.

```
> momentum:=m[a]*v[ai]=m[a]*v[af]+m[b]*v[bf];
```
$$momentum := m_a\, v_{ai} = m_a\, v_{af} + x\, m_a\, v_{bf}$$

The mass m_a is eliminated by dividing *momentum* by m_a and simplifying.

```
> momentum2:=simplify(momentum/m[a]);
```
$$momentum2 := v_{ai} = v_{af} + x\, v_{bf}$$

For an elastic collision, $E = (1/2)\, m\, v^2$, so $v = \sqrt{2}\, \sqrt{E/m}$. Jennifer enters this relation in v_{ai}, v_{af}, and v_{bf}. The expressions v_{ai}, etc., will be automatically substituted into *momentum2*.

```
> v[ai]:=sqrt(2*E[ai]/m[a]); v[af]:=sqrt(2*E[af]/m[a]);
> v[bf]:=sqrt(2*E[bf]/m[b]);
```
$$v_{ai} := \sqrt{2}\,\sqrt{\frac{E_{ai}}{m_a}}$$

Multiplying *momentum2* by $\sqrt{m_a/2}$ and symbolically simplifying, the momentum conservation relation will now be expressed in terms of the initial and final energies of particles a and b and the still-to-be-determined parameter x.

```
> momentum3:=simplify(sqrt(m[a]/2)*momentum2,symbolic);
```
$$momentum3 := \sqrt{E_{ai}} = \sqrt{E_{af}} + \sqrt{E_{bf}}\,\sqrt{x}$$

Jennifer analytically solves the *energy* and *momentum3* equations for the final energies E_{af} and E_{bf}, generating two answers in *sol*.

```
> sol:=solve({energy,momentum3},{E[af],E[bf]});
```
$$sol := \{E_{bf} = 0,\ E_{af} = E_{ai}\},\ \{E_{bf} = \frac{4\,E_{ai}\,x}{(1+x)^2},\ E_{af} = \frac{E_{ai}\,(1 - 2\,x + x^2)}{(1+x)^2}\}$$

Jennifer is interested in the answer for which the final energy E_{bf} of particle b depends on x. In *sol1*, she uses the command syntax select(has,e,x), where the expression e is {sol[1],sol[2]}, to select the desired answer. The operand(s) in e which has x is selected. The output would be in the form of a set of sets, {{ }}. The first, and only, member of the outer set is extracted by entering [1] after the select command.

```
> sol1:=select(has,{sol[1],sol[2]},x)[1];
```
$$sol1 := \{E_{bf} = \frac{4\,E_{ai}\,x}{(1+x)^2},\ E_{af} = \frac{E_{ai}\,(1 - 2\,x + x^2)}{(1+x)^2}\}$$

E_{bf} and E_{af} can be similarly selected from *sol1*. Applying the right-hand side (rhs) command, the results are assigned the names E_{bf1} and E_{af1}.

```
> E[bf1]:=rhs((select(has,sol1,E[bf])[1]));
```
$$E_{bf1} := \frac{4\,E_{ai}\,x}{(1+x)^2}$$

```
> E[af1]:=rhs(select(has,sol1,E[af])[1]);
```
$$E_{af1} := \frac{E_{ai}\,(1 - 2\,x + x^2)}{(1+x)^2}$$

To graphically determine the x value which maximizes the kinetic energy transfer from a to b, Jennifer plots the ratios E_{bf1}/E_{ai} and E_{af1}/E_{ai} in Fig. 2.7, choosing different colors and line styles for the two curves.

```
> plot([E[bf1]/E[ai],E[af1]/E[ai]],x=0..3,color=[red,blue],
> linestyle=[1,3],labels=["x","ratio"],thickness=2);
```

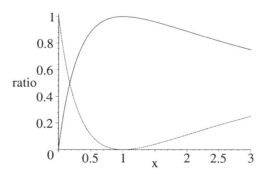

Figure 2.7: Top curve: E_{bf1}/E_{ai}. Bottom curve: E_{af1}/E_{ai}.

From the above figure, it appears that particle a transfers all of its kinetic energy to b when $x = 1$. Jennifer now confirms this by differentiating E_{bf1} with

2.2. ENERGY AND MOMENTUM

respect to x, setting the result equal to zero, and solving *eq1* for x.

> `eq1:=diff(E[bf1],x)=0;`

$$eq1 := \frac{4\,E_{ai}}{(1+x)^2} - \frac{8\,E_{ai}\,x}{(1+x)^3} = 0$$

> `xmax:=solve(eq1,x);`

$$xmax := 1$$

Indeed the value of x which maximizes the kinetic energy transfer is $xmax = 1$.

Now Jennifer examines the situation when a stationary particle c is inserted between a and b. In the analysis, she considers a two-collision process, a with c and then c with b. For the first collision, she sets $y = m_c/m_a$ as the ratio of the masses of particles c and a. She can use the earlier result for E_{bf1} by substituting $x=y$ in E_{bf1} and labeling the "final" energy of c as E_{cf}.

> `E[cf]:=subs(x=y,E[bf1]);`

$$E_{cf} := \frac{4\,E_{ai}\,y}{(1+y)^2}$$

In the subsequent collision of c with b, E_{cf} will be the "initial" kinetic energy of c. Letting $z = m_b/m_c$ and replacing x with z and E_{ai} with E_{cf} in E_{bf1}, then E_{bf2} gives the final energy of b in terms of the initial kinetic energy of a.

> `E[bf2]:=subs({x=z,E[ai]=E[cf]},E[bf1]);`

$$E_{bf2} := \frac{16\,E_{ai}\,y\,z}{(1+y)^2\,(1+z)^2}$$

But the product $y\,z = (m_c/m_a)(m_b/m_c) = m_b/m_a = x$, where x is the ratio to be determined. Jennifer therefore substitutes $z=x/y$ into E_{bf2}.

> `E[bf2]:=subs(z=x/y,E[bf2]);`

$$E_{bf2} := \frac{16\,E_{ai}\,x}{(1+y)^2\,(1+\frac{x}{y})^2}$$

In *eq2* she now differentiates E_{bf2} with respect to y. For maximum energy transfer to b, this result must be equal to zero.

> `eq2:=diff(E[bf2],y)=0;`

$$eq2 := -\frac{32\,E_{ai}\,x}{(1+y)^3\,(1+\frac{x}{y})^2} + \frac{32\,E_{ai}\,x^2}{(1+y)^2\,(1+\frac{x}{y})^3\,y^2} = 0$$

Solving *eq2* for y yields three answers.

> `eq3:=solve(eq2,y);`

$$eq3 := 0,\ \sqrt{x},\ -\sqrt{x}$$

On physical grounds, the second solution ($\sqrt{x}$) in *eq3* must be selected.

> `eq4:=y=eq3[2];`

$$eq4 := y = \sqrt{x}$$

To express this result in terms of the actual masses, the mass m_b is unassigned and the substitutions $y = m_c/m_a$ and $x = m_b/m_a$ are made in eq4.

> `unassign('m[b]'): eq5:=subs(y=m[c]/m[a],x=m[b]/m[a],eq4);`

$$eq5 := \frac{m_c}{m_a} = \sqrt{\frac{m_b}{m_a}}$$

Then eq5 is solved for m_c using the `isolate` and `radsimp` commands.

> `eq6:=isolate(eq5,m[c]);`

$$eq6 := m_c = \sqrt{\frac{m_b}{m_a}}\, m_a$$

> `eq7:=radsimp(eq6);`

$$eq7 := m_c = \sqrt{m_b\, m_a}$$

So, Jennifer finds that if particle c has the mass $m_c = \sqrt{m_a m_b}$, the kinetic energy transfer from a to b is maximized. To demonstrate that the energy transfer is larger than when only two particles are involved, she substitutes eq4 into E_{bf2} and forms and simplifies the ratio E_{bf2}/E_{bf1}.

> `E[bf2]:=subs(eq4,E[bf2]);`

$$E_{bf2} := \frac{16\, E_{ai}\, x}{(1+\sqrt{x})^4}$$

> `ratio:=simplify(E[bf2]/E[bf1]);`

$$ratio := \frac{4\,(1+x)^2}{(1+\sqrt{x})^4}$$

Jennifer plots the above ratio, obtaining Figure 2.8.

> `plot(ratio,x=0..3,view=[0..3,0.9..1.2],tickmarks=[3,3],`
> `thickness=2,labels=["x","ratio"]);`

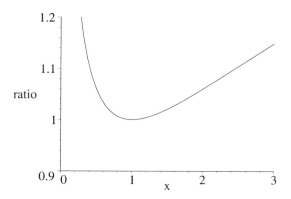

Figure 2.8: The ratio E_{bf2}/E_{bf1} as a function of x.

2.2. ENERGY AND MOMENTUM

The curve appears to have a minimum value of 1 at about $x = 1$. To confirm this, Jennifer differentiates *ratio* with respect to x, sets the result equal to zero, and numerically solves *eq8* for x over the range $x = 0..3$.

> `eq8:=diff(ratio,x)=0;`

$$eq8 := -\frac{8(1+x)^2}{(1+\sqrt{x})^5 \sqrt{x}} + \frac{8(1+x)}{(1+\sqrt{x})^4} = 0$$

> `xmin:=fsolve(eq8,x=0..3);`

$$xmin := 1.000000000$$

The minimum in the curve occurs at $xmin = 1$. Evaluating the ratio at this value of x, the minimum ratio is 1.

> `minratio:=eval(ratio,x=xmin);`

$$minratio := 1.000000000$$

Thus, provided $x \neq 1$, inserting a third particle (c) increases the kinetic energy transfer from a to b. With her recipe complete, Jennifer hopes that her students will be at least appreciative of, if not amazed by, these elastic collision results.

2.2.2 How Arthur Won the Nobel Prize

When you are courting a nice girl an hour seems like a second.
When you sit on a red-hot cinder a second seems like an hour.
That's relativity.
Albert Einstein, Nobel physics laureate, *News Chronicle* (14 March 1949)

In 1923, Arthur Holly Compton measured the wavelength shift of X-rays scattered by free electrons. The experimental results could be readily understood in terms of a particle (photon) model of light, but not in terms of a wave model, thus lending strong support to the "reality" of photons. For this pioneering work, now referred to as the "Compton effect", Arthur won the Nobel prize in

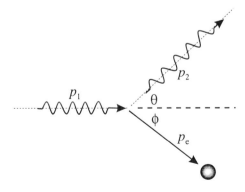

Figure 2.9: Compton effect: X-ray photon scattered by electron.

physics in 1927. We have asked Vectoria to derive the analytic expression for the wavelength shift of a photon incident on an electron initially at rest as well as an expression for the kinetic energy of the recoiling electron.

Referring to Figure 2.9, Vectoria considers an X-ray photon with initial momentum p_1 being scattered with final momentum p_2 in the direction θ by the initially stationary electron of mass m_e. The electron recoils in the direction ϕ with momentum p_e. If the incident X-ray wavelength is λ_1 and the scattered X-ray wavelength is λ_2, the incident photon momentum then is $p_1 = h/\lambda_1$ and scattered photon momentum is $p_2 = h/\lambda_2$, where h is Planck's constant. Vectoria enters these momentum expressions,

> `restart:`
> `p[1]:=h/lambda[1]; p[2]:=h/lambda[2];`

along with the momentum conservation relations in the x (horizontal) and y (vertical) directions. For convenience in solving the problem, she multiplies each term in the Px and Py equations by c, the speed of light.

> `Px:=p[1]*c=p[2]*c*cos(theta)+p[e]*c*cos(phi);`

$$Px := \frac{h\,c}{\lambda_1} = \frac{h\,c\cos(\theta)}{\lambda_2} + p_e\,c\cos(\phi)$$

> `Py:=0=p[2]*c*sin(theta)-p[e]*c*sin(phi);`

$$Py := 0 = \frac{h\,c\sin(\theta)}{\lambda_2} - p_e\,c\sin(\phi)$$

Because of the high speeds involved, the relativistic expression for the energy E of a particle of momentum p and rest mass m, namely $E = \sqrt{p^2c^2 + m^2c^4}$, must be used. Since a photon has no rest mass, the incident photon has energy $p_1 c$ and the scattered photon energy $p_2 c$. The electron initially has no momentum, so its initial energy is $m_e c^2$, while its final energy is $\sqrt{p_e^2 c^2 + m_e^2 c^4}$. Vectoria enters the energy conservation relation Et for the photon-electron collision and solves the three equations Px, Py, and Et for λ_2, p_e, and ϕ. The lengthy analytic output, involving `RootOf` expressions, is suppressed here in the text. The function `RootOf` is a place holder for representing all the roots of an equation in one variable. Consult Maple's Help for more information on this function.

> `Et:=p[1]*c+m[e]*c^2=p[2]*c+sqrt(p[e]^2*c^2+m[e]^2*c^4);`

$$Et := \frac{h\,c}{\lambda_1} + m_e\,c^2 = \frac{h\,c}{\lambda_2} + \sqrt{p_e^2\,c^2 + m_e^2\,c^4}$$

> `sol:=solve({Px,Py,Et},{lambda[2],p[e],phi});`

On assigning the solution, she extracts the expression for λ_2 and expands it.

> `assign(sol);`
> `lambda_2:=expand(lambda[2]);`

$$lambda\_2 := -\frac{h\cos(\theta)}{m_e\,c} + \frac{h}{m_e\,c} + \lambda_1$$

The wavelength shift follows on subtracting λ_1 from λ_2, and successively collecting the coefficients of h, $1/m_e$, and $1/c$ in eq.

2.2. ENERGY AND MOMENTUM

```
> eq:=lambda2-lambda1=lambda_2-lambda[1];
```
$$eq := \lambda 2 - \lambda 1 = -\frac{h\cos(\theta)}{m_e\,c} + \frac{h}{m_e\,c}$$
```
> Compton_formula:=collect(eq,[h,1/m[e],1/c]);
```
$$Compton\_formula := \lambda 2 - \lambda 1 = \frac{(1-\cos(\theta))\,h}{c\,m_e}$$

The Compton wavelength shift is then given by the above Compton formula.

Next, Vectoria extracts the expression for the momentum of the recoiling electron. Since the answer given in the *sol* involved the RootOf expression, she applies the `allvalues` command in *Pe* to obtain the analytic forms.

```
> Pe:=allvalues(p[e]);
```

Two expressions are given in the output of *Pe*, involving a positive and negative square root. Vectoria selects the positive (first) answer. The kinetic energy K_e of the recoil electron is obtained by subtracting its rest energy from its total energy, and simplifying the lengthy output with the `symbolic` option.

```
> K[e]:=sqrt(Pe[1]^2*c^2+m[e]^2*c^4)-m[e]*c^2;
> K[e]:=simplify(K[e],symbolic);
```
$$K_e := -\frac{c(-1+\cos(\theta))\,h^2}{(-h\cos(\theta)+h+m_e\,c\,\lambda_1)\,\lambda_1}$$

The incident photon has an initial frequency ν_1 related to the wavelength by $\lambda_1 = c/\nu_1$. She makes this substitution as well as the trigonometric half-angle identity $\cos\theta = 1 - 2\sin(\theta/2)^2$ in K_e.

```
> K[e]:=subs(lambda[1]=c/nu[1],
>    cos(theta)=(1-2*sin(theta/2)^2),K[e]);
```
$$K_e := \frac{2\sin(\frac{\theta}{2})^2\,h^2\,\nu_1}{-h(1-2\sin(\frac{\theta}{2})^2)+h+\frac{m_e\,c^2}{\nu_1}}$$

Vectoria calculates the ratio of the recoil electron kinetic energy to the incident photon energy, simplifying the result with the `expand` command. She assigns a name Ke/E, which is enclosed in back quotes to make it a valid assignment.

```
> 'Ke/E':=expand(K[e]/(h*nu[1]));
```
$$Ke/E := \frac{2\sin(\frac{\theta}{2})^2\,h}{2h\sin(\frac{\theta}{2})^2+\frac{m_e\,c^2}{\nu_1}}$$

Being curious about how big the wavelength shift was in Compton's experiment, Vectoria needs the values (in SI units) of Planck's constant h, the electron rest mass m_e, and the vacuum speed of light c. Rather than consulting a physics text, she loads the `ScientificConstants` package and uses the command `evalf(Constant())` to display the numerical values of h, m_e, and c.

```
> with(ScientificConstants): h:=evalf(Constant(h));
```

```
> m[e]:=evalf(Constant(m[e])); c:=evalf(Constant(c));
```
$$h := 0.662606876\,10^{-33}$$
$$m_e := 0.910938188\,10^{-30}$$
$$c := 0.299792458\,10^9$$

According to Tipler [Tip91], in Compton's experiment the incident X-rays had a wavelength $\lambda_1 = 0.0711 \times 10^{-9}$ m. She enters this number and calculates the corresponding frequency ν_1.

```
> lambda[1]:=0.0711*10^(-9): nu[1]:=c/lambda[1];#x rays
```
$$\nu_1 := 0.4216490267\,10^{19}$$

Since the wavelength shift is small, she divides the Compton formula by the incident wavelength to form a ratio and converts the result to a percentage.

```
> percent_wavelength_shift:=(rhs(Compton_formula)/lambda[1])*100;
```
$$percent\_wavelength\_shift := 3.412531948 - 3.412531948\cos(\theta)$$

Converting the photon scattering angle to degrees,

```
> theta:=Theta*Pi/180:
```

Vectoria plots the percentage wavelength shift over the angular range $\Theta = 0$ to $180°$ (backscattering). She also plots the electron kinetic energy ratio in the same figure. She colors the former curve blue, the latter curve green, and adds a title to the figure. The resulting plot is shown in Figure 2.10. The upper curve is the wavelength shift, the lower one the kinetic energy ratio. The maximum shifts in each curve are about 6%.

```
> plot([percent_wavelength_shift,'Ke/E'*100],Theta=0..180,
> color=[blue,green],tickmarks=[3,3],labels=["scattering angle",
> "percent"],title="blue=wavelength shift,green=Ke/E");
```

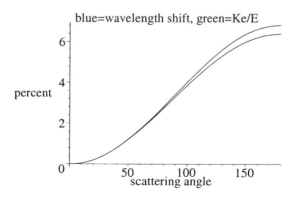

Figure 2.10: Top curve: Percentage wavelength shift. Bottom: Ke/E ratio.

2.2. ENERGY AND MOMENTUM

2.2.3 A Heck of a Wreck

If you don't like the way I drive, get off the sidewalk.
A bumper sticker

Greg Arious Nerd has been called in as an expert consultant in the police investigation of a three-car pileup at an intersection of three streets in Wen Kroy, the capital city of Erehwon. Evidently, each car arrived at the intersection simultaneously, each traveling along a different street. According to eyewitnesses, a Drof was traveling due east, a Callidac was traveling 30° north of east, and an Adnoh was traveling due north prior to the collision. After the collision, the wrecked cars remained locked together and their tires made skid marks 18 m long in a direction 45° north of east before coming to rest. According to the car manufacturers, the Drof, Callidac, and Adnoh have masses of 1400, 1500, and 1300 kg, respectively. The coefficient of kinetic friction between the locked wheels and the dry pavement was about $\mu = 0.8$. One last vital piece of information is that a reliable eyewitness estimated that the Drof was traveling about 30% faster than the Callidac just before the collision. The police have asked Greg to determine the speed of each car just before the collision. The legal speed limit is 90 km/h (25 m/s). Were any of the cars exceeding the speed limit prior to the crash? Although not requested by the police, we have asked Greg to calculate the amount of kinetic energy lost in the inelastic collision.

Being a mathematician, Greg likes to initially keep his analysis quite general, so that he can change the masses, velocities, or angles, if so desired. He takes the Drof to be car number 1, the Callidac to be number 2, and the Adnoh to be number 3. Labeling the car masses as $m1$, $m2$, $m3$, their speeds before the collision as $v1$, $v2$, $v3$, and their angles with the east (positive x) direction to be $\phi1$, $\phi2$, and $\phi3$, Greg loads the VectorCalculus package, so that he can see the momenta displayed as vectors. Greg then enters the momenta magnitudes $p1$, $p2$, $p3$ for each car and forms the momentum vectors $P1$, $P2$, $P3$.

```
>    restart: with(VectorCalculus):
>    p1:=m1*v1: p2:=m2*v2: p3:=m3*v3:
>    P1:=<p1*cos(phi1),p1*sin(phi1)>;
>    P2:=<p2*cos(phi2),p2*sin(phi2)>;
>    P3:=<p3*cos(phi3),p3*sin(phi3)>;
```

$$P1 := m1\,v1\cos(\phi1)\,\mathrm{e_x} + m1\,v1\sin(\phi1)\,\mathrm{e_y}$$
$$P2 := m2\,v2\cos(\phi2)\,\mathrm{e_x} + m2\,v2\sin(\phi2)\,\mathrm{e_y}$$
$$P3 := m3\,v3\cos(\phi3)\,\mathrm{e_x} + m3\,v3\sin(\phi3)\,\mathrm{e_y}$$

The total initial momentum Pin before the collision is calculated.

```
>    Pin:=P1+P2+P3;
```

$$Pin := (m1\,v1\cos(\phi1) + m2\,v2\cos(\phi2) + m3\,v3\cos(\phi3))\,\mathrm{e_x}$$
$$+(m1\,v1\sin(\phi1) + m2\,v2\sin(\phi2) + m3\,v3\sin(\phi3))\,\mathrm{e_y}$$

Letting vf be the speed of the joined vehicles just after the collision and θ the angular direction in which they move, the final momentum Pf is entered.

> `Pf:=(m1+m2+m3)*<vf*cos(theta),vf*sin(theta)>;`

$$Pf := (m1 + m2 + m3)\, vf \cos(\theta)\, e_x + (m1 + m2 + m3)\, vf \sin(\theta)\, e_y$$

Greg forms the momentum difference $Pf - Pin$.

> `eq:=Pf-Pin;`

$$\begin{aligned}eq := &\,((m1 + m2 + m3)\, vf \cos(\theta) - m1\, v1 \cos(\phi 1) - m2\, v2 \cos(\phi 2) \\ &- m3\, v3 \cos(\phi 3))\, e_x + ((m1 + m2 + m3)\, vf \sin(\theta) - m1\, v1 \sin(\phi 1) \\ &- m2\, v2 \sin(\phi 2) - m3\, v3 \sin(\phi 3))\, e_y\end{aligned}$$

Since the momentum is conserved in the collision, the momentum difference is zero. Greg extracts the momentum differences in the x and y directions from eq and sets them equal to zero in $eq2$ and $eq3$.

> `eq2:=eq[1]=0; eq3:=eq[2]=0;`

The final speed vf can be related to the length L of the skid marks as follows. The normal force on the joined vehicles is $N = (m1 + m2 + m3)\, g$, where g is the acceleration due to gravity. The frictional force $Fr = \mu N$.

> `N:=(m1+m2+m3)*g: Fr:=mu*N;`

$$Fr := \mu\, (m1 + m2 + m3)\, g$$

By the conservation of energy, the kinetic energy immediately after the collision must be equal to the work $Fr\, L$ done by Fr over the distance L. The energy conservation relation is entered in Ef.

> `Ef:=1/2*(m1+m2+m3)*vf^2=Fr*L;`

$$Ef := \left(\frac{m1}{2} + \frac{m2}{2} + \frac{m3}{2}\right) vf^2 = \mu\, (m1 + m2 + m3)\, g\, L$$

The final energy equation Ef is solved for vf. Application of the `radsimp` command yields the positive square root.

> `vf:=radsimp(solve(Ef,vf));`

$$vf := \sqrt{2}\,\sqrt{\mu\, g\, L}$$

Since all the parameter values in vf are known ($g = 10$ m/s^2 on Erehwon), the final speed vf just after the collision can be calculated. Greg takes the speed $v1$ to be r times the speed $v2$, where in the present case, $r = 1.30$.

> `v1:=r*v2;`

Greg then solves the set of equations $eq2$ and $eq3$ for $v2$ and $v3$ and assigns the lengthy solution sol (we suppress the output here).

> `sol:=solve({eq2,eq3},{v2,v3});`
> `assign(sol):`

Examining sol, Greg decides to create a functional operator V which will `combine` the terms in the velocity v with the trigonometric option. This result will then be factored. Then forming $V(v2)$ and $V(v3)$, the analytic forms of the velocities $v2$ and $v3$ are calculated. The velocity $v1$ is also displayed.

2.2. ENERGY AND MOMENTUM

```
> V:=v->factor(combine(v,trig)):
> v2:=V(v2); v3:=V(v3); v1:=v1;
```

$$v2 := -\frac{\sqrt{2}\sqrt{\mu g L}\sin(\phi 3-\theta)(m1+m2+m3)}{m1\,r\sin(-\phi 3+\phi 1)+m2\sin(-\phi 3+\phi 2)}$$

$$v3 := \frac{\sqrt{2}\sqrt{\mu g L}(m1+m2+m3)(m1\,r\sin(-\theta+\phi 1)+m2\sin(-\theta+\phi 2))}{(m1\,r\sin(-\phi 3+\phi 1)+m2\sin(-\phi 3+\phi 2))\,m3}$$

$$v1 := -\frac{r\sqrt{2}\sqrt{\mu g L}\sin(\phi 3-\theta)(m1+m2+m3)}{m1\,r\sin(-\phi 3+\phi 1)+m2\sin(-\phi 3+\phi 2)}$$

To determine the amount of kinetic energy lost in the inelastic collision, Greg calculates the initial kinetic energy Ki of the three cars prior to the collision.

```
> Ki:=1/2*m1*v1^2+1/2*m2*v2^2+1/2*m3*v3^2:
```

The kinetic energy change is calculated in *KChange* by subtracting Ki from the lhs of *Ef*. As authors of this text we have again overruled an outraged Greg and suppressed the lengthy output here.

```
> Kchange:=lhs(Ef)-Ki;
```

After regaining his composure, Greg continues with his calculation, now entering the given parameter values, and evaluating the speeds (to four figures) in m/s of the three cars before the collision.

```
> L:=18: mu:=0.8: g:=10: m1:=1400: m2:=1500: m3:=1300: r:=1.30:
> phi1:=0: phi2:=Pi/6: phi3:=Pi/2: theta:=Pi/4:
> v1:=evalf(v1,4); v2:=evalf(v2,4); v3:=evalf(v3,4);
```

$$v1 := 21.01$$
$$v2 := 16.16$$
$$v3 := 29.43$$

From the output, Greg notes that the third vehicle, the Adnoh, was exceeding the 25 m/s speed limit prior to the collision, while the Drof ($v1$) and Callidac ($v2$) were not speeding. Greg calculates the kinetic energy change to six figures,

```
> KEchange:=evalf(KEchange,6);
```

$$KEchange := -463546.$$

and finds that there was a loss of 463,546 Joules in the inelastic collision.

2.2.4 G Forces

Gravity is a contributing factor in nearly 73 percent of all accidents involving falling objects.
Dave Barry, American Pulitzer Prize winner (1988)

Mr. X wants to redesign his roller coaster ride and has gone to his mathematician friend Mike for help in determining the G forces that passengers would experience on the ride. The G force is the effective weight that a rider would feel and is usually expressed as a multiple of his or her normal weight. A rider experiencing, say, 5 G would have an apparent weight five times normal. If the G force becomes too large, the rider will suffer a "blackout" (a momentary lack of consciousness). Although X muses that this might be a good way of keeping the usually noisy young riders quiet, he realizes that he must keep his customers happy and also avoid lawsuits so the G forces must be kept below this threshold. According to X, a portion of the proposed roller coaster track, spanning the horizontal distance $x = -100$ to $x = 100$ meters, is described by the height function $h(x) = 0.25 \times 10^{-5} x^4 - .02 x^2 + 40$.

Let's listen in on what Mike has to say to Mr. X about the G forces. He has opened his laptop computer and is about to develop a relevant recipe for calculating the G forces, given any height function $h(x)$.

"To start out with, I will completely neglect all frictional forces. We can always create a more realistic model later. My goal here is to show you how the G forces can be calculated, given $h(x)$ for the roller coaster track. It goes without saying that the roller coaster car is assumed to stay on the track at all times. Let's create a height operator h which will calculate the height if the horizontal position x is specified. I will use the form of h that you specified."

```
> restart: with(plots):
> h:=(x)->0.25*10^(-5)*x^4-0.02*x^2+40;
```

$$h := x \to 0.2500000000\, 10^{-5}\, x^4 - 0.02\, x^2 + 40$$

"Let's first create a graph `gr1` of $h(x)$ over the given range $x = -100...100$, keeping the scaling constrained and producing a thick curve. We will display the plot later, after I have developed a formula for the curvature and calculated the radius of curvature for the curve as a function of x."

```
> gr1:=plot(h(x),x=-100..100,scaling=constrained,thickness=2):
```

"Why do we have to determine the radius of curvature, Mike? And what is the formula for calculating it?", Mr. X inquires.

"Well X, in order to determine the G force, we have to calculate the centripetal force $F = m v^2 / r$ on a passenger of mass m traveling with speed v at any point on the track. For motion in a circle, r would of course be the radius of the circle. The radius of curvature in this case would be equal to the fixed value of r. But for your $h(x)$ the radius of curvature will vary from point to point along the track and therefore so will the centripetal force on the rider. First, I will develop the formula for calculating the radius of curvature and then apply it to your $h(x)$. If ds is an element of infinitesimal length as you move along a curve and $d\theta$ is the rotation angle of the tangent to the curve in moving along ds, the radius of curvature $rc = |R|$, where $R = ds/d\theta$. The curvature is given by $1/R$ and is measure of the rate of turning of the tangent to the curve relative to the arclength s. The curvature has a positive sign when the curve is concave up and a negative sign when it is concave down. Let me enter R."

```
> R:=ds/d[theta];
```

$$R := \frac{ds}{d_\theta}$$

2.2. ENERGY AND MOMENTUM

"The arclength, $ds = \sqrt{(dx)^2 + (dy)^2} = \sqrt{1 + (dy/dx)^2}\, dx$."

```
>  ds:=sqrt(1+diff(y(x),x)^2)*dx;
```

$$ds := \sqrt{1 + (\frac{d}{dx}\mathrm{y}(x))^2}\, dx$$

"The curve's slope at a point x is related to the angle θ by $\tan\theta(x) = dy/dx$."

```
>  eq1:=tan(theta(x))=diff(y(x),x);
```

$$eq1 := \tan(\theta(x)) = \frac{d}{dx}\mathrm{y}(x)$$

"We need to determine $d\theta(x)$. So let's differentiate $eq1$ with respect to x in $eq2$, and isolate the derivative $d\theta(x)/dx$ to the left-hand side in $eq3$."

```
>  eq2:=diff(eq1,x);
```

$$eq2 := (1 + \tan(\theta(x))^2)(\frac{d}{dx}\theta(x)) = \frac{d^2}{dx^2}\mathrm{y}(x)$$

```
>  eq3:=isolate(eq2,diff(theta(x),x));
```

$$eq3 := \frac{d}{dx}\theta(x) = \frac{\frac{d^2}{dx^2}\mathrm{y}(x)}{1 + \tan(\theta(x))^2}$$

"Then the relation of $d\theta$ to dx is obtained by substituting $eq1$ into $eq3$, taking the rhs of the result, and multiplying by dx. This relation will be automatically substituted into R."

```
>  d[theta]:=rhs(subs(eq1,eq3))*dx;
```

$$d_\theta := \frac{(\frac{d^2}{dx^2}\mathrm{y}(x))\, dx}{1 + (\frac{d}{dx}\mathrm{y}(x))^2}$$

"The general mathematical expression for the curvature then is given by $1/R$."

```
>  curvature:=1/R;
```

$$curvature := \frac{\frac{d^2}{dx^2}\mathrm{y}(x)}{(1 + (\frac{d}{dx}\mathrm{y}(x))^2)^{(3/2)}}$$

"The unapply command can be used to turn an expression, such as *curvature* into a functional operator. Subsequently entering curv(y), where y is the height, the curvature at a point x on the curve will be calculated."

```
>  curv:=unapply(curvature,y);
```

$$curv := y \rightarrow \frac{\frac{d^2}{dx^2}y(x)}{(1 + (\frac{d}{dx}y(x))^2)^{(3/2)}}$$

"Taking your expression for h as the argument in *curv* produces the curvature *curv2* for your proposed roller coaster track."

```
> curv2:=curv(h);
```

$$curv2 := \frac{0.00003000000000\, x^2 - 0.04}{(1 + (0.00001000000000\, x^3 - 0.04\, x)^2)^{(3/2)}}$$

"Let's create a graph of *curv2*. Actually I will multiply *curv2* by 1000, so that it will not be too small when superimposed on the plot of $h(x)$. To distinguish it from the latter, let's color the curve green and use a thick dashed linestyle."

```
> gr2:=plot(1000*curv2,x=-100..100,color=green,linestyle=3,
> thickness=2):
```

"If you now look at my laptop screen, you will see a plot containing $h(x)$ and the corresponding curvature when I execute the following `display` command line." (For the reader's benefit, a black and white version of what Mike and Mr. X observe is shown in Figure 2.11.)

```
> display(gr1,gr2,tickmarks=[4,4]);
```

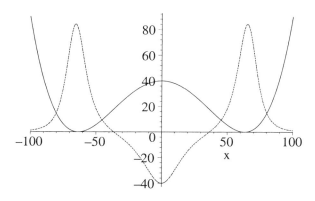

Figure 2.11: Solid curve: $h(x)$. Dashed curve: 1000 × curvature.

"The solid red curve is the $h(x)$ that I gave you, Mike. I believe it is usually referred to as a double well potential, that is to say two potential wells separated by an intervening hill. But, how do we interpret the green dashed curve describing the curvature?"

"As I mentioned earlier, where the curvature is negative (e.g., around $x = 0$), the $h(x)$ curve is concave downwards, while in the two valleys, the curvature is positive because $h(x)$ is concave upwards. The maximum negative curvature occurs at $x = 0$ and the maximum positive curvature occurs at the valley bottoms. The curvature is zero at the two inflection points of the height curve. At these inflection points, the radius of curvature will be infinite, and the centripetal force equal to zero. The radius of curvature is obtained by forming the

2.2. ENERGY AND MOMENTUM

reciprocal of the absolute value of *curv2*."

```
>   rc:=1/abs(curv2):    #radius of curvature
```

"Now that we have determined the radius of curvature at any point on the height curve, we can begin to look at the G forces that might be experienced by a typical rider. My young niece Ruth has a mass m of about 25 kg. Since this is only a preliminary calculation, I will take $g=10$ m/s^2."

```
>   m:=25: g:=10:
```

"As a simple example, let's start the roller coaster car from rest at a height slightly less than the central maximum height of 40 m at $x=0$. I will take the initial height to be, say, $y0=39.999$ m. Recall that the curvature is largest at the bottom of each valley so the radius of curvature is smallest here and therefore the centripetal force the largest. The maximum G force will occur at a point where the height is zero. Let me label this as the "final" height yf for this part of the calculation."

```
>   y0:=39.999: yf:=0:
```

"The initial speed $v0=0$ is entered. I must now find the initial x value, $x0$ at which the car starts."

```
>   v0:=0:
```

"Setting $h(x)=y0$ and numerically solving for x in a range which covers the right-hand valley yields two possible x values for the starting point $x0$."

```
>   x0:=fsolve(h(x)=y0,x,0..100);
```

$$x0 := 0.2236074965, 89.44243959$$

"The first answer, near $x=0$, is selected."

```
>   x0:=x0[1];
```

$$x0 := 0.2236074965$$

"We also need to know the final x coordinate, xf, at the bottom of the valley. Setting $h(x)=yf$ and numerically solving for x, yields two identical values."

```
>   xf:=fsolve(h(x)=yf,x,1..90);
```

$$xf := 63.24555320, 63.24555320$$

"It doesn't matter which xf value we choose, so let's take the first solution."

```
>   xf:=xf[1];
```

$$xf := 63.24555320$$

"To calculate the speed v at a position x, we shall use the conservation of mechanical (kinetic plus potential) energy. Let's create an operator relation for calculating the total mechanical energy Et, given the position x and speed v."

```
>   Et:=(x,v)->1/2*m*v^2+m*g*h(x); # mechanical energy
```

$$Et := (x,\, v) \to \frac{1}{2}\, m\, v^2 + m\, g\, \mathrm{h}(x)$$

"Now, the mechanical energy must be conserved since all frictional forces are being neglected. The total mechanical energy at $x=xf$ must be equal to the mechanical energy at $x0$."

```
> Et1:=Et(x0,v0)=Et(xf,v); # conservation of energy
```
$$Et1 := 9999.750000 = \frac{25\,v^2}{2}$$

"Numerically solving $Et1$ for v over the range 0..100 yields the speed vf at xf."
```
> vf:=fsolve(Et1,v,0..100);
```
$$vf := 28.28391769$$

"Noting that 25 m/s corresponds to 90 km/h, a speed of 28.3 m/s is quite fast. Of course, frictional forces will reduce this speed. The radius of curvature, rf, at xf is obtained by evaluating rc at $x=xf$."
```
> rf:=eval(rc,x=xf);
```
$$rf := 12.50000000$$

"The radius of curvature at $x = xf$ is 12.5 m. Let's create a functional operator F for calculating the centripetal force, when the mass m, the speed v, and the radius of curvature r are supplied as arguments."
```
> F:=(m,v,r)->m*v^2/r;
```
$$F := (m,\,v,\,r) \to \frac{m\,v^2}{r}$$

"Taking Ruth's mass and the calculated values of vf and rf, the centripetal force Ff acting on Ruth at the bottom of the valley,"
```
> Ff:=F(m,vf,rf);
```
$$Ff := 1599.960000$$

"is about 1600 Newtons. The force Ff is equal to the difference between the normal force N exerted upwards on Ruth by the car seat and Ruth's 'true' weight $m\,g$ downwards. So the normal force $N = m\,g + Ff$,"
```
> N:=m*g+Ff;
```
$$N := 1849.960000$$

"exerted on Ruth is about 1850 N. By Newton's third law, Ruth will be pressing down on the seat with a force of 1850 N. This will be her 'effective' weight. The number of Gs that she will feel is equal to N divided by her true weight."
```
> gees:=N/(m*g);
```
$$gees := 7.399840000$$

"Ruth would experience a maximum force of about 7.4 G at the bottom of the valley. On searching the Internet, I have read that most roller coaster operators like to keep the G force below 5, and that for an average healthy person the blackout limit is around 8 G. So Ruth would probably be O.K., but you probably will want to readjust your heights somewhat to lower the G force at the bottom of the valley. If 8 G is the maximum tolerable G force, we can ask, for curiosity's sake, what is the maximum height that Ruth can start from without exceeding 8 G? Further, if she starts at this height, what is her speed as she passes over the central hump at $x = 0$ and what G force does she experience there? In the next line, I will calculate the maximum speed $vmax$ that Ruth can have as she passes through xf and experiences 8 G."

2.2. ENERGY AND MOMENTUM

```
>  vmax:=fsolve(m*g+F(m,v,rf)=8*m*g,v=0..100);
```
$$vmax := 29.58039892$$

"This speed is about 29.6 m/s. We can find the new starting position xs, the car starting from rest, which will give this speed. Again applying the conservation of mechanical energy, the energy at xf is equated to the energy at xs."

```
>  Et2:=Et(xf,vmax)=Et(xs,0); # conservation of energy
```
$$Et2 := 10937.50000 = 0.0006250000000\, xs^4 - 5.00\, xs^2 + 10000$$

"Numerically solving $Et2$ for xs, yields a new starting position $xs \approx 90.5$ m."

```
>  xs:=fsolve(Et2,xs,0..200);
```
$$xs := 90.46159479$$

"The corresponding starting height ys is obtained by calculating $h(xs)$,"

```
>  ys:=h(xs);
```
$$ys := 43.7500000$$

"yielding $ys = 43.75$ m. The speed $vtop$ at the top of the hill at $x = 0$ is obtained by again applying energy conservation and solving for $vtop$."

```
>  Et3:=Et(xs,0)=Et(0,vtop);
```
$$Et3 := 10937.50000 = \frac{25\, vtop^2}{2} + 10000$$

```
>  vtop:=fsolve(Et3,vtop,0..100);
```
$$vtop := 8.660254038$$

"At $x = 0$, the speed is about 9 m/s. Evaluating rc at $x=0$,"

```
>  r0:=eval(rc,x=0);
```
$$r0 := 25.00000000$$

"yields a radius of curvature of 25 m when the car is at the top of the hill. At $x = 0$, the centripetal force acts downwards and is equal to the difference between Ruth's true weight downwards and the normal force $N2$ exerted upwards on Ruth by the seat. So, $N2$ is calculated as follows,"

```
>  N2:=m*g-F(m,vtop,r0);
```
$$N2 := 175.0000000$$

"yielding $N2 = 175$ N, or about 0.7 G."

```
>  gees2:=N2/(m*g);
```
$$gees2 := 0.7000000000$$

"Ruth will feel slightly lighter than normal as she passes through $x = 0$. Oh, oh, I didn't realize what time it is. I have to leave now as Vectoria and I are meeting at the Greek Taverna for supper. You're welcome to join us."

"Thanks for the invite, Mike, but I have other plans for tonight. Thanks for all your help, and perhaps at a later date you can develop a more realistic recipe which includes frictional forces."

2.3 Rotational Dynamics

To this point in the chapter, all of the recipes have dealt with the translational motion of rigid bodies. Now three examples of rotational dynamics are considered involving (a) pure rotation about a fixed axis, (b) a generalization of Atwood's machine, and (c) the motion of a billiard ball given backspin.

2.3.1 The Case of the Falling Pencil

When you have eliminated the impossible, whatever remains, however improbable, must be the truth.
Sherlock Holmes, in *The Sign of Four, ch. 6* (1889)

In his diary, a modern-day Dr. Watson has chronicled the latest exploits of his detective friend, Sherlock Holmes. Watson has entitled this diary entry *The Case of the Falling Pencil*. Let us peek at what Watson has written.

Shortly after midnight, I received a call from Holmes, who was at the scene of a heinous murder committed in the drawing room of the old Abercrombie mansion. Driving across the lonely moors through the swirling mist, I wondered what new adventure Holmes and I were to embark on. What ingenious piece of computer code would he use this time to help him solve the case? On arriving at the mansion, I found Holmes in the library and already working furiously on his ever present laptop computer.

"Ah, there you are Watson. Having already interviewed the sole witness, Jeeves the butler, I have been toying with an idea that arose from our discussion. I think that the key aspect to his testimony involves a pencil, to be more precise, a falling pencil. According to Jeeves, his master was still alive when he discovered him, lying on the floor in the drawing room with a pencil clutched in his hand. He was evidently trying to write something but, before he could do so, he died and the pencil, which had been in a nearly vertical position with one end touching the floor, fell out of his hand. Jeeves further claimed that the pencil underwent pure rotation without slipping. This statement puzzled me because, when I arrived, the body was lying on a smooth uncarpeted floor. After Jeeves had retired to the pantry to calm the cook down, I attempted to duplicate what Jeeves had claimed. However, I noticed that the pencil began to slip when the angle θ with the vertical was about 30°. Clearly, Jeeves was mistaken or he had inadvertently given me a clue. Before interrogating him further, and waiting for you to arrive, I was playing around with the rotational dynamics of a falling pencil on my laptop. Let me explain what I have done."

"As long as the pencil doesn't slip, this problem involves pure rotation of the pencil about a fixed axis through its contact point with the floor. I will be using Newton's second law, $\tau = I\alpha$, for rotation about a fixed axis. Here τ is the torque on the pencil about the axis, I is the moment of inertia, and α is the angular acceleration. In the Maple system that I am using, I is usually reserved for the square root of minus one. The following `interface` command

2.3. ROTATIONAL DYNAMICS

frees up I for use as the moment of inertia by letting $j = \sqrt{-1}$."

> `restart: with(plots): interface(imaginaryunit=j):`

"For convenience, I have taken the pencil length $L = 2\,d$. I will ignore the pencil's small sharpened end and consider it to be a thin uniform rod of mass m and linear density $\rho = m/L$."

> `L:=2*d; rho:=m/L;`

$$L := 2\,d$$
$$\rho := \frac{m}{2\,d}$$

"The moment of inertia of the pencil around the axis of rotation is obtained by evaluating the integral relation $I = \int_0^L \rho r^2 \, dr$."

> `I:=int(rho*r^2,r=0..L);`

$$I := \frac{4\,m\,d^2}{3}$$

"Next, I thought it was worthwhile to create a free body diagram, indicating all the forces acting on the falling pencil, as well as including the acceleration vector of the center of mass, located half-way along the pencil. A functional operator `gr` was formed for producing an arrow graph, when the arrow base coordinates b, the vector coordinates v, and the color c are provided."

> `gr:=(b,v,c)->arrow(b,v,color=c):`

"The following base and vector coordinates were chosen."

> `b1:=[0,0]: b2:=[1.5,1.5]: v1:=[0,1]: v2:=[1,0]: v3:=[0.7,-0.7]:`

"The graphs gr||1 and gr||2 produce red arrows schematically representing the vertical normal force N and the horizontal frictional force Fr exerted by the floor on the bottom (base coordinates b1) of the pencil."

> `gr||1:=gr(b1,v1,red): gr||2:=gr(b1,v2,red):`

"The third and fourth graphs produce red and greens arrows to schematically represent the weight mg of the pencil and the linear acceleration a perpendicular to the rod, both arrows having base coordinates b2 at the center of mass."

> `gr||3:=gr(b2,-v1,red): gr||4:=gr(b2,v3,green):`

"The rod was schematically plotted as a thick blue line in gr||5 and text added in gr||6 and gr||7 with the `textplot` commands. Note how I used the `font=[SYMBOL,12]` option and the entry q in gr||7 to generate the size 12 Greek symbol θ, schematically representing the angle that the pencil makes with vertical. In a later figure, I have used the same font option, entering m to produce the Greek symbol μ for the coefficient of friction."

> `gr||5:=plot([b1,2*b2],color=blue,thickness=3):`
> `gr||6:=textplot({[.2,1,"N"],[1.1,.15,"Fr"],[1.7,.5,"mg"],`
> `[.75,.9,"d"],[2.2,2.35,"d"],[2.3,.8,"a"]}):`
> `gr||7:=textplot([.28,.6,"q"],font=[SYMBOL,12]):`

"The seven graphs were then superimposed with the sequence (seq) and display command to produce the free body diagram shown in Figure 2.12. The scaling was constrained, no axes or tickmarks were included and a title was added."

```
> display({seq(gr||i,i=1..7)},scaling=constrained,
> axes=none,tickmarks=[0,0],title="free body diagram");
```

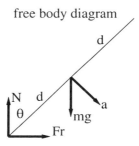

Figure 2.12: Forces on the falling pencil undergoing linear acceleration a.

"The normal (N) and frictional force (Fr) both pass through the axis of rotation so have zero moment arm and therefore do not contribute to the torque on the pencil. The perpendicular distance from the line of action of the weight vector to the rotation axis is $d\sin(\theta)$, so the weight produces a torque $\tau = m\,g\,d\sin(\theta)$."

```
> tau:=m*g*d*sin(theta);
```

$$\tau := m\,g\,d\sin(\theta)$$

"Next, I entered Newton's second law for rotational motion, the torque and moment of inertia being automatically substituted, and solved eq1 for α."

```
> eq1:=tau=I*alpha;
```

$$eq1 := m\,g\,d\sin(\theta) = \frac{4\,m\,d^2\,\alpha}{3}$$

```
> alpha:=solve(eq1,alpha);
```

$$\alpha := \frac{3}{4}\frac{g\sin(\theta)}{d}$$

"Now, by definition, the angular velocity $\omega = d\theta/dt$, so $\omega\,d\omega = (d\theta/dt)\,d\omega = \alpha\,d\theta$. So the angular velocity Ω of the pencil when it has fallen through an angle Θ from the vertical can be obtained by integrating over this last relation."

```
> eq2:=int(omega,omega=0..Omega)=int(alpha,theta=0..Theta);
```

$$eq2 := \frac{\Omega^2}{2} = -\frac{3}{4}\frac{g\,(\cos(\Theta)-1)}{d}$$

"Actually, the output in eq2 is simply a statement about the conservation of mechanical energy, as is easily confirmed in the next two command lines. Initially, the vertical pencil has the potential energy $m\,g\,d$ and zero kinetic energy, assuming that it starts falling from rest. At an angle Θ its potential energy is $m\,g\,d\cos(\Theta)$ and its kinetic energy is $(1/2)\,I\,\omega^2$. In eq2b, I entered

2.3. ROTATIONAL DYNAMICS

the conservation of energy statement and solved for Ω^2 in *eq2c*. On dividing by 2, *eq2c* yields the same result as in *eq2*."

```
>  eq2b:=(1/2)*I*Omega^2+m*g*d*cos(Theta)=0+m*g*d;
```
$$eq2b := \frac{2\,m\,d^2\,\Omega^2}{3} + m\,g\,d\cos(\Theta) = m\,g\,d$$

```
>  eq2c:=(1/2)*Omega^2=solve(eq2b,Omega^2)/2;
```
$$eq2c := \frac{\Omega^2}{2} = -\frac{3}{4}\frac{g\,(\cos(\Theta) - 1)}{d}$$

"The half-angle trigonometric identity $\cos(\Theta) = 1 - 2\sin^2(\Theta/2)$ was then substituted into *eq2* and *eq3* solved for Ω, the radsimp command being applied to extract the positive square root."

```
>  eq3:=subs(cos(Theta)=1-2*sin(Theta/2)^2,eq2);
```
$$eq3 := \frac{\Omega^2}{2} = \frac{3}{2}\frac{g\sin(\frac{\Theta}{2})^2}{d}$$

```
>  Omega:=radsimp(solve(eq3,Omega));
```
$$\Omega := \frac{\sqrt{3}\,\sqrt{d\,g}\sin(\frac{\Theta}{2})}{d}$$

"The angular velocity of the pencil when it has fallen through an angle Θ is then determined. Although intellectually interesting, I realized that a more relevant objective as far as this case is concerned was to determine the relation between the coefficient of friction μ of the floor and the angle that the pencil makes with the vertical. As long as μ remains below the maximum value μ_S, the coefficient of static friction, the pencil will not slip at its bottom end. To determine an expression for μ, the linear acceleration of the center of mass, $a = d\alpha$ was entered. The direction of a can be seen in Figure 2.12."

```
>  a:=d*alpha;
```
$$a := \frac{3}{4}g\sin(\theta)$$

"The frictional force is related to the normal force by the relation $Fr = \mu N$."

```
>  Fr:=mu*N;
```

"The linear acceleration vector was resolved into horizontal and vertical components, $a\cos\theta$ and $a\sin\theta$. In the horizontal direction, there is only one force causing the acceleration of the pencil's center of mass, namely Fr. So Newton's second law for linear motion of the center of mass in the horizontal direction gives $Fr = m\,a\,\cos(\theta)$."

```
>  eq4:=Fr=m*a*cos(theta);
```
$$eq4 := \mu\,N = \frac{3}{4}m\,g\sin(\theta)\cos(\theta)$$

"In the vertical direction, the net force downwards on the pencil is $mg - N$. Newton's second law in this direction is entered in *eq5*."

```
> eq5:=m*g-N=m*a*sin(theta);
```
$$eq5 := mg - N = \frac{3}{4} mg \sin(\theta)^2$$

"Equations eq4 and eq5 are easily solved for the two unknowns μ and N. On assigning the solution, the expression for μ was extracted and plotted over the angular range $\theta = 0$ to $\pi/2$."

```
> sol:=solve({eq4,eq5},{mu,N});
```
$$sol := \{\mu = -\frac{3\sin(\theta)\cos(\theta)}{-4 + 3\sin(\theta)^2}, \; N = mg - \frac{3}{4} mg \sin(\theta)^2\}$$

```
> assign(sol): mu:=mu;
```
$$\mu := -\frac{3\sin(\theta)\cos(\theta)}{-4 + 3\sin(\theta)^2}$$

```
> plot(mu,theta=0..Pi/2,labels=["q","m"],font=[SYMBOL,12]);
```

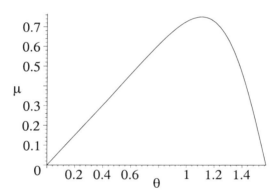

Figure 2.13: Dependence of μ on θ.

"The dependence of μ on the angle θ is shown in Figure 2.13. The frictional force, and therefore μ, is zero when $\theta = 0$ and $\pi/2$. The curve rises to a maximum at an intermediate value of θ, which is easily determined by differentiating μ with respect to θ, setting the result equal to zero, and numerically solving for θ over the range $\theta = 0.8$ to 1.2."

```
> eq6:=diff(mu,theta)=0;
```
$$eq6 := -\frac{3\cos(\theta)^2}{-4 + 3\sin(\theta)^2} + \frac{3\sin(\theta)^2}{-4 + 3\sin(\theta)^2} + \frac{18\sin(\theta)^2 \cos(\theta)^2}{(-4 + 3\sin(\theta)^2)^2} = 0$$

```
> ang:=fsolve(eq6,theta=0.8..1.2);
```
$$ang := 1.107148718$$

"The maximum value of μ occurs at about 1.1 rads, or on converting to degrees,
```
> ang2:=convert(ang,units,radian,degree);
```

2.3. ROTATIONAL DYNAMICS

$$ang2 := 63.43494881$$

at 63.4°. The maximum value of $\mu \simeq 0.75$ follows on evaluating μ at $\theta = ang$."

> `max_mu:=eval(mu,theta=ang);`

$$max\_mu := 0.7500000003$$

"So what does this mean as far as this case is concerned, Holmes?"

"Well, my friend, remember that I observed the pencil to slip at about 30°. Evaluating μ at $\theta = \pi/6$ radians, yields $\mu \approx 0.4$."

> `mu[s]:=eval(mu,theta=evalf(Pi/6));`

$$\mu_s := 0.3997040328$$

"Since slipping occurs, this is the value of the coefficient of static friction (μ_s). But Jeeves claimed that the pencil did not slip at any angle with the vertical, so this implies that $\mu_s > 0.75$. Either Jeeves is wrong or Jeeves is lying and the floor was covered with a rug of some sort which prevented the falling pencil from slipping. I did notice that there was surprisingly little blood on the floor. Why the rug was removed is a puzzle and I think that now that you have arrived we should interrogate Jeeves together. I am not saying that the butler did it, but we can't rule out that possibility."

"O.K. Holmes, but what is the relevance of the remaining lines of your computer code?"

"You hadn't arrived yet, and remember that I had calculated the angular velocity Ω of the pencil at any angle Θ with the vertical assuming that slipping doesn't occur. The pencil is about 15 cm long, so $d = 0.075$ meters. Taking $\Theta = \pi/2$ radians and $g = 9.8$ m/s^2,"

> `g:=9.8: d:=0.075: Theta:=Pi/2:`

"the pencil would hit the floor with an angular velocity of 14 radians/second."

> `ang_vel:=evalf(Omega);`

$$ang\_vel := 14.00000000$$

At this stage, Dr. Watson's diary entry stops. Presumably, he will finish it later and perhaps let us know how the case turned out.

2.3.2 The Atwood Supreme

The machine does not isolate man from the great problems of nature but plunges him more deeply into them.
Antoine de Saint-Exupéry, French aviator/author, *Wind, Sand, and Stars* (1939)

In 1784 George Atwood developed an experimental configuration, now commonly referred to as Atwood's "machine", to measure the acceleration of gravity g. With the crude timing devices available at that time, a direct accurate measurement of g was difficult. Atwood's machine consisted of two masses m_1 and $m_2 > m_1$ tied together by a length of very light rope slung over an essentially massless free-turning pulley. As shown in most first year physics texts, for this device the tension in the rope is given by $T = 2 m_1 m_2 g/(m_1 + m_2)$

and g is related to the acceleration a of either mass by the relation $g = (m_2 + m_1)a/(m_2 - m_1)$. For $m_2 \approx m_1$, the acceleration $a \ll g$ and could be readily measured. With a determined and m_1 and m_2 known, then the value of g is easily calculated.

In this recipe, we create a generalized Atwood machine, a device not to measure g, but to illustrate how a complex system is easily handled using a computer algebra system. Our "Atwood supreme" is as depicted in Figure 2.14. The two masses m_1 and m_2, which are initially at rest on the floor, are connected

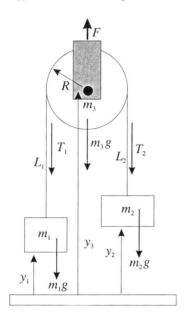

Figure 2.14: Richard and George's supreme Atwood's machine.

by a very light inextensible rope passing over a pulley of radius R and mass m_3. The pulley, also initially at rest, is given an upwards acceleration a_3 by an applied force F, while m_1 and m_2 have accelerations a_1 and a_2, respectively. At some instant in time, m_1, m_2, and m_3 are located at vertical distances y_1, y_2, and y_3 above a reference plane. The tensions in the portions of string of length L_1, connected to m_1, and of length L_2, connected to m_2, are T_1 and T_2. We pose the following questions for our device:

(a) What are the analytic expressions for a_1, a_2, a_3, T_1, and T_2?

(b) Do these expressions correctly reduce to those for Atwood's machine?

(c) Given $m_1 = m_3 = 2\,m_2 = 2$ kg, and $g = 10$ m/s^2, what F is required so that $a_2 = 0$ (m_2 remains at rest)? What are a_1, a_2, a_3, T_1, and T_2?

To free the symbol I for use as the moment of inertia, let's set $j = \sqrt{-1}$.

2.3. ROTATIONAL DYNAMICS

```
> restart: interface(imaginaryunit=j):
```
In eq_1, eq_2, and eq_3, the net force on each mass, m_1, m_2, and m_3 is set equal to the mass times the linear acceleration.

```
> eq[1]:=T[1]-m[1]*g=m[1]*a[1];
```
$$eq_1 := T_1 - m_1 g = m_1 a_1$$

```
> eq[2]:=T[2]-m[2]*g=m[2]*a[2];
```
$$eq_2 := T_2 - m_2 g = m_2 a_2$$

```
> eq[3]:=F-T[1]-T[2]-m[3]*g=m[3]*a[3];
```
$$eq_3 := F - T_1 - T_2 - m_3 g = m_3 a_3$$

For rotational motion of the pulley, the net torque on the pulley, $RT_1 - RT_2$, is equal to the pulley's moment of inertia times its angular acceleration α.

```
> eq[4]:=R*T[1]-R*T[2]=I*alpha;
```
$$eq_4 := RT_1 - RT_2 = I\alpha$$

Treating the pulley as a disc of radius R with the axis of rotation through its center, its moment of inertia is $I = (1/2)\, m_3\, R^2$. Assuming that the rope doesn't slip on the pulley, $\alpha = a_{L1}/R$, where a_{L1} is the linear acceleration of the string segment of length L_1.

```
> I:=1/2*m[3]*R^2; alpha:=a[L1]/R;
```
$$I := \frac{1}{2} m_3 R^2$$

$$\alpha := \frac{a_{L1}}{R}$$

There are three constraint equations for the accelerations. First, since $y_3 = y_1 + L_1$, then on differentiating twice with respect to time, one has $a_3 = a_1 + a_{L1}$ which is entered in eq_5.

```
> eq[5]:=a[3]=a[1]+a[L1];
```
$$eq_5 := a_3 = a_1 + a_{L1}$$

Similarly, since $y_3 = y_2 + L_2$, then $a_3 = a_2 + a_{L2}$ which is entered in eq_6.

```
> eq[6]:=a[3]=a[2]+a[L2];
```
$$eq_6 := a_3 = a_2 + a_{L2}$$

Finally, since $L_1 + L_2$ is constant, then $a_{L1} = -a_{L2}$ which is entered in eq_7.

```
> eq[7]:=a[L1]=-a[L2];
```
$$eq_7 := a_{L1} = -a_{L2}$$

Solving seven equations in seven unknowns (a_1, a_2, a_3, a_{L1}, a_{L2}, T_1, and T_2) is a tedious task to do by hand but trivial with the analytic solve command.

```
> sol:=solve({seq(eq[i],i=1..7)},{a[1],a[2],a[3],a[L1],a[L2],
> T[1],T[2]});
```

$$sol := \{a_{L2} = -\frac{2F(m_1 - m_2)}{3m_3 m_1 + 3m_3 m_2 + m_3{}^2 + 8m_2 m_1},$$

$$a_3 = \frac{2m_2 F - 8m_2 m_1 g - 3m_3 m_2 g - m_3{}^2 g + F m_3 + 2m_1 F - 3m_3 m_1 g}{3m_3 m_1 + 3m_3 m_2 + m_3{}^2 + 8m_2 m_1},$$

$$T_2 = \frac{m_2 F (m_3 + 4m_1)}{3m_3 m_1 + 3m_3 m_2 + m_3{}^2 + 8m_2 m_1},$$

$$T_1 = \frac{F m_1 (m_3 + 4m_2)}{3m_3 m_1 + 3m_3 m_2 + m_3{}^2 + 8m_2 m_1},$$

$$a_1 = -\frac{-4m_2 F + 8m_2 m_1 g + 3m_3 m_2 g + m_3{}^2 g - F m_3 + 3m_3 m_1 g}{3m_3 m_1 + 3m_3 m_2 + m_3{}^2 + 8m_2 m_1},$$

$$a_2 = -\frac{8m_2 m_1 g + 3m_3 m_2 g + m_3{}^2 g - F m_3 - 4m_1 F + 3m_3 m_1 g}{3m_3 m_1 + 3m_3 m_2 + m_3{}^2 + 8m_2 m_1},$$

$$a_{L1} = \frac{2F(m_1 - m_2)}{3m_3 m_1 + 3m_3 m_2 + m_3{}^2 + 8m_2 m_1}\}$$

The analytic expressions shown in the above output are quite formidable. Having answered our first question, let's now assign this solution.

> `assign(sol):`

Next, let's check to see if we can regain the acceleration and tension expressions for Atwood's original machine. For his device, $a_3 = m_3 = 0$. In the following command line, we determine the force F_check needed to give $a_3 = 0$ and also set $m_3 = 0$ in F_check.

> `F_check:=solve(a[3]=0,F): F_check:=subs(m[3]=0,F_check):`

Substituting $F = F\_check$ and $m_3 = 0$ into a_1 (or into a_2) yields the correct form of the acceleration for Atwood's machine. Similarly, the correct form for the rope's tension results if the same substitutions are made into T_1 (or T_2).

> `accel_check:=simplify(subs({F=F_check,m[3]=0},a[1]));`

$$accel\_check := \frac{g(m_2 - m_1)}{m_2 + m_1}$$

> `tension_check:=simplify(subs({F=F_check,m[3]=0},T[1]));`

$$tension\_check := 2\frac{m_2 m_1 g}{m_2 + m_1}$$

Finally, to answer the third question, we enter $m_1 = 2$, $m_2 = 1$, $m_3 = 2$, $g = 10$.

> `m[1]:=2: m[2]:=1: m[3]:=2: g:=10:`

The force F required to produce $a_2 = 0$ is found to be 38 Newtons.

> `F:=solve(a[2]=0,F);`

$$F := 38$$

The accelerations (in m/s^2) and tensions (in N) are then obtained.

> `a[1]:=a[1]; a[2]:=a[2]; a[3]:=a[3]; T[1]:=T[1]; T[2]:=T[2];`

2.3. ROTATIONAL DYNAMICS

$$a_1 := -4$$
$$a_2 := 0$$
$$a_3 := -2$$
$$T_1 := 12$$
$$T_2 := 10$$

The minus signs in the accelerations indicate that the masses m_1 and m_3 (the pulley) are not accelerating upwards as originally assumed in the formulation, but are accelerating downwards. For example, in eq_3 the net force on m_3 is $F - T_1 - T_2 - m_3 g = 38 - 12 - 10 - (2)(10) = -4$ N, i.e., acts downwards.

2.3.3 Fast Freddie's Trick Shot

Playing snooker gives you firm hands and helps to build up character. It is the ideal recreation for dedicated nuns.
Archbishop Luigi Barbarito, Italian cleric, *Daily Telegraph* (15 Nov. 1989)

Fast Freddie is not only an excellent pool player, but is also a talented physics student. When a software company, GAMECO, asked Freddie to work on its new computer pool game he jumped at the chance to make some money doing both things he enjoyed. Referring to Figure 2.15, GAMECO want Freddie to

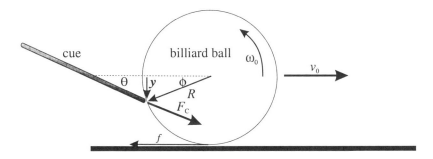

Figure 2.15: Geometry for cue striking billiard ball.

analyze and animate the following "trick shot". A billiard ball (the cue ball) is struck with the cue in such a way as to give the ball an initial linear velocity $v0$ and backspin with initial angular velocity $\omega 0$. In the animation, the ball is to travel to the right (slipping as it moves), momentarily come to rest, and reverse direction. When the ball crosses its starting position, its slipping is to cease and the ball is to start rolling with a final linear velocity $vf = -v0$, i.e., the same magnitude as the initial velocity but traveling to the left. In addition, Freddie has been asked to find the relationship between the cue's angle θ and the vertical distance y at which the ball must be struck to produce this trick shot. In particular what is the value of θ if $y = R/2$, where R is the radius of the ball? Freddie is told to assume that, for his trick shot, the cue ball does not strike any other ball or strike the bumper rails of the pool table.

In response to GAMECO's request, Fast Freddie has created the following animated recipe. After loading the `plots` and `plottools` (needed for the `circle` command) packages, he uses the following `interface` command to replace Maple's protected symbol I for $\sqrt{-1}$ with j. This frees the symbol I for use as the moment of inertia of the billiard ball.

> `restart: with(plots): with(plottools):`

> `interface(imaginaryunit=j):`

Freddie employs the arrow operator to enter the two relevant kinematical formulas for translation of the center of mass and the two needed expressions for angular motion of the cue ball about its axis of rotation. Here s, a, and v are the linear distance, velocity, and acceleration at time t, while θ, α, and ω are the angle through which the ball rotates, the angular acceleration, and angular velocity at time t. Freddie will have to unassign θ later so it can be used to represent the cue angle.

> `s:=(v0,a,t)->v0*t+1/2*a*t^2;`

$$s := (v0,\, a,\, t) \to v0\, t + \frac{1}{2}\, a\, t^2$$

> `v:=(v0,a,t)->v0+a*t;`

$$v := (v0,\, a,\, t) \to v0 + a\, t$$

> `omega:=(omega0,alpha,t)->omega0+alpha*t;`

$$\omega := (\omega 0,\, \alpha,\, t) \to \omega 0 + \alpha\, t$$

> `theta:=(omega0,alpha,t)->omega0*t+1/2*alpha*t^2;`

$$\theta := (\omega 0,\, \alpha,\, t) \to \omega 0\, t + \frac{1}{2}\, \alpha\, t^2$$

As the ball (of mass m) begins to move to the right, it is rotating counterclockwise and slipping at its contact point with the surface of the pool table. The linear velocity of the center of mass decreases due to the frictional force f exerted on the ball's bottom by the table surface. Assuming that the frictional force is constant (i.e., independent of the relative slipping velocity), the ball's acceleration is given by $a = -f/m$, the minus sign indicating deceleration.

> `a:=-f/m;`

The time To at which the ball momentarily comes to rest (has zero velocity) and the distance So that its center of mass has moved from the starting point are calculated.

> `To:=solve(v(v0,a,t)=0,t); So:=s(v0,a,To);`

$$To := \frac{v0\, m}{f}$$

$$So := \frac{v0^2\, m}{2\, f}$$

Freddie enters the final linear velocity of the ball, i.e., $vf = -v0$.

2.3. ROTATIONAL DYNAMICS

```
>   vf:=-v0;
```
The time T to achieve the final velocity vf and the distance $S1$ of the ball's center of mass from the starting point at this time are calculated.
```
>   T:=solve(v(v0,a,t)=vf,t);
```
$$T := \frac{2\,v0\,m}{f}$$
```
>   S1:=s(v0,a,T);
```
$$S1 := 0$$

The ball acquires its final velocity as it passes back through the starting point at a time $T=2To$. Since the ball is of radius R, the frictional force f exerts a torque $\tau = Rf$ on the ball which causes a decrease in the ball's angular speed.
```
>   tau:=R*f;
```
The ball is spherical so its moment of inertia is $I = (2/5)\,m\,R^2$.
```
>   I:=2/5*m*R^2;
```
By Newton's second law for rotational motion, the angular acceleration is $\alpha = -\tau/I$, the minus sign again indicating deceleration.
```
>   alpha:=-tau/I;
```
$$\alpha := -\frac{5\,f}{2\,R\,m}$$

The ball is to begin to roll without slipping when it crosses its starting position. The condition for rolling without slipping is that the linear velocity of the center of mass is equal to the radius times the angular rotation velocity. Using this rolling condition, which occurs at time T, $eq1$ establishes the relation needed to determine the initial angular velocity $w0$. Then $eq1$ is solved for $w0$.
```
>   eq1:=-vf/R=omega(omega0,alpha,T);
```
$$eq1 := \frac{v0}{R} = w0 - \frac{5\,v0}{R}$$
```
>   omega0:=solve(eq1,omega0);
```
$$w0 := \frac{6\,v0}{R}$$

The expression for the frictional force f on the billiard ball (mass m) is entered. Here μ is the frictional coefficient and g the gravitational acceleration.
```
>   f:=mu*m*g;
```
$$f := \mu\,m\,g$$

Then the times To and T, the distance So, and the distance S and velocity V of the ball's center of mass at time t are calculated.
```
>   To:=To; So:=So; T:= T; S:=s(v0,a,t); V:=v(v0,a,t);
```
$$To := \frac{v0}{\mu\,g}$$

$$So := \frac{v0^2}{2\,\mu\,g}$$

$$T := \frac{2\,v0}{\mu\,g}$$

$$S := v0\,t - \frac{\mu\,g\,t^2}{2}$$

$$V := v0 - \mu\,g\,t$$

The angular rotation Θ and angular velocity Ω are similarly calculated at t.

> Theta:=theta(omega0,alpha,t); Omega:=omega(omega0,alpha,t);

$$\Theta := \frac{6\,v0\,t}{R} - \frac{5\,\mu\,g\,t^2}{4\,R}$$

$$\Omega := \frac{6\,v0}{R} - \frac{5\,\mu\,g\,t}{2\,R}$$

In order to animate the motion of the cue ball, Freddie considers a billiard ball of mass $m = 0.3$ kg and radius $R = 0.03$ m. He takes $g = 9.8$ m/s^2, the initial velocity $v0 = 2$ m/s, and the frictional coefficient $\mu = 0.6$.

> m:=0.3: R:=0.03: g:=9.8: v0:=2: mu:=0.6:

The time To for the ball to come momentarily to rest is found to be 0.34 s, the time T to pass back through the starting point to be 0.68 s, and the distance So at time To to be 0.34 m.

> To:=To; T:=T; So:=So;

$$To := 0.3401360544$$
$$T := 0.6802721088$$
$$So := 0.3401360544$$

At arbitrary time t, the expressions for S, V, Θ, and Ω are as follows.

> S:=S; V:=V; Theta:=Theta; Omega:=Omega;

$$S := 2\,t - 2.940000000\,t^2$$
$$V := 2 - 5.88\,t$$
$$\Theta := 400.0000000\,t - 245.0000000\,t^2$$
$$\Omega := 400.0000000 - 490.0000000\,t$$

At time T, the slipping stops, and the ball's linear position is given by $S1=0$ and its velocity by $vf=-v0=-2$ m/s. The values of Θ and Ω at this instant are calculated.

> Theta1:=eval(Theta,t=T); Omega1:=eval(Omega,t=T);

$$\Theta1 := 158.7301587$$
$$\Omega1 := 66.6666667$$

To simulate the translational plus rotational motion of the cue ball, Freddie generates $N=500$ graphs. So that the ball will not look too small on the computer screen if the input parameter values are changed, he has introduced a magnification factor M ($M=1.5$ here) to make the ball look larger. The entry sp controls the spacing of the graphs. Here every graph will be used, but if for example Freddie took $sp=5$, then only every fifth graph would be plotted in the animation. The entry ex controls the "extra" time that will be included in the

2.3. ROTATIONAL DYNAMICS

simulation after the ball has returned to its starting position. The ball will be allowed to roll (without slipping) an extra distance to the left.

>   N:=500: M:=1.5: sp:=1: ex:=0.25:

Using the pointplot command, Freddie creates a graph function P to plot a small circle on the ith time step at the ball's center of mass (axis of rotation) and on the rim of the ball to give a sense of rotation. The color cc will change from blue for $t < T$ (slipping occurs) to red when $t > T$ (rolling without slipping).

>   P:=(i,A,B)->pointplot([A,B],symbol=circle,color=cc):

Taking artistic licence, since cue balls are normally white, Freddie employs the circle command to plot the ball's rim as a thick green circle. The coordinates of the circle's center on the ith time step are entered as a list in the first argument. The second argument is the (magnified) radius of the ball.

>   C:=i->circle([s,M*R],M*R,scaling=constrained,thickness=3,
>   color=green):

The following do loop generates the N graphs. The total time is divided into N time intervals, so t is the time on the ith time step.

>   for i from 0 to N do
>   t:=(T+ex*T)/N*i;

For $t < T$, slipping occurs, and the expressions derived above for S and Θ are used. The small circles at the center of mass and on the rim are colored blue.

>   if t<T then s:=S;theta:=Theta;cc:=blue:

The following else statement is used for $t > T$. The linear coordinate is then given by $S1 + vf(t - T)$ and the angular coordinate by $\Theta 1 + \Omega 1, (t - T)$. The color of the two small circles changes to red.

>   else s:=S1+vf*(t-T);theta:=Theta1+Omega1*(t-T);cc:=red:
>   end if;

On ending the if statement, all the relevant graphs on the ith time step are superimposed in d[i]. The first graph entry in the display command is for the small colored circle at the center of mass which moves with the linear velocity of the ball. The second entry plots a colored circle on the rim of the cue ball which rotates as the ball moves. The third entry is for the rim of the ball.

>   d[i]:=display(P(i,s,M*R),P(i,s+M*R*cos(theta),M*R+
>   M*R*sin(theta)),C(i));
>   end do:

On completion of the do loop, Freddie animates the graphs with the display command and the insequence=true option. The reader should execute the following command line and click on the play arrow to see the cue ball's motion.

>   display(seq(d[sp*i],i=0..N/sp),insequence=true,
>   tickmarks=[3,0],scaling=constrained);

To determine the angle θ that the cue must make for the trick shot, Freddie unassigns θ as well as m, R, $v0$, and ω. Instead of using `unassign`, he achieves the same goal by enclosing the entries in single quotes.

> `m:='m': R:='R': v0:='v0': theta:='theta': omega:='omega':`

He has been asked to relate the cue angle θ to the vertical distance y below the ball's center at which contact of the cue with the ball takes place. If contact takes place for the (short) time interval δ, and the cue exerts an average force F_c on the ball, the impulse delivered to the ball is $F_c \delta$.

> `impulse:=F[c]*delta;`

$$impulse := F_c \, \delta$$

The linear momentum $m \, v0$ acquired by the ball is equal to the component of the impulse in the direction of the motion of the center of mass.

> `eq2:=m*v0=impulse*cos(theta);`

$$eq2 := m \, v0 = F_c \, \delta \cos(\theta)$$

The contact time interval δ is obtained by solving $eq2$.

> `delta:=solve(eq2,delta);`

$$\delta := \frac{m \, v0}{F_c \cos(\theta)}$$

The cue delivers a torque to the ball of amount $R \, F_c \sin(\theta + \phi)$.

> `torque:=R*F[c]*sin(theta+phi);`

$$torque := R \, F_c \sin(\theta + \phi)$$

The ball acquires an angular momentum $I\omega$ equal to the angular impulse (torque times the contact time interval) imparted to the ball by the cue.

> `eq3:=I*omega=torque*delta;`

$$eq3 := \frac{2 \, m \, R^2 \, \omega}{5} = \frac{R \sin(\theta + \phi) \, m \, v0}{\cos(\theta)}$$

The angular velocity ω, obtained by solving $eq3$, must be equal to the initial angular velocity $\omega 0$ derived earlier.

> `eq4:=omega0=solve(eq3,omega);`

$$eq4 := \frac{6 \, v0}{R} = \frac{5}{2} \frac{\sin(\theta + \phi) \, v0}{R \cos(\theta)}$$

The cue angle needed to produce the trick shot is found by solving $eq4$ for θ and simplifying with the `trig` option.

> `cue_angle:=simplify(solve(eq4,theta),trig);`

$$cue\_angle := -\arctan\left(\frac{1}{5} \frac{-12 + 5 \sin(\phi)}{\cos(\phi)}\right)$$

To relate the cue angle to y, Freddie has noted that $\phi = \arcsin(y/R)$ from Figure 2.15. He substitutes this relation into cue_angle and simplifies the resulting cue angle expression (Θ) with the `radical` option.

> `Theta:=subs(phi=arcsin(y/R),cue_angle);`

```
> Theta:=simplify(Theta,radical);
```

$$\Theta := \arctan\left(\frac{12\,R - 5\,y}{5\,R\sqrt{\frac{R^2 - y^2}{R^2}}}\right)$$

Freddie calculates the cue angle (in degrees) when the ball is struck at $y = R/2$. In this case, the necessary angle to produce the trick shot is about 65.5°.

```
> "theta"=evalf(subs(y=R/2,Theta*180/Pi));
```

"theta" $= 65.49636652$

Having completed this trick shot analysis and animation, Freddie is eager to learn what other pool game simulations GAMECO might have in mind.

2.4 Supplementary Recipes

02-S01: Gabrielle's Toy Car
Gabrielle has a toy car of mass m, initially at rest at the origin. A constant force F_0 is applied to it for the first T seconds, $2F_0$ applied for the next NT seconds, $4F_0$ applied for the following $N^2 T$ seconds, $8F_0$ for the next $N^3 T$ seconds, and so on. Here N is a constant. Develop a recipe that calculates the total distance traveled by the car and its speed after an arbitrary (finite) number of time intervals. The total distance traveled by Gabrielle's car is r times the distance traveled in the first time interval. What is the value of N if 10 time intervals elapse and $r = 21.5$?

02-S02: Mike's Race
While attending a mathematics conference in a downtown Metropolis hotel, Mike takes an elevator from the lobby to the top floor (98 m higher), where the conference room is located. If Mike does not experience any discomfort unless his apparent weight increases by more than 15% or decreases by more than 10%, what is the shortest time it takes to transport him comfortably from the lobby to the top floor? Take $g=9.8$ m/s^2.

Several overhead projectors and related equipment needed for the conference are sent up in the freight elevator. The conditions for this elevator are that all the equipment must remain in contact with the floor and the elevator cable can safely support a load equal to five times the combined weight of the elevator car and equipment. If both elevators start off at the same time, who will win the "race" to the top floor, Mike or the equipment? What is the time difference?

Plot Mike's velocity and the velocity of the equipment in the same figure. By integrating over the piecewise velocity functions, show that the area under each curve is equal to 98 m.

02-S03: The Dirty Bird Window Washer
One summer, while earning money to pay her way through university, Jennifer's sister Heather worked for the Dirty Bird window washing company. On one

particular job, she leaned her ladder against the side of a house which had the same coefficient μ of static friction as the level surface on which the bottom end of the ladder rested. Make a labeled free body diagram indicating all of the forces acting on the ladder when Heather was standing part way up it. Before climbing the ladder to clean the house windows, Heather leaned the ladder at different angles θ to the vertical and found that the ladder began to slip when $\theta = \Theta$ degrees. Determine how μ depends on Θ. Then Heather went up the ladder a fraction f of the ladder's total length. If she weighed r times the ladder, what was the maximum angle that the ladder could make with the vertical to avoid slipping? Express this angle in terms of r, f, Θ, and simplify your expression as much as possible. If Heather found that $\Theta = 60°$ and she weighed three times as much as the ladder, what is the maximum angle that the ladder could have without slipping if she had wanted to go 3/4 of the way up the ladder?

02-S04: Suspension Bridge

While updating the teaching skills of physics instructors in the eastern Indonesian territories of Irian Jaya, Sulawesi, and the "spice island" of Ambon, Richard had the opportunity to hike in remote jungle regions and see many types of suspension bridges. In the spirit of this experience, here is a bridge problem that is very tedious to solve with pen and paper as it involves solving eight simultaneous equations in eight unknowns. A suspension bridge (Fig. 2.16) spans a deep river gorge 54 m wide. The bridge deck is a steel truss of 48,000 kg weight,

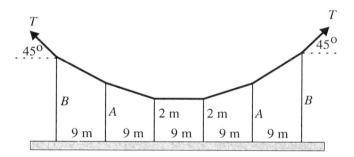

Figure 2.16: Suspension bridge.

supported by six pairs of vertical cables spaced 9 m apart and each carrying an equal amount of the weight. The two central pairs of vertical cables are 2 m in length. The end cables of the suspension arc make an angle of 45° with the horizontal. Determine the lengths A and B of the remaining vertical cables and the tension T in the end cables of the arc. Neglect the weights of all cables.

02-S05: Erehwonese Serving Platter

An ancient, thin, flat Erehwonese serving platter is bounded by the x-axis, y-axis, and the curve $y = 1 - x^2$ and has a density $\rho = 1 + 3x^2 e^{-x} + y^3 \sin y$.

2.4 SUPPLEMENTARY RECIPES

(a) Determine the mass m of the platter and locate its center of mass. The center of mass coordinates are given by $\vec{r}_{cm} = (1/m) \int \int \vec{r} \rho(x,y) \, dy \, dx$.

(b) Create a labeled picture of the green-shaded platter with black trim on its edge. Indicate the center of mass by a red circle.

(c) Calculate the moment of inertia $I = \int \int r^2 \rho(x,y) \, dy \, dx$ of the platter about an axis perpendicular to the platter and through the origin.

(d) Calculate the moment of inertia I_{cm} about an axis perpendicular to the platter through the center of mass.

(e) Use parts (c) and (d) to confirm the parallel-axis theorem that states that $I = I_{cm} + m \, d^2$, where d is the distance between the two axes.

02-S06: A Sticky Encounter

Justine and Kevin, who have sticky putty balls of mass m_a and $m_b = r \, m_a$, are standing on the viewing platforms of adjacent apartment buildings. The balls have the initial positions $\vec{r}_a(0) = 7\,\hat{e}_x + 100\,\hat{e}_z$ and $\vec{r}_b(0) = 49\,\hat{e}_x + 100\,\hat{e}_z$ meters and are thrown with initial velocities $\vec{v}_a(0) = 7\hat{e}_x + 3\hat{e}_y$ and $\vec{v}_b(0) = -7\hat{e}_x + 3\hat{e}_y$ m/s. Taking $g = 9.8$ m/s^2 and $r = 2$, determine the positions of the balls as a function of time until they hit the ground. If a collision takes place, assume that it is totally inelastic. Animate the motion of the balls over the entire time interval, superimposing the animation on a plot of the complete trajectories.

02-S07: Rockets Away

After visiting an army testing ground in the Nevada desert where small rockets are being fired, an engineer, Russell, is confronted with the following rocket problem. A rocket of initial mass (including fuel) m_i is fired from the ground ($y = 0$) at an angle θ degrees with the horizontal (x-direction). Its initial velocity is zero, but it acclerates as it burns the fuel and expels the burnt gases out the back end of the rocket with an exhaust speed U relative to the rocket. The mass of the rocket decreases at a steady rate r for a burn time tb to a final mass m_f. After the burning of fuel ceases, the rocket undergoes "free fall" motion until it hits the ground. Russell has provided a recipe which answers the following questions:

(a) Derive analytic expressions for the x and y coordinates of the rocket over its entire flight until it hits the ground again. For simplicity, neglect air resistance and the variation of g with altitude. Hint: Since the mass of the rocket varies, the more general form of Newton's second law (i.e., external force equals the rate of change of linear momentum) must be used.

(b) Derive analytic expressions for the time $tmax$ at which the rocket reaches its maximum elevation, the maximum elevation $hmax$, the time for the rocket to hit the ground again, and the horizontal range.

(c) Given $m_i = 28.5$kg, $m_f = 0.27\,m_i$, $tb = 20$ seconds, $g = 9.8$ m/s^2, $\theta = 60°$, and that the exhaust gases exert a thrust force of 600 N, determine the values of all the quantities in part (b). Note that the exhaust speed U is equal to the thrust force divided by the rate of mass decrease.

(d) Animate the motion of the rocket for its entire flight. Represent the rocket as a size 16, orange, circle. In the same figure, plot the portion of the trajectory during the burn stage as a thick blue curve and the subsequent free fall portion as a thick red curve.

02-S08: Erehwon Space Probe Explosion

While in flight to distant Earth, an Erehwon space probe (mass M) explodes into three fragments, each with mass $M/3$. One fragment continues along the original flight path, while the other two fly off in directions inclined at 60° to the original path. The energy released in the explosion and given to the fragments is R times as much as the kinetic energy possessed by the probe at the time of explosion. Determine the velocity and kinetic energy of each fragment immediately after the explosion. Plot the velocities and determine the range of R for which the results are physically meaningful. Taking $R = 3$, evaluate the ratios of the kinetic energies of the fragments to the original kinetic energy of the probe. Make a labeled vector diagram showing the velocity vectors.

02-S09: Gabrielle's Slippery Blocks

Reaching into her toy box, Gabrielle places a frictionless cube of mass M and length a on each edge on a frictionless horizontal table. She leans a thin uniform rod of mass $M/3$ and length $4a$ against a vertical face of the cube at its upper edge, with the other end of the rod touching the table. The rod is oriented parallel to the two adjacent vertical sides of the block and in the plane passing through the center of the cube. Gabrielle then releases the rod from rest.

(a) If the rod makes an angle $\theta(t)$ with the table at time t, derive the angular velocity $\dot{\theta}$ of the rod while it is still in contact with the cube.

(b) Show that the rod loses contact with the cube at an angle Θ which satisfies the transcendental equation $(3/2)\sin(4\Theta) - 32\cos(\Theta) + 93\sin(2\Theta) = 0$.

(c) Solve the transcendental equation for Θ. Through what angle does the rod rotate before losing contact with the cube?

02-S10: Another Toybox Problem

Reaching into her overflowing toybox once again, Gabrielle places a wedge of mass M and angle α on top of a fixed frictionless inclined plane of inclination α, the top of the wedge being horizontal and rough. A uniform spherical croquet ball of mass m is placed on the top face of the wedge, and the whole system is released from rest. As the wedge slides down the inclined plane, the croquet ball rolls without slipping on the top face of the wedge. Prove that as the wedge slides down a distance h from rest parallel to the inclined plane, the croquet ball moves through a distance $(5\,h\,\cos\alpha)/7$ along the top of the wedge, provided that the ball remains on the wedge. Also prove that the coefficient μ of friction between the wedge and the ball must satisfy $\mu \geq 2(M+m)\tan\alpha/(7M+2m)$.

Part II
THE ENTREES

*The universe is full of magical things,
patiently waiting for our wits to grow sharper.*
    Eden Philpotts, British author (1862–1960)

*Nature and nature's laws lay hid in night;
God said "Let Newton be!" and all was light.*
    Alexander Pope, English poet (1688–1744)

*It did not last: the Devil, howling "Ho
Let Einstein be!" restored the status quo.*
    John Squire, British author (1884–1958)

Chapter 3

Vector Calculus

3.1 Curvilinear Coordinates

In the Appetizers, the recipes made use of Cartesian coordinates. Depending on the underlying spatial symmetry, other orthogonal curvilinear coordinate systems may prove more useful in trying to solve certain mechanics problems. In the following three recipes, we illustrate kinematics in plane polar and spherical polar coordinates and how to generate "scale factors" for calculating area and volume elements, gradients, Laplacians, etc., in toroidal coordinates.

3.1.1 The Case of the Artistic Slug

Elementary, my dear Watson, elementary.
Attributed to the fictional detective Sherlock Holmes but not found in this form in any of Conan Doyle's books.

After interviewing several scientifically trained witnesses, Dr. Watson has informed Sherlock Holmes of the movements of Slimy slug, prior to Slimy's demise at the hands of a rival gang. On leaving his hiding place beneath a large oak tree at time $t = 0$, Slimy had been observed moving along a 2-dimensional planar path with radial coordinate $r(t) = t^2 e^{-t}$ and angular coordinate $\theta(t) = \frac{3}{5}t^2 - 2t + 12$. All quantities are expressed in "scientific slug" units. Evidently, Slimy was attacked by the rival gang at the time that the maximum angle between his velocity and acceleration vectors occured. He managed to escape unscathed and travel along his original path until he was attacked by the same gang a second time. This happened when his speed was a maximum. Mortally wounded through a vicious salt attack, he still managed to continue along his original path and eventually return to his hiding place where he succumbed from his injuries. Little did the rival gang know that Slimy was an artistic slug who, anticipating an attack, had cunningly designed his trail of slime to leave a clue as to the gang's identity.

As in the earlier Case of the Falling Pencil, where Sherlock Holmes eventually proved that the butler did it, let us peek at what Dr. Watson has recorded in his diary for the Case of the Artistic Slug.

On noting that Slimy's path was given in terms of r and θ, Holmes decided to work with these polar coordinates. On his trusty laptop computer, code-named the Brain, he began his file by loading the `plots` and `LinearAlgebra` packages. He has chosen to work with this latter package, instead of the `VectorCalculus` package for dealing with vectors. To make the time-dependent results simpler in appearance, Holmes has also entered the command `assume(t>0)`.[1]

```
> restart: with(plots): with(LinearAlgebra):
> assume(t>0):
```

The well-known relations between Cartesian (X, Y) and polar coordinates were entered, as well as a Cartesian position vector $\vec{R} = X\,\hat{e}_x + Y\,\hat{e}_y$. The syntax `<X|Y>` produces a "row" vector, while `<X,Y>` would produce a "column" vector.

```
> X:=r*cos(theta); Y:=r*sin(theta); R:=<X|Y>;
```

$$X := r\cos(\theta)$$
$$Y := r\sin(\theta)$$
$$R := [r\cos(\theta),\ r\sin(\theta)]$$

On remarking to Holmes that the position vector looked like a Maple list to me, he used the `type` command to verify that $\vec{R}$ was a vector, rather than a list.

```
> type(R,Vector); type(R,list);
```

$$true$$
$$false$$

Holmes then pointed out to me that the unit vectors $\hat{e}_r$, $\hat{e}_\theta$ in the r and θ directions are related to the unit vectors $\hat{e}_x$, $\hat{e}_y$ in the x and y directions thus:

$$\hat{e}_r = \frac{\partial \vec{R}/\partial r}{|\partial \vec{R}/\partial r|} = \frac{(\partial x/\partial r)\,\hat{e}_x + (\partial y/\partial r)\,\hat{e}_y}{\sqrt{(\partial x/\partial r)^2 + (\partial y/\partial r)^2}},$$
$$\hat{e}_\theta = \frac{\partial \vec{R}/\partial \theta}{|\partial \vec{R}/\partial \theta|} = \frac{(\partial x/\partial \theta)\,\hat{e}_x + (\partial y/\partial \theta)\,\hat{e}_y}{\sqrt{(\partial x/\partial \theta)^2 + (\partial y/\partial \theta)^2}}.$$
(3.1)

To evaluate the numerators $\partial \vec{R}/\partial r$ and $\partial \vec{R}/\partial \theta$, Holmes used the `map` command to differentiate the position vector with respect to r in $n1$ and θ in $n2$.

```
> n1:=map(diff,R,r); n2:=map(diff,R,theta);
```

$$n1 := [\cos(\theta),\ \sin(\theta)]$$
$$n2 := [-r\sin(\theta),\ r\cos(\theta)]$$

---

[1] When an assumption is made about a variable, it normally appears in the output with a trailing tilde (∼). Trailing tildes can be removed by preceding the `assume` command with the command `interface(showassumed=0)`. Alternately, one can remove the trailing tildes from all Maple sessions by clicking on File in Maple's toolbar, then on Preferences, I/O Display, and then under Assumed Variables on No Annotation and on Apply Globally.

3.1. CURVILINEAR COORDINATES

The map command applies diff to each component of $\vec{R}$. Just using, e.g., the command structure diff(R,r) does not work here. To evaluate the magnitudes $|\partial \vec{R}/\partial r|$ and $|\partial \vec{R}/\partial \theta|$, he applied the Norm(,2) command to $n1$ and $n2$.

> d1:=Norm(n1,2); d2:=Norm(n2,2);

$$d1 := \sqrt{|\cos(\theta)|^2 + |\sin(\theta)|^2}$$

$$d2 := \sqrt{|r\sin(\theta)|^2 + |r\cos(\theta)|^2}$$

To complete the evaluation of Eq. (3.1), Holmes formed the ratios $n1/d1$ and $n2/d2$ in er, et and applied the simplify command with the symbolic option to remove the modulus signs which appeared in $d1$ and $d2$.

> er:=simplify(n1/d1,symbolic); et:=simplify(n2/d2,symbolic);

$$er := [\cos(\theta), \sin(\theta)]$$

$$et := [-\sin(\theta), \cos(\theta)]$$

A function f was formed for substituting the time dependence into r and θ.

> f:=v->subs(r=r(t),theta=theta(t),v):

In req and teq, Holmes has used f to derive the (time-dependent) relations between the polar unit vectors $\hat{e}_r$, $\hat{e}_\theta$ and the Cartesian unit vectors $\hat{e}_x$, $\hat{e}_y$.

> req:=e[r]=f(er[1])*e[x]+f(er[2])*e[y];

$$req := e_r = \cos(\theta(t))\,e_x + \sin(\theta(t))\,e_y$$

> teq:=e[theta]=f(et[1])*e[x]+f(et[2])*e[y];

$$teq := e_\theta = -\sin(\theta(t))\,e_x + \cos(\theta(t))\,e_y$$

These relations could have also been deduced by making a sketch of the two coordinate systems and using simple geometry. However, the above approach has the advantage of being quite general and can be applied to any 2- or 3-dimensional curvilinear coordinate system, no matter how complex. Now the *position* vector was formed in terms of $\hat{e}_x$, $\hat{e}_y$. These unit vectors are constants with respect to time, whereas $\hat{e}_r$ and $\hat{e}_\theta$ are not. The velocity and acceleration were calculated by differentiating *position* and *vel* with respect to time.

> position:=f(R[1])*e[x]+f(R[2])*e[y];

$$position := r(t)\cos(\theta(t))\,e_x + r(t)\sin(\theta(t))\,e_y$$

> vel:=diff(position,t); accel:=diff(vel,t);

$$vel := (\frac{d}{dt}r(t))\cos(\theta(t))\,e_x - r(t)\sin(\theta(t))(\frac{d}{dt}\theta(t))\,e_x$$
$$+(\frac{d}{dt}r(t))\sin(\theta(t))\,e_y + r(t)\cos(\theta(t))(\frac{d}{dt}\theta(t))\,e_y$$

To express the velocity and acceleration completely in polar coordinates, Holmes solved the two equations, req and teq, for e_x and e_y and assigned the solution.

> sol:=solve({req,teq},{e[x],e[y]}); assign(sol):

$$sol := \{e_y = \cos(\theta(t))\,e_\theta + e_r\sin(\theta(t)),\; e_x = -e_\theta\sin(\theta(t)) + \cos(\theta(t))\,e_r\}$$

The desired velocity and acceleration expressions then followed on applying the `simplify` command, and collecting the coefficients of e_r, e_θ in *accel*.

> `vel:=simplify(vel); accel:=simplify(accel);`

$$vel := r(t)\left(\frac{d}{dt}\theta(t)\right)e_\theta + \left(\frac{d}{dt}r(t)\right)e_r$$

> `accel:=collect(accel,[e[r],e[theta]]);`

$$accel := \left(\left(\frac{d^2}{dt^2}r(t)\right) - r(t)\left(\frac{d}{dt}\theta(t)\right)^2\right)e_r$$
$$+ \left(2\left(\frac{d}{dt}r(t)\right)\left(\frac{d}{dt}\theta(t)\right) + r(t)\left(\frac{d^2}{dt^2}\theta(t)\right)\right)e_\theta$$

Holmes assured me that the expressions *vel* and *accel* are the standard forms of the velocity and acceleration in plane polar coordinates. By this stage, I was beginning to wonder when Holmes was going to tackle the actual case, but I kept my mouth shut. Fortunately, he next entered Slimy's radial and angular coordinates at time t and I eagerly waited for him to solve the case.

> `r(t):=t^2*exp(-t); theta(t):=(3/5)*t^2-2*t+12;`

$$r(t) := t^2 e^{(-t)}$$

$$\theta(t) := \frac{3}{5}t^2 - 2t + 12$$

Using the coefficient (`coeff`) command to extract the radial and angular components and forming Maple row vectors, Slimy's velocity *vel2* and acceleration *accel2* at time t were then determined.

> `vel2:=<coeff(vel,e[r])|coeff(vel,e[theta])>;`

$$vel2 := \left[2te^{(-t)} - t^2 e^{(-t)},\ t^2 e^{(-t)}\left(\frac{6t}{5} - 2\right)\right]$$

> `accel2:=<coeff(accel,e[r])|coeff(accel,e[theta])>;`

$$accel2 := \left[2e^{(-t)} - 4te^{(-t)} + t^2 e^{(-t)} - t^2 e^{(-t)}\left(\frac{6t}{5} - 2\right)^2,\right.$$
$$\left. 2\left(2te^{(-t)} - t^2 e^{(-t)}\right)\left(\frac{6t}{5} - 2\right) + \frac{6}{5}t^2 e^{(-t)}\right]$$

Slimy's *speed* and acceleration magnitude (*a_mag*) immediately followed on applying `Norm` to *vel2* and *accel2* and simplifying.

> `speed:=simplify(Norm(vel2,2));`

$$speed := \frac{1}{5}te^{(-t)}\sqrt{100 - 100t + 125t^2 + 36t^4 - 120t^3}$$

> `a_mag:=simplify(Norm(accel2,2));`

Recalling that the second attack on Slimy occured when his speed was a maximum, Holmes decided to plot Slimy's speed to see at what time this attack took place. For completeness sake, he also plotted the acceleration magnitude

3.1. CURVILINEAR COORDINATES

in the same figure. On the computer screen, the speed was represented by a solid red curve, the acceleration magnitude by a blue dashed line. He added a title to this effect to his graph. The resulting plot is shown in Figure 3.1.

```
> plot([speed,a_mag],t=0..15,color=[red,blue],linestyle=[1,3],
> title="solid-speed, dashed-acceleration");
```

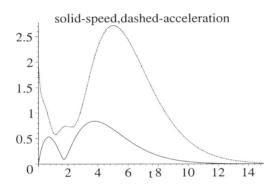

Figure 3.1: Slimy's speed and acceleration magnitude as a function of time.

By clicking on the computer plot and on the highest point of the speed curve, Holmes was able to deduce that the second attack took place at about 3.8 time units after Slimy left his hiding place. To determine the time of the first attack, Holmes formed the dot product of the *accel2* and *vel2* vectors, divided by the magnitudes, and took the arccosine of the result. The lengthy output was simplified somewhat with the simplify(symbolic) command.

```
> phi:=arccos(DotProduct(accel2,vel2)/(speed*a_mag));
> phi:=simplify(%,symbolic);
```

Holmes plotted the angle ϕ as a function of time, coloring the curve red. He also plotted horizontal lines corresponding to the angles π and $\pi/2$, coloring them blue and green, respectively. The resulting plot is shown in Figure 3.2.

```
> plot([phi,Pi,Pi/2],t=0..15,color=[red,blue,green],
> numpoints=200);
```

By clicking on the plot and the angle curve, Holmes concluded that the maximum angle between the velocity and acceleration vectors occured about 1.3 time units after Slimy left his hiding place. This was the time of the first unsuccessful attack. While impressed by Holmes's conclusions, I could no longer refrain from speaking and said, "But Holmes, who did it?"

Whereupon, Holmes calmly proceeded to create an animated picture of the path traced out by Slimy on his ill-fated journey. He formed a functional operator p to use the polarplot command to plot Slimey's trail for the time

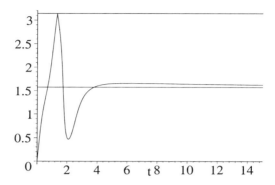

Figure 3.2: Angle between Slimy's acceleration and velocity vectors versus t.

range $t = 0$ to $0.05i$. He then entered the display command with the sequence command seq(p(i),i=1..200) and included the option insequence=true.

```
> p:=i->polarplot([r(t),theta(t),t=0..0.05*i]):
> display(seq(p(i),i=1..200),insequence=true,thickness=2);
```

When he executed this last command line, I could see the coordinate axes on the computer screen but nothing else. Glancing at me and chuckling in a slightly sadistic manner, Holmes clicked on the plot and then on the start arrow to initiate the animation. And Slimy's last journey began to evolve on the screen. When the animation was complete, a picture similar to that shown in Figure 3.3 resulted. I stared at the final picture blankly, but to Holmes's non-sluggish imagination the identity of the evil gang was immediately obvious.

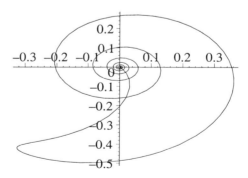

Figure 3.3: The trail of slime revealed the culprits' identity.

"The answer is elementary, my dear Watson," he said. "This is clearly a stylized drawing of a snail. So, the notorious Snail gang did the foul deed."

3.1.2 Abigail Ant Roams the Beach Ball

*It is the supreme art of the teacher to awaken joy in
creative expression and knowledge.*
Albert Einstein, Nobel laureate in physics (1879–1955)

Being an avid reader of mystery novels, particularly those that have a scientific slant to them, Jennifer has just finished reading the Case of the Artistic Slug. In an attempt to breathe some life into what might appear to her mechanics students as a dry mathematical topic, she has created a recipe which is not only a generalization of Sherlock Holmes' kinematic example but displays a simpler approach using the VectorCalculus package. Specifically, Jennifer has derived the general analytic forms of the velocity and acceleration vectors in spherical polar coordinates for Ms. Abigail (Abby) Ant and then applied the results to Abby's motion along a prescribed path on the surface of a large beach ball.

To derive general expressions for the velocity and acceleration in spherical polar coordinates, Jennifer considers Abby to be located at some instant in time at a radial distance r from the origin, the radius vector making an angle θ with the z-axis, and its projection in the x-y plane making an angle ϕ with the x-axis. The radial distance $r \geq 0$, θ ranges from 0 to π, and ϕ from 0 to 2π. After loading the plots and VectorCalculus packages, Jennifer enters the relationships between spherical polar and Cartesian coordinates (X, Y, Z).

```
>  restart: with(plots): with(VectorCalculus):
>  X:=r(t)*cos(phi(t))*sin(theta(t));
```
$$X := \mathrm{r}(t)\cos(\phi(t))\sin(\theta(t))$$
```
>  Y:=r(t)*sin(phi(t))*sin(theta(t));
```
$$Y := \mathrm{r}(t)\sin(\phi(t))\sin(\theta(t))$$
```
>  Z:=r(t)*cos(theta(t));
```
$$Z := \mathrm{r}(t)\cos(\theta(t))$$

Although she can easily sketch the surfaces corresponding to holding each of the coordinates, r, θ, and ϕ, equal to a constant value, Jennifer knows that this task is not so easy for some other 3-dimensional coordinate systems. She wonders if Maple has a command to plot these surfaces. Changing the colon to a semi-colon in the with(plots) command and re-executing the opening command line, she finds in the lengthy output list of plotting command structures the words coordplot and coordplot3d. These sound promising. Since she is working in three dimensions, Jennifer decides to select the latter. To find out what this command can do, she clicks her cursor on coordplot3d and then on Maple's Help at the top of her computer screen. The phrase Help on "coordplot3d" appears. Clicking on this entry opens an information window on coordplot3d.

Reading the description, this command appears to be just what she wants for plotting surfaces in three dimensions, holding each coordinate fixed. For two dimensions, the command `coordplot` can be used. Closing the information window and returning to her worksheet, she now enters `coordplot3d` with the spherical option. The grid option controls the smoothness of the surfaces drawn, the default values being [22,22]. Jennifer chooses to use a finer grid even though this will take more computer time to create the plot.

```
>   coordplot3d(spherical,axes=FRAME,labels=["x","y","z"],
>   grid=[40,40],tickmarks=[2,2,2],orientation=[-20,50]);
```

The resulting picture is shown in Figure 3.4. The sphere, cone, and half-plane correspond to holding r, θ, and ϕ equal to some particular constant values.

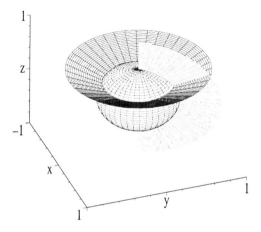

Figure 3.4: Surfaces of constant r, constant θ, and constant ϕ.

Jennifer now uses the `VectorField` command to enter a general position vector $\vec{R}$ for Abby in terms of the Cartesian unit vectors, the coordinate transformations being automatically substituted. A vector field object is created, this being indicated by the overbars appearing above the unit (basis) vectors. For general curvilinear coordinate systems, such as the spherical polar system, the directions of the unit vectors will vary from point to point in space. Only Cartesian unit vectors are independent of position.

```
>   R:=VectorField(<X,Y,Z>,'cartesian'[x,y,z]);
```

$R := \mathrm{r}(t)\cos(\phi(t))\sin(\theta(t))\,\overline{e}_x + \mathrm{r}(t)\sin(\phi(t))\sin(\theta(t))\,\overline{e}_y + \mathrm{r}(t)\cos(\theta(t))\,\overline{e}_z$

Jennifer calculates the velocity $\vec{v}$ and acceleration $\vec{a}$ (lengthy outputs not shown here) by differentiating $\vec{R}$ once, and then twice, with respect to time t.

```
>   v:=diff(R,t); a:=diff(R,t,t);
```

3.1. CURVILINEAR COORDINATES

The outputs are still in terms of the Cartesian basis vectors $\bar{e}_x, \bar{e}_y, \bar{e}_z$. Jennifer converts to the spherical polar basis vectors $\bar{e}_r, \bar{e}_\theta, \bar{e}_\phi$ by creating the following MapToBasis function f and then forming $f(v)$ in $v2$ and $f(a)$ in $a2$.

```
> f:=V->MapToBasis(V,'spherical'[r,theta,phi]):
> v2:=f(v); a2:=f(a);
```

The above two command lines are equivalent to the many steps carried out by Sherlock Holmes in establishing and using the relationships between the unit vectors in Cartesian and polar coordinates. Again, Jennifer advises the reader to execute the recipe to see the lengthy output which has been suppressed here. In the output, the time-independent terms $\cos(\theta), \sin(\theta), \sin(\phi), \cos(\phi)$ appear, arising from the transformation of the basis vectors. The following function F restores the time dependence to such terms.

```
> F:=V->simplify(subs({cos(theta)=cos(theta(t)),sin(theta)=
> sin(theta(t)),cos(phi)=cos(phi(t)),sin(phi)=sin(phi(t))},V)):
```

Then $F(v2)$ and $F(a2)$ generate the velocity and acceleration expressions in spherical polar coordinates. To obtain the standard form for the acceleration, the identity $\cos^2\theta(t) = 1 - \sin^2\theta(t)$ is substituted with the algsubs command.

```
> v3:=F(v2); a3:=algsubs(cos(theta(t))^2=1-sin(theta(t))^2,F(a2));
```

$$v3 := \left(\frac{d}{dt}r(t)\right)\bar{e}_r + r(t)\left(\frac{d}{dt}\theta(t)\right)\bar{e}_\theta + r(t)\left(\frac{d}{dt}\phi(t)\right)\sin(\theta(t))\bar{e}_\phi$$

$$a3 := \left(\left(\frac{d^2}{dt^2}r(t)\right) - r(t)\sin(\theta(t))^2\left(\frac{d}{dt}\phi(t)\right)^2 - r(t)\left(\frac{d}{dt}\theta(t)\right)^2\right)\bar{e}_r +$$

$$\left(r(t)\left(\frac{d^2}{dt^2}\theta(t)\right) + 2\left(\frac{d}{dt}r(t)\right)\left(\frac{d}{dt}\theta(t)\right) - r(t)\sin(\theta(t))\cos(\theta(t))\left(\frac{d}{dt}\phi(t)\right)^2\right)\bar{e}_\theta$$

$$+ \left(r(t)\left(\frac{d^2}{dt^2}\phi(t)\right)\sin(\theta(t)) + 2\left(\frac{d}{dt}r(t)\right)\left(\frac{d}{dt}\phi(t)\right)\sin(\theta(t))\right.$$

$$\left. + 2r(t)\left(\frac{d}{dt}\theta(t)\right)\left(\frac{d}{dt}\phi(t)\right)\cos(\theta(t))\right)\bar{e}_\phi$$

Jennifer confirms that the above spherical polar expressions for the general velocity and acceleration are correct by comparing them with those given in the mechanics text by Fowles and Cassiday [FC99]. The general expression for the speed follows on taking the square root of the dot product of $v3$ with itself.

```
> speed:=sqrt(v3 . v3);
```

$$speed := \sqrt{(\frac{d}{dt}r(t))^2 + r(t)^2(\frac{d}{dt}\theta(t))^2 + r(t)^2(\frac{d}{dt}\phi(t))^2\sin(\theta(t))^2}$$

As a simple application of the above results, Jennifer now considers Abby to be moving on the surface of a beach ball of radius $r = A$ in such a way that Abby's angular coordinates at time t are $\theta(t) = (\pi/2)(1 + \cos(B\omega t)/4)$ and $\phi(t) = \omega t$. After deriving the analytic forms of Abby's position coordinates, velocity, and speed, she will choose specific parameter values and plot the speed as a function of time and create a 3-dimensional picture of Abby's motion on the surface of the ball. Jennifer enters the analytic forms of $r(t)$, $\theta(t)$, and $\phi(t)$.

> `r(t):=A; theta(t):=(Pi/2)*(1+cos(B*omega*t)/4);`
> `phi(t):=omega*t;`

These expressions are automatically substituted into X, Y, Z, the velocity *vel*, and the *speed*. The velocity and speed outputs are displayed here.

> `X:=X; Y:=Y; Z:=Z; vel:=v3; speed:=speed;`

$$vel := -\frac{1}{8} A \pi \sin(B\omega t) B\omega \, \overline{e}_\theta + A\omega \sin(\frac{1}{2}\pi(\frac{1}{4}\cos(B\omega t)+1))\overline{e}_\phi$$

$$speed := \sqrt{\frac{1}{64} A^2 \pi^2 \sin(B\omega t)^2 B^2 \omega^2 + A^2 \omega^2 \sin(\frac{1}{2}\pi(\frac{1}{4}\cos(B\omega t)+1))^2}$$

Jennifer takes the beach ball to be of radius $A = 1$ m, sets $B = 4$, and takes $\omega = 1/10$ rad/s. The time $T = 2\pi/\omega$ for Abby to complete one revolution of the ball is calculated and then rounded off to the nearest integer TT with the round command. This latter number will determine the number of graphs to be used in the animation of Abby's motion.

> `A:=1: B:=4: omega:=1/10: T:=evalf(2*Pi/omega); TT:=round(T);`

$$T := 62.83185308$$
$$TT := 63$$

Abby's speed is now plotted for one revolution, the result being shown in Figure 3.5. Abby's speed varies periodically with time.

> `plot(speed,t=0..T,tickmarks=[4,4],labels=["t","speed"],`
> `color=blue,view=[0..T,0..0.2]);`

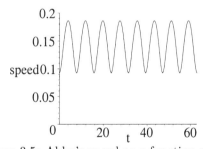

Figure 3.5: Abby's speed as a function of time.

Next, Jennifer creates a 3-dimensional plot of Abby's trajectory in `gr1` with the spacecurve command, the path being represented by a thick red curve.

> `gr1:=spacecurve([X,Y,Z],t=0..T,color=red,thickness=2):`

The surface of the beach ball is plotted in `gr2` with the `plot3d` command with the option `coords=spherical`. The choice of a `wireframe` style allows Abby's entire trajectory to be seen, even on the opposite side of the ball.

> `gr2:=plot3d(A,theta=0..2*Pi,phi=0..Pi,coords=spherical,`
> `style=wireframe):`

3.1. CURVILINEAR COORDINATES

A graph function gr3 is created, using the `pointplot3d` command, for plotting Abby's position at times $t = i$, where $i=0, 1, 2, ...TT$. Abby's position is represented by a size 20 black circle. The graphs gr1, gr2, gr3(i) are then superimposed at times $t=i$ with the operator pl.

```
> gr3:=i->pointplot3d([eval([X,Y,Z],t=i),symbol=circle,
> symbolsize=20,color=black):
> pl:=i->display({gr1,gr2,gr3(i)}):
```

The sequence of plots pl(i) from $i=0$ to TT are displayed, the Maple option `insequence =true` producing animation when the `display` command is executed, and one clicks on the plot and on the start arrow in the tool bar.

```
> display([seq(pl(i),i=0..TT)],insequence=true,labels=["x","y",
> "z"],axes=FRAME,view=[-1..1,-1..1,-1..1],tickmarks=[2,2,2]);
```

The opening frame of the animation shows a large black point superimposed on the entire trajectory, Abby's trajectory being shown in Figure 3.6.

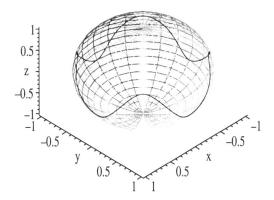

Figure 3.6: Abigail ant's path on the surface of the beach ball.

Jennifer is pleased with her example and her approach for a number of reasons. Taking her laptop to the lecture and projecting the executed recipe on a screen, she will be able to show the students what is possible with a computer algebra system. She can choose some other non-Cartesian coordinates besides spherical polar and easily modify the recipe to show the students what the velocity and acceleration vectors look like in these coordinate systems. More importantly, she can get the students involved and perhaps excited by asking how the ant's motion is altered when the parameter values are altered. They will be able to see that once a fairly general recipe has been developed, "what if...?" questions are easily and quickly answered without further tedious calculations. Jennifer feels that one of the most important things that she can do as a teacher is to introduce the students to the joys of discovery and provide them with the analytic and computer tools to accomplish their scientific goals.

3.1.3 This Doughnut Isn't for Eating

'Twixt the optimist and pessimist, The difference is droll:
The optimist sees the doughnut, But the pessimist sees the hole.
McLandburgh Wilson, *Optimist and Pessimist* (1915)

On October 31, 1952, the first hydrogen bomb was exploded in a test at the Eniwetok Atoll in the Pacific ocean. The fusion of the hydrogen nuclei led to an uncontrolled release of energy equivalent to 10 million tons of TNT. A current goal of experimental physics is to achieve a sustained and controlled source of fusion power in a fusion reactor. To alleviate possible future power shortages, the production of such fusion power is important because there is a great abundance of the fuel and a fusion reactor has less inherent dangers than a fission reactor. As an example of the latter, one might recall the disastrous accident at Chernobyl in the U.S.S.R in 1986.

The "Tokamak", an acronym in the Russian language for "toroidal magnetic chamber", is an experimental fusion reactor that was originally developed in the U.S.S.R., but has since been built and operated in several other countries. In a Tokamak, the charged particles that make up the high temperature plasma are confined by toroidal and poloidal magnetic fields inside a torus or doughnut. The magnetic fields exert forces that keep the plasma from touching the walls of the doughnut chamber. In order to model the motion of the charged particles inside a Tokamak, a toroidal coordinate system should prove useful. Figure 3.7 shows a cross section of a torus, the radial distance r being measured from the central circular axis of the doughnut, while the angles ϕ and θ are measured from the x- and z-axis, respectively. From the figure, it is easy to see that $x = (R + r\sin(\theta))\cos(\phi)$, $y = (R + r\sin(\theta))\sin(\phi)$, and $z = r\cos(\theta)$.

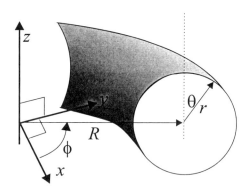

Figure 3.7: Relation of toroidal coordinates to the rectangular system.

For a general 3-dimensional orthogonal coordinate system, the differential elements of length in, say, the u, v, and w directions can be written in the form $ds_u = h_u du$, $ds_v = h_v dv$, and $ds_w = h_w dw$, where the "scale factor"

3.1. CURVILINEAR COORDINATES

$h_u = [(\partial x/\partial u)^2 + (\partial y/\partial u)^2 + (\partial z/\partial u)^2]^{1/2}$, and the other two scale factors, h_v and h_w, are obtained by replacing u with v and w. The relations of the unit vectors $\hat{e}_u$, $\hat{e}_v$, and $\hat{e}_w$, to the Cartesian unit vectors may be obtained by calculating $\hat{e}_u = \partial \vec{r}/\partial s_u = \partial \vec{r}/(h_u \, \partial u)$, etc., where $\vec{r} = x\hat{e}_x + y\hat{e}_y + z\hat{e}_z$ is the position vector with $x = x(u,v,w)$, etc.. The element of area on, say, a surface of constant u is $dA_u = ds_v \, ds_w = h_v h_w du \, dv$ while the volume element is $dV = ds_u ds_v ds_w = h_u h_v h_w du \, dv \, dw$. The gradient (grad or ∇), divergence (div or $\nabla \cdot$), curl ($\nabla \times$), and Laplacian (div grad or $\nabla \cdot \nabla \equiv \nabla^2$) operators are calculated as follows [Gri99], ψ and $\vec{A}$ being scalar and vector functions:

$$\nabla \psi = \frac{\hat{e}_u}{h_u}\frac{\partial \psi}{\partial u} + \frac{\hat{e}_v}{h_v}\frac{\partial \psi}{\partial v} + \frac{\hat{e}_w}{h_w}\frac{\partial \psi}{\partial w},$$

$$\nabla \cdot \vec{A} = \frac{1}{h_u h_v h_w}\left[\frac{\partial}{\partial u}(A_u h_v h_w) + \frac{\partial}{\partial v}(A_v h_u h_w) + \frac{\partial}{\partial w}(A_w h_u h_v)\right],$$

$$\nabla \times \vec{A} = \frac{\hat{e}_u}{h_v h_w}\left[\frac{\partial}{\partial v}(h_w A_w) - \frac{\partial}{\partial w}(h_v A_v)\right] + \frac{\hat{e}_v}{h_u h_w}\left[\frac{\partial}{\partial w}(h_u A_u) - \frac{\partial}{\partial u}(h_w A_w)\right]$$
$$+ \frac{\hat{e}_w}{h_u h_v}\left[\frac{\partial}{\partial u}(h_v A_v) - \frac{\partial}{\partial v}(h_u A_u)\right],$$

$$\nabla^2 f = \frac{1}{h_u h_v h_w}\left[\frac{\partial}{\partial u}\left(\frac{h_v h_w}{h_u}\frac{\partial f}{\partial u}\right) + \frac{\partial}{\partial v}\left(\frac{h_u h_w}{h_v}\frac{\partial f}{\partial v}\right) + \frac{\partial}{\partial w}\left(\frac{h_u h_v}{h_w}\frac{\partial f}{\partial w}\right)\right].$$

In this recipe, we will not look at the operation of a tokomak, but instead derive some important analytic results for the toroidal coordinate system. Although the toroidal coordinates that we have introduced seem "natural", we shall discover that they are not the same as those built into the Maple system. So we shall introduce "our" toroidal coordinate system and plot surfaces of constant r, θ, and ϕ to see if we do indeed get a doughnut. Then we will calculate the scale factors, h_r, h_θ, h_ϕ, and perform the integrations $\int dA_r$ and $\int dV$ to determine the doughnut's total surface area and volume. We shall also determine the toroidal unit vectors $\hat{e}_r$, $\hat{e}_\theta$, $\hat{e}_\phi$, and check that they are indeed orthogonal and have unit magnitude. Finally, with the scale factors known, we will calculate the gradient, divergence, and curl for the toroidal system.

The plots and LinearAlgebra packages are loaded, the latter being needed for calculating the norm and the dot product. We then query Maple on what coordinate systems it knows. On executing the ?coords command, a Help

```
>   restart: with(plots): with(LinearAlgebra):
>   ?coords;
```

window on the 2- and 3-dimensional coordinate systems supported by Maple appears. Scanning down the list, the toroidal coordinate system defined there differs in structure from the one introduced in Figure 3.7. Closing the Help window, let us enter the three relations that define "our" toroidal system. We shall refer to this coordinate system, which is not built in to Maple, as "ourtoroid".

```
>   x:=(R+r*sin(theta))*cos(phi);
```

$$x := (R + r\sin(\theta))\cos(\phi)$$

```
>   y:=(R+r*sin(theta))*sin(phi);
```
$$y := (R + r\sin(\theta))\sin(\phi)$$
```
>   z:=r*cos(theta);
```
$$z := r\cos(\theta)$$
For plotting purposes, we shall take $R=3$. Later, R will be unassigned.
```
>   R:=3:
```
The command addcoords can be used to add a new coordinate system to Maple.
```
>   addcoords(ourtoroid,[r,theta,phi],[x,y,z],[r,theta,phi],
>   [[2],[0],[0],[0..2,0..2*Pi,0..2*Pi],[-5..5,0..5,-3..3]]);
```
The name ourtoroid is inserted as the first argument. The second argument in the above command line is a list of the variables (r, θ, ϕ) in this new toroidal system. The third argument is a list of expressions (x, y, z) relating the Cartesian coordinates to the toroidal coordinates. These relations have already been entered. The fourth argument is a list of the quantities which are regarded as constants in the Cartesian system. The last argument is a list of lists which gives default values to be used in coordplot3d, the command for plotting surfaces of constant r, θ, and ϕ. The first three entries in the list of lists indicate that the surfaces $r = 2$, $\theta = 0$, and $\phi = 0$ are to be drawn. More surfaces can be included if so desired. The fourth entry gives the ranges of r (from 0 to 2), θ (from 0 to 2π), and ϕ (from 0 to 2π). The last entry gives the range of the viewing box ($x = -5$ to 5, $y = 0$ to 5, $z = -3$ to 3). The y range has been chosen to show half a doughnut. If a full doughnut is desired, take $y = -5$ to 5. The half-doughnut is now drawn by using the coordplot3d command, the
```
>   coordplot3d(ourtoroid,axes=boxed,scaling=constrained,
>   labels=["x","y","z"],orientation=[-40,40]);
```

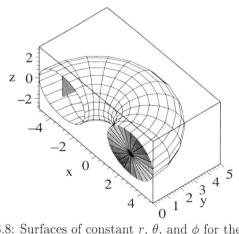

Figure 3.8: Surfaces of constant r, θ, and ϕ for the half-doughnut.

3.1. CURVILINEAR COORDINATES

result being shown in Fig. 3.8. In the figure, the half-doughnut with radius $r = 2$ is clearly seen. The darkly shaded right end corresponds to the surface $\phi = 0$, while a small portion of the $\theta = 0$ surface can be seen in the left end. The latter surface may be more clearly viewed on the computer screen by rotating the doughnut with your cursor.

R is now unassigned and, after assuming that $r > 0$, $R > r$, $\theta > 0$, and $\phi > 0$, a general *position* (row) vector is entered.

> unassign('R'): assume(r>0,R>r,theta>0,phi>0):
> position:=<x|y|z>;

$$position := [(R + r\sin(\theta))\cos(\phi),\ (R + r\sin(\theta))\sin(\phi),\ r\cos(\theta)]$$

A function f is formed for differentiating *position* with respect to a variable u, and a function hh for generating the scale factors for our toroidal system.

> f:=u->map(diff,position,u):
> hh:=f->simplify(Norm(f,2)):

Choosing u to be r, θ, and ϕ, the scale factors h_r, h_θ, h_ϕ follow on calculating $hh(f(r))$, $hh(f(\theta))$, and $hh(f(\phi))$, respectively.

> h[r]:=hh(f(r)); h[theta]:=hh(f(theta)); h[phi]:=hh(f(phi));

$$h_r := 1$$
$$h_\theta := r$$
$$h_\phi := R + r\sin(\theta)$$

With all the scale factors known, a functional operator F for calculating the ratio $f(v)/h[v]$ and simplifying is created. This ratio will generate the unit vectors $\hat{e}_r$, $\hat{e}_\theta$, $\hat{e}_\phi$. The unit vectors for our toroidal system follow on subsequently calculating $F(r)$, $F(\theta)$, and $F(\phi)$. They are assigned the names er, et, ep.

> F:=v->simplify(f(v)/h[v]):
> er:=F(r); et:=F(theta); ep:=F(phi);

$$er := [\sin(\theta)\cos(\phi),\ \sin(\theta)\sin(\phi),\ \cos(\theta)]$$
$$et := [\cos(\theta)\cos(\phi),\ \cos(\theta)\sin(\phi),\ -\sin(\theta)]$$
$$ep := [-\sin(\phi),\ \cos(\phi),\ 0]$$

That these are indeed orthogonal unit vectors is readily checked. We create a function G for calculating the dot product between two vectors $\vec{a}$ and $\vec{b}$.

> G:=(a,b)->simplify(DotProduct(a,b)):

In $m1$, $m2$, and $m3$, we verify that $\hat{e}_r \cdot \hat{e}_r = 1$, $\hat{e}_\theta \cdot \hat{e}_\theta = 1$, $\hat{e}_\phi \cdot \hat{e}_\phi = 1$,

> m1:=G(er,er); m2:=G(et,et); m3:=G(ep,ep);

$$m1 := 1$$
$$m2 := 1$$
$$m3 := 1$$

while in $m4$, $m5$, and $m6$, one has $\hat{e}_r \cdot \hat{e}_\theta = 0$, $\hat{e}_r \cdot \hat{e}_\phi = 0$, and $\hat{e}_\phi \cdot \hat{e}_\theta = 0$.

> m4:=G(er,et); m5:=G(er,ep); m6:=G(ep,et);

$$m4 := 0$$

$$m5 := 0$$
$$m6 := 0$$

The surface area $\int_0^{2\pi} \int_0^{2\pi} h_\theta h_\phi d\theta\, d\phi$ of a doughnut of radius $r = b$ is displayed by using the Int command twice and then is evaluated with the value command.

```
> Area:=Int(Int(eval(h[theta]*h[phi],r=b),theta=0..2*Pi),
> phi=0..2*Pi);
```

$$Area := \int_0^{2\pi} \int_0^{2\pi} b\,(R + b\sin(\theta))\, d\theta\, d\phi$$

```
> Area:=value(Area);
```

$$Area := 4\,b\,R\,\pi^2$$

The surface area of our toroid is $4\pi^2 bR$. Similarly, the volume of the doughnut is calculated by performing a triple integration. The area and volume expressions agree with the results quoted in standard calculus texts (e.g., Stewart[Ste87]).

```
> Volume:=Int(Int(Int(h[r]*h[theta]*h[phi],r=0..b),
> theta=0..2*Pi),phi=0..2*Pi);
```

$$Volume := \int_0^{2\pi} \int_0^{2\pi} \int_0^b r\,(R + r\sin(\theta))\, dr\, d\theta\, d\phi$$

```
> Volume:=value(Volume);
```

$$Volume := 2\,R\,b^2\,\pi^2$$

Now functional operators g, d, and c are formed for calculating the gradient, divergence, and curl, when the coordinates u, v, w are specified.

```
> g:=(u,v,w)->e[u]*diff(psi(u,v,w),u)/h[u]+
> e[v]*diff(psi(u,v,w),v)/h[v]+e[w]*diff(psi(u,v,w),w)/h[w];
```

$$g := (u,\,v,\,w) \to \frac{e_u \operatorname{diff}(\psi(u,\,v,\,w),\,u)}{h_u}$$
$$+ \frac{e_v \operatorname{diff}(\psi(u,\,v,\,w),\,v)}{h_v} + \frac{e_w \operatorname{diff}(\psi(u,\,v,\,w),\,w)}{h_w}$$

```
> d:=(u,v,w)->diff(A[u](u,v,w)*h[v]*h[w],u)/(h[u]*h[v]*h[w])
> +diff(A[v](u,v,w)*h[u]*h[w],v)/(h[u]*h[v]*h[w])
> +diff(A[w](u,v,w)*h[u]*h[v],w)/(h[u]*h[v]*h[w]);
```

$$d := (u,\,v,\,w) \to \frac{\operatorname{diff}(A_u(u,\,v,\,w)\,h_v\,h_w,\,u)}{h_u\,h_v\,h_w}$$
$$+ \frac{\operatorname{diff}(A_v(u,\,v,\,w)\,h_u\,h_w,\,v)}{h_u\,h_v\,h_w} + \frac{\operatorname{diff}(A_w(u,\,v,\,w)\,h_u\,h_v,\,w)}{h_u\,h_v\,h_w}$$

```
> c:=(u,v,w)->e[u]*(diff(h[w]*A[w](u,v,w),v)-diff(h[v]*
> A[v](u,v,w),w))/(h[v]*h[w])+e[v]*(diff(h[u]*A[u](u,v,w),w)
> -diff(h[w]*A[w](u,v,w),u))/(h[u]*h[w])+e[w](diff(h[v]*
> A[v](u,v,w),u)-diff(h[u]*A[u](u,v,w),v))/(h[u]*h[v]);
```

3.2. VECTOR OPERATORS

$$c := (u, v, w) \to \frac{e_u \left(\text{diff}(h_w A_w(u, v, w), v) - \text{diff}(h_v A_v(u, v, w), w)\right)}{h_v h_w}$$
$$+ \frac{e_v \left(\text{diff}(h_u A_u(u, v, w), w) - \text{diff}(h_w A_w(u, v, w), u)\right)}{h_u h_w}$$
$$+ \frac{e_w \left(\text{diff}(h_v A_v(u, v, w), u) - \text{diff}(h_u A_u(u, v, w), v)\right)}{h_u h_v}$$

The gradient of ψ, the divergence of $\vec{A}$, and the curl of $\vec{A}$ are now calculated,

> `gradpsi:=g(r,theta,phi); divA:=d(r,theta,phi);`
> `curlA:=c(r,theta,phi);`

$$gradpsi := \mathrm{e}_r \left(\frac{\partial}{\partial r} \psi(r, \theta, \phi)\right) + \frac{\mathrm{e}_\theta \left(\frac{\partial}{\partial \theta} \psi(r, \theta, \phi)\right)}{r} + \frac{\mathrm{e}_\phi \left(\frac{\partial}{\partial \phi} \psi(r, \theta, \phi)\right)}{R + r \sin(\theta)}$$

the lengthy outputs for the divergence and curl being suppressed here.

3.2 Vector Operators

Clearly, the procedure of the last recipe could be used to derive the scale factors and then the vector operators for other coordinate systems, but this is not necessary. In the `VectorCalculus` library package, the commands `Curl`, `Divergence`, `Gradient`, and `Laplacian` are available in the common coordinate systems with the scale factors already worked out for us. Even for coordinate systems unknown to Maple (such as "our" toroidal system), the scale factors do not have to be calculated. If you know the relations between the Cartesian coordinates and those of the "new" system, the new system can be added to Maple's repertoire by using the `AddCoordinates` command. The vector operators for the new system are then available for use. Our first recipe in this section involves calculating the gradient in Cartesian coordinates (here $h_x=h_y=h_z=1$) and making use of the idea that the gradient at a point P represents the maximum slope at P and is perpendicular to the equipotential through P.

3.2.1 Enon on the Hill

And Noah he often said to his wife when he sat down to dine,
'I don't care where the water goes if it doesn't get into the wine.'
G. K. Chesterton, English essayist, novelist, and poet (1874–1936)

The all-season resort (featuring skiing in the winter and hiking in the summer) of Enon is located near the top of a tall hill in the Southern Alps of Erehwon. Eoj Ig and his wife Enaj, whom we met earlier, are attending a short summer course on mathematical applications of computer algebra at the resort and have the afternoon free from lectures. They had wanted to go hiking, but it has begun to rain heavily so they are spending the time working on their computer algebra assignment. They have been told that the hilly region surrounding Enon has a height profile $h(x,y)$ given approximately by

$$h := e^{(-2(x-.01)^2 - 2(y-1)^2)} + 2xy\, e^{(-2(x-.002)^2 - 2(y+.75)^2)} + \cos(.5x)\cos(.4y)$$

and that Enon is located at $xe = 0.02$, $ye = 0.9$. A small lake is positioned at $xl = 0.55$, $yl = -1.05$. All distances are in km with positive x pointing to the east and positive y to the north. The assignment is as follows:

(a) Determine the elevation of Enon and the lake.

(b) Produce a 3-dimensional contour plot of the region over the range $x = -2..2$, $y = -2..2$, and use colored circles to represent Enon and the lake.

(c) Determine the location and height of the hill on which Enon is located.

(d) Determine the direction and magnitude of the maximum downhill slope at Enon. Assuming that rain water flows in this direction, what is the acceleration of rain water at Enon? Neglect friction.

(e) Assuming that the rain water from Enon follows the path of steepest descent down the hill, calculate this path and show (by superimposing it on the 3-d contour plot) that the water flows into the lake.

(f) Produce a constrained 2-d contour plot of the region with arrows indicating the magnitude and direction of steepest downward slope. Place Enon, the lake, and the path of the rain water from Enon on this plot.

Let us eavesdrop on Enaj and Eoj as they tackle this assignment. Enaj is at the keyboard of their laptop and begins by loading the `plots` and `VectorCalculus` packages. The latter contains the `Gradient` command which they will be using.

```
>  restart: with(plots): with(VectorCalculus):
```

"Enaj," Eoj says, "We could enter the given hill function $h(x,y)$ directly, but I have noticed that the first two terms involve Gaussian shapes which will produce two hills while the third term, involving the product of two cosines, generates a rolling background. Let's be a bit more general and create functions with adjustable parameters which can produce these shapes. This will enable us to tweak the overall topography, if necessary, by choosing different parameter values than those that Professors Snne and Eriugcm have given us."

"Good point, Eoj. I will create a Gaussian function $h1$ which produces a hill of height $H1$ centered at $x = a1$, $y = b1$, with a width parameter $c1$, and then a cosine product function $h2$ of height $H2$ with $a2$ and $b2$ determining the wavelength of the undulations in the x and y directions. The hill function h is then given by $h = h1(1, 0.01, 1, 2) + xy\, h1(2, 0.002, -0.75, 2) + h2(1, 0.5, 0.4)$. While we are in a function-creating mood, I might as well introduce a function hi for evaluating the height, when the x and y coordinates are specified."

```
>  h1:=(H1,a1,b1,c1)->H1*exp(-c1*((x-a1)^2+(y-b1)^2)):
>  h2:=(H2,a2,b2)->H2*cos(a2*x)*cos(b2*y):
>  h:=h1(1,0.01,1,2)+x*y*h1(2,0.002,-0.75,2)+h2(1,0.5,0.4);
```

3.2. VECTOR OPERATORS

$$h := e^{(-2(x-.01)^2 - 2(y-1)^2)} + 2\,x\,y\,e^{(-2(x-.002)^2 - 2(y+.75)^2)} + \cos(.5\,x)\cos(.4\,y)$$

> `hi:=(xi,yi)->eval(h,{x=xi,y=yi}):`

"O.K., Enaj. Why don't we now answer the first part of the assignment by entering the coordinates $xe = 0.02$, $ye = 0.9$ of Enon and $xl = 0.55$, $yl = -1.05$ for the lake. Enon's elevation ze is then given by calculating $hi(xe, ye)$ and the lake's elevation zl is similarly found."

> `xe:=0.02: ye:=0.9: xl:=0.55: yl:=-1.05:`

> `ze:=hi(xe,ye); zl:=hi(xl,yl);`

$$ze := 1.916008024$$
$$zl := 0.3497675486$$

"So, Enon's elevation is 1.916 km, or 1916 meters, above sea level, while the lake is at an elevation of about 350 meters. To answer part (b) of the assignment, you can use the `plot3d` command in `hplot`, with a `patchcontour` style and 20 contours, to plot h and the `pointplot3d` command in `elplot` to produce size 20 red circles indicating the positions of Enon and the lake."

> `hplot:=plot3d(h,x=-2..2,y=-2..2,axes=framed,orientation=`
> `[-20,40],style=patchcontour,contours=20,labels=["x","y","z"]):`
> `elplot:=pointplot3d([[xe,ye,ze],[xl,yl,zl]],symbol=circle,`
> `symbolsize=20,color=red):`

Having entered the relevant plotting commands, Enaj superimposes the two graphs to create a 3-dimensional contour plot of the region in Figure 3.9.

> `display({hplot,elplot});`

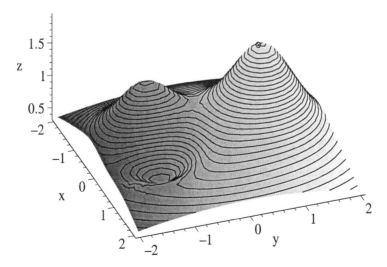

Figure 3.9: 3-dimensional contour plot of the region around Enon.

Clicking on the computer plot and dragging with the mouse, Enaj rotates the 3-dimensional figure to look at it from different angles. "It looks pretty good, Eoj. There's Enon, colored red, near the top of the higher hill and the lake at the bottom of the dip. What do we have to do next?"

"In part (c), we are supposed to locate the top of the hill on which Enon is located and the elevation of this peak. Since this is a 3-dimensional plot, you can't simply put your cursor at the top of the hill and click to find the elevation. We can use the gradient, however. At the top of the hill, the gradient must be zero. We are also asked in part (d) to find the maximum downhill slope at Enon, so why don't you enter minus the gradient of h in terms of the Cartesian coordinates x and y. If we computed the positive gradient, we would obtain the maximum uphill slope."

Enaj does this, labeling the result as *gradeq*. Viewing the lengthy output (not shown here) on her computer screen, she notes that it is given as a vector expressed in terms of the Cartesian basis vectors $\bar{e}_x$, $\bar{e}_y$.

> `gradeq:=-Gradient(h,[x,y]);`

She then sets the components of *gradeq* equal to zero and numerically solves for x and y over the range $x = -2$ to 2 and $y = 0$ to 2. The y range is chosen so as to pick out the top of the taller of the two hills.

> `peak:=fsolve({gradeq[1]=0,gradeq[2]=0},{x=-2..2,y=0..2});`

$$peak := \{x = 0.01074572229, y = 0.9622607909\}$$

"The top of the hill is at about $x = 0.011$, $y = 0.96$," remarks Eoj. "You can assign *peak* and apply the function hi to find the height of the peak."

> `assign(peak): hpeak:=hi(x,y);`

$$hpeak := 1.924034174$$

"So, the top of the hill on which Enon is located is about 1924 meters above sea level. Enon is only about eight meters lower. You had now better **unassign** x and y, so that they can be used again as general variables."

> `unassign('x','y'):`

Having done this, Enaj creates a slope function s for evaluating the x and y components of the maximum slope at an arbitrary point xi, yi. In addition to xi, yi, the component n of *gradeq* must also be specified.

> `s:=(xi,yi,n)->eval(gradeq[n],{x=xi,y=yi}):`

"With this slope function, I can now determine the maximum downhill slope components at Enon by calculating $s(xe, ye, 1)$ and $s(xe, ye, 2)$," she remarks.

> `sxe:=s(xe,ye,1); sye:=s(xe,ye,2);`

$$sxe := 0.03612361953$$
$$sye := -0.2502457609$$

"The x and y components of the maximum slope at Enon are $sxe = 0.036$ and $sye = -0.25$. By taking the arctangent of sye/sxe and converting to degrees, I can determine the angular direction of the maximum downhill slope at Enon."

> `angulardirection:=evalf(arctan(sye/sxe)*180/Pi)*degrees;`

3.2. VECTOR OPERATORS

$$angulardirection := -81.78594702\ degrees$$

"The answer is -81.8 degrees, i.e., remembering the meaning of our coordinates, this is 81.8 degrees south of east. I can now calculate the maximum downhill slope at Enon by forming $\sqrt{sxe^2 + sye^2}$."

> `max_slope:=sqrt(sxe^2+sye^2);`

$$max\_slope := 0.2528395870$$

"The magnitude of the downhill slope at Enon is about 0.25. We can convert this to an angle in radians, by taking the arctangent of this maximum slope, and then into degrees using the `convert` command."

> `angle:=arctan(max_slope);#in radians`

$$angle := 0.2476494248$$

> `angle2:=convert(angle,units,radian,degree)*degrees;`

$$angle2 := 14.18926683\ degrees$$

"The downhill slope at Enon makes an angle of 14.2 degrees with the horizontal. We were also asked to find the acceleration of rain water at Enon, neglecting friction. Let's take the acceleration due to gravity $g \approx 10$ m/s^2. Using Newton's second law, the acceleration of the rain water down the slope is equal to g times the sine of the angle (in radians)."

> `accel:=10*sin(angle);`

$$accel := 2.451257764$$

"So, at Enon the rain water has an acceleration down the hill of 2.45 m/s^2. Assuming that this water follows the path of steepest descent, we are asked to determine its path and show that the rain water flows into the lake. What approach would you suggest, Eoj?"

"Move over and let me at the keyboard. I will construct a simple do loop that will iterate the change in position of the rain water as it descends from Enon along the path of maximum slope. I will take a sufficient number N of steps of step size d to move from Enon to the lake. The starting coordinates $(x[0], y[0], z[0])$ for the loop are, of course, those for Enon. Since we will superimpose the path on the 3-dimensional contour plot made earlier and also on a 2-d contour plot, we enter the starting points $pt[0] = [x[0], y[0], z[0]]$ and $pt2[0] = [x[0], y[0]]$."

> `N:=800: d:=0.005: x[0]:=xe: y[0]:=ye: z[0]:=ze:`
> `pt[0]:=[x[0],y[0],z[0]]: pt2[0]:=[x[0],y[0]]:`

"In the do loop, the x-coordinate of the rain water on step $i+1$ is related to the value on the ith step by the relation $x[i+1] = x[i] + s(x[i], y[i], 1)\, d$. The y coordinate change is of a similar form. The z-coordinate on step $i+1$ is then calculated and the points $pt[i+1]$ and $pt2[i+1]$ formed."

> `for i from 0 to N do`
> `x[i+1]:=x[i]+s(x[i],y[i],1)*d:`
> `y[i+1]:=y[i]+s(x[i],y[i],2)*d:`

```
>   z[i+1]:=hi(x[i+1],y[i+1]):
>   pt[i+1]:=[x[i+1],y[i+1],z[i+1]];
>   pt2[i+1]:=[x[i+1],y[i+1]];
>   end do:
```
"On completion of the do loop, we can use the `spacecurve` command to plot the sequence of points $pt[i]$ as a 3-dimensional thick blue line."
```
>   path:=spacecurve([seq(pt[i],i=0..N)],style=line,color=blue,
>   thickness=3):
```
"If we now include the `path` plot in the `display` command along with `hplot` and `elplot`, we will see the 3-d path of the rain water flowing from Enon to the lake." (The reader will have to execute the recipe to see this picture.)
```
>   display({hplot,elplot,path});
```
"That's a neat picture, Eoj. I can see other applications of what you have done. For example, we could start a daredevil skier somewhere else on the hill and find his path of steepest descent to the bottom. Of course, this doesn't take into account such minor impediments as trees, large boulders, etc. Well, we are almost done. Only the last question on the assignment remains to be answered, and I see that it has quit raining and the sun is out. Maybe we can get in a short hike before supper."

"You're right. We have more lectures after supper, so let's quickly finish and get in that hike before going to eat. I will use the `fieldplot` command to plot the 2-dimensional vector field produced in *gradeq*. The arrows are taken to be thick and red and a `grid=[15,15]` is chosen. The default grid is [20,20]. Here, the arrows will point in the direction of the negative gradient, their length being an indication of magnitude. Longer arrows correspond to a steeper slope."
```
>   fplot:=fieldplot(gradeq,x=-2..2,y=-2..2,arrows=THICK,
>   grid=[15,15],color=red):
```
"In `cplot`, a 2-dimensional contour plot with 20 contours is created, while in `elplot2` the positions of Enon and the lake are represented by red circles."
```
>   cplot:=contourplot(h,x=-2..2,y=-2..2,colour=green,
>   grid=[40,40],contours=20,numpoints=5000,thickness=1):
>   elplot2:=plot([[xe,ye],[xl,yl]],style=point,symbol=circle,
>   symbolsize=20,color=red):
```
"Using the sequence of points $pt2[i]$ generated earlier in the do loop, we can plot the path of the rainwater in two dimensions. Let's use a line style in the `pointplot` command and take the curve to be thick and colored blue."
```
>   path2:=pointplot([seq(pt2[i],i=0..N)],style=line,color=blue,
>   thickness=3):
```
"The four graphs are then superimposed to produce Figure 3.10."
```
>   display({fplot,cplot,elplot2,path2},scaling=constrained);
```

3.2. VECTOR OPERATORS

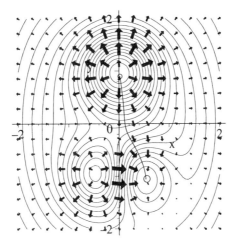

Figure 3.10: Path of rain water superimposed on 2-d contour/vector field plot.

"The contours of constant height are clearly seen. The arrows indicate the direction and magnitude of the steepest downhill slope. The path of the rain water follows the direction of the negative gradient from Enon to the lake. I guess that I am biased, but that's one mighty fine picture. Let's remove the output from our file and save it. Now where did I put my hiking boots?"

3.2.2 Are You Conservative, Mr. Vector Field?

To live your life is not as simple as to cross a field.
Boris Pasternak, Russian novelist and poet (1890–1960)

Eoj and Enaj have been given another assignment in their summer short course, this one involving the curl operator and conservative vector fields. For an irrotational (conservative) vector field $\vec{F}$, curl$\vec{F} = 0$ everywhere, or equivalently the line integral $\oint \vec{F} \cdot d\vec{s} = 0$ for any closed loop. For a conservative field, $\vec{F}$ is the gradient of some scalar potential function ϕ, $\vec{F} = -\nabla\phi$. Specifically, Eoj and Enaj have been asked to answer the following questions:

(a) Is the following velocity field (in Cartesian coordinates) conservative?

$$\vec{V} := [2x\sin(y) + 2y^2 - 12x^3y^3 + 2xyz - 3z + 4xe^z]\hat{e}_x +$$

$$[(1+x^2)\cos(y) - 3y^2 + 4xy - 9x^4y^2 + (1+x^2)z]\hat{e}_y + [(1+x^2)y - 3x + 2x^2e^z]\hat{e}_z$$

(b) If so, determine ϕ. Evaluate ϕ at the point $x = 1.0$, $y = 1.8$, $z = -1.2$. Rounding this value of ϕ off to the nearest integer Φ, plot $\vec{V}$ and the two equipotentials corresponding to $\phi = \pm\Phi$.

(c) Is the following velocity field (in spherical coordinates) conservative?
$$\vec{V}2 := (r\cos^2\theta)\,\hat{e}_r - (r\cos\theta\sin\theta)\,\hat{e}_\theta + 3r\,\hat{e}_\phi$$
If not, confirm that $\vec{V}2$ satisfies the identity (valid for any vector field) div(curl($\vec{V}2$))=$\nabla\cdot(\nabla\times\vec{V}2)$=0.

(d) Evaluate the curl of $\vec{V}2$ at the point ($r=1$, $\theta=\pi/2$, $\phi=0$). Also express the answer in terms of Cartesian coordinates.

(e) Support your conclusion that $\vec{V}2$ is not conservative by computing $\oint \vec{V}2 \cdot d\vec{s}$ for the following closed loop which has four "legs": Leg 1: Along x-axis from origin O to $A = (x=1, y=0, z=0)$; Leg 2: Along an arc of radius 1 in the x-y plane from A to $B = (0,1,0)$; Leg 3: Along a vertical path in the y-z plane from B to $C = (0,1,2)$; Leg 4: Along a straight line in the y-z plane from C back to O. Plot the loop before doing the line integral.

(f) By integrating curl $\vec{V}2$ over the surface area S enclosed by the above contour Γ, verify Stokes's theorem which states that
$$\int_S (\nabla\times\vec{F})\cdot d\vec{A} = \oint_\Gamma \vec{F}\cdot d\vec{s}.$$
In applying this theorem, the following right-hand rule applies: If the fingers of your right hand point in the direction of the line integral, then your thumb points in the direction of $d\vec{A}$.

Enaj has gone into town to celebrate a girlfriend's recent engagement, leaving Eoj on his own. He decides to tackle the assignment, leaving it to Enaj to make comments and suggest improvements when she returns.

After loading the plots and VectorCalculus packages, Eoj enters the formidable-looking velocity field $\vec{V}$, using the VectorField command with the option that the Cartesian coordinates x, y, z are to be used.

```
>   restart: with(plots): with(VectorCalculus):
>   V:=VectorField(<2*x*sin(y)+2*y^2-12*x^3*y^3+2*x*y*z-3*z+
>   4*x*exp(z),(1+x^2)*cos(y)-3*y^2+4*x*y-9*x^4*y^2+(1+x^2)*z,
>   (1+x^2)*y-3*x+2*x^2*exp(z)>,'cartesian'[x,y,z]);
```

$$V := (2x\sin(y) + 2y^2 - 12x^3y^3 + 2xyz - 3z + 4xe^z)\,\bar{e}_x$$
$$+((x^2+1)\cos(y) - 3y^2 + 4xy - 9x^4y^2 + (x^2+1)z)\,\bar{e}_y$$
$$+((x^2+1)y - 3x + 2x^2 e^z)\,\bar{e}_z$$

He then uses the Curl command to take the curl of $\vec{V}$.

```
>   Curl1:=Curl(V);
```

$$Curl1 := 0\,\bar{e}_x$$

The result is a zero vector, so $\vec{V}$ is a conservative field. The potential ϕ can be determined from $\phi = -\int \vec{V}\cdot d\vec{s}$, the integration being from, say, the origin

3.2. VECTOR OPERATORS

along a straight line to an arbitrary point (x,y,z). This can be accomplished by using the LineInt command with the option Line(<0,0,0>,<x,y,z>).

> phi:=-LineInt(V,Line(<0,0,0>,<x,y,z>));

$\phi := -yz + 3xz - x^2 \sin(y) + y^3 - \sin(y) - 2xy^2 + 3x^4 y^3 - x^2 yz - 2x^2 e^z$

The potential $\phi(x,y,z)$ then is given by the above output. Eoj is struck by how easy it was to obtain the answer using computer algebra. The hand calculation would have taken much longer, with the possibility of producing an algebraic error. He now evaluates the potential ϕ at $x=1.0$, $y=1.8$, $z=-1.2$ and then uses the round command to round the answer off to the nearest integer in Φ.

> phi1:=eval(phi,{x=1.0,y=1.8,z=-1.2});

$$\phi 1 := 15.01791632$$

> Phi:=round(phi1);

$$\Phi := 15$$

The 3-dimensional implicit plotting command, implicitplot3d, is used in gr1 to plot the two equipotential surfaces corresponding to $\phi = \pm\Phi = \pm 15$,

> gr1:=implicitplot3d({phi=Phi,phi=-Phi},x=-3..3,y=-3..3,
> z=-3..3,grid=[20,20,20],style=patchcontour,shading=zhue):

while in gr2 the 3-d velocity field is represented by thick red arrows by using the fieldplot3d command. To produce a better final figure, the default grids [10,10,10] for implicitplot3d and [8,8,8] for fieldplot3d have been overruled.

> gr2:=fieldplot3d(V,x=-3..3,y=-3..3,z=-3..3,grid=[7,7,7],
> arrows=THICK,color=red):

The graphs gr1 and gr2 are then superimposed to yield Figure 3.11.

> display({gr1,gr2},axes=boxed,orientation=[40,40],
> tickmarks=[3,3,3],labels=["x","y","z"]);

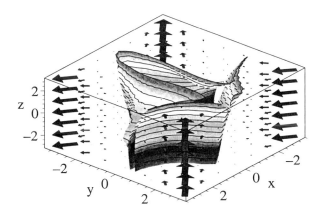

Figure 3.11: Two equipotential surfaces with the velocity field arrows.

To free ϕ for use as a spherical angular coordinate, Eoj unassigns it.

> unassign('phi'):

Although Eoj could enter the second velocity field with the `VectorField` command with the option 'spherical'[r,theta,phi], he decides for variety to use the following `SetCoordinates` command to set the new coordinates. The default coordinate system is now the spherical polar coordinate system until a new coordinate system is specified or the `restart` command is entered.

> SetCoordinates('spherical'[r,theta,phi]):

The second velocity field is now entered and the curl taken.

> V2:=VectorField(<r*cos(theta)^2,-r*cos(theta)*sin(theta),3*r>);

$$V2 := r\cos(\theta)^2\,\bar{e}_r - r\cos(\theta)\sin(\theta)\,\bar{e}_\theta + 3r\,\bar{e}_\phi$$

> Curl2:=Curl(V2);

$$Curl2 := \frac{3\cos(\theta)}{\sin(\theta)}\,\bar{e}_r - 6\,\bar{e}_\theta$$

Since $Curl2$ is not equal to zero, Eoj concludes that $\vec{V}2$ is not a conservative vector field. He confirms the identity div(curl($\vec{V}$))=0 for $\vec{V}=\vec{V}2$.

> Divergence(Curl2);

$$0$$

An alternate way to confirm the identity is to calculate $\nabla \cdot (\nabla \times \vec{V}2)$ working with the ∇ operator (command `Del` or `Nabla`) and the dot and cross products.

> Del.(Del &x Curl2);

$$0$$

To evaluate $Curl2$ at the point $(r=1, \theta=\pi/2, \phi=0)$, this point is entered as the vector $\vec{R}$. Because this is not a vector field, the unit vectors do not have bars over them.

> R:=<1,Pi/2,0>;

$$R := e_r + \frac{\pi}{2}\,e_\theta$$

In Cartesian coordinates, this point should be $(x=1, y=0, z=0)$. Eoj checks this out by using `MapToBasis` to convert $\vec{R}$ into Cartesian coordinates.

> R2:=MapToBasis(R,'cartesian'[x,y,z]);

$$R2 := e_x$$

The command `evalVF(Curl2,R)` evaluates the Vector Field $Curl2$ at the point $\vec{R}$. The result is then simplified in $value1$.

> value1:=simplify(evalVF(Curl2,R));

$$value1 := 6\,e_r$$

$Curl2$ points in the positive radial direction at $\vec{R}$, having a magnitude 6. Again using `MapToBasis`, $value1$ is converted to Cartesian coordinates in $value2$.

> value2:=MapToBasis(value1,'cartesian'[x,y,z]);

$$value2 := 6\,e_z$$

3.2. VECTOR OPERATORS

Curl2 points radially outwards along the positive z-axis.

Since $\vec{V}2$ is not a conservative vector field, the line integral of $\vec{V}2$ around the closed loop specified in part (e) of the assignment will not be zero. Eoj first plots the four legs of the loop. He introduces the graph function A to plot arrows for the three straight-line segments of the loop. The arrow(u,v,color=c) function plots at each base point in u a colored (specify c) arrow in the direction of each point in v. Eoj enters u and v as 3-dimensional Cartesian lists. The shape of the arrow head is taken to be a cylindrical arrow and the other arrow parameter values are arrived at by trial and error.

```
>   A:=(u,v,c)->arrow(u,v,color=c,width=0.02,head_length=0.2,
>   head_width=0.07,shape=cylindrical_arrow):
```

In gr||1, the first leg of the loop between the origin [0,0,0] and the (Cartesian) point [1,0,0] is graphed as a red arrow.

```
>   gr||1:=A([0,0,0],[1,0,0],red):
```

The second leg is in the x-y plane along a circular arc of radius 1 from the point [1,0,0] to the point [0,1,0]. This leg cannot be plotted with the arrow command. Instead, Eoj uses the spacecurve command with the list argument [cos(t),sin(t),0] to plot the arc through the angular range $t=0$ to $\pi/2$, coloring the curve as a thick orange line.

```
>   gr||2:=spacecurve([cos(t),sin(t),0],t=0..Pi/2,color=orange,
>   thickness=3):
```

The third and fourth legs again involve straight line segments which are graphed in gr||3 and gr||4 as green and blue arrows, respectively.

```
>   gr||3:=A([0,1,0],[0,0,2],green):
>   gr||4:=A([0,1,2],[0,-1,-2],blue):
```

Using the textplot3d command, Eoj adds the Cartesian coordinates of the end points of each leg and labels the axes as X, Y, and Z. The entries will be colored blue in the final plot.

```
>   gr||5:=textplot3d({[.9,0.1,0,"(1,0,0)"],[0,1.1,0.1,
>   "(0,1,0)"],[0,1,2.1,"(0,1,2)"],[.5,0,0.2,"X"],[0,.5,.1,"Y"],
>   [0,.05,1,"Z"]},color=blue):
```

The sequence of five graphs are then superimposed to produce the closed loop shown in Figure 3.12. Eoj has not bothered to use constrained scaling.

```
>   display({seq(gr||i,i=1..5)},axes=normal,
>   tickmarks=[0,0,0],orientation=[40,60]);
```

Now Eoj is ready to calculate the line integral of $\vec{V}2$ for each leg of the loop. He finds that it is more convenient to work in Cartesian coordinates, rather than spherical coordinates. So he uses MapToBasis in *V2b* to convert $\vec{V}2$ into 3-dimensional Cartesian coordinates.

```
>   V2b:=simplify(MapToBasis(V2,'cartesian'[x,y,z]));
```

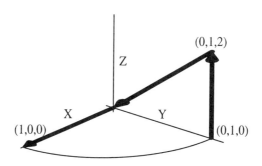

Figure 3.12: Path for performing the closed line integral of $\vec{V}2$.

$$V2b := -\frac{3\sqrt{x^2+y^2+z^2}\,y}{\sqrt{x^2+y^2}}\,\bar{e}_x + \frac{3\sqrt{x^2+y^2+z^2}\,x}{\sqrt{x^2+y^2}}\,\bar{e}_y + z\,\bar{e}_z$$

He tackles the curved second leg first. Since the arc of a circle in a plane is involved, he sets the coordinates to the 2-dimensional Cartesian system. He then forms the vector field $\vec{V}3$ from the first two components of $V2b$.

> SetCoordinates('cartesian'[x,y]):
> V3:=VectorField(<V2b[1],V2b[2]>);

$$V3 := -\frac{3\sqrt{x^2+y^2+z^2}\,y}{\sqrt{x^2+y^2}}\,\bar{e}_x + \frac{3\sqrt{x^2+y^2+z^2}\,x}{\sqrt{x^2+y^2}}\,\bar{e}_y$$

He uses the line integral command LineInt to evaluate the line integral along an arc in a plane at a fixed value z. The angular range varies from 0 to $\pi/2$. The arc is part of a circle of radius 1 centered at the origin.

> LineInt(V3,Arc(Circle(<0,0>,1),0,Pi/2));

$$\frac{3\pi\sqrt{1+z^2}}{2}$$

Evaluating the previous output at $z=0$, the line integral contribution

> Int2:=eval(%,z=0);

$$Int2 := \frac{3\pi}{2}$$

from the second leg is $Int2=3\pi/2$. To determine the line integral contributions from the other three legs, which lie in different planes, he now sets the coordinates to the 3-dimensional Cartesian system.

> SetCoordinates('cartesian'[x,y,z]):

The line integral contribution from the first leg is now determined using $V2b$, the Line option being entered with the end points of the line entered as the vectors $<0,0,0>$ and $<1,0,0>$.

> Int1:=LineInt(V2b,Line(<0,0,0>,<1,0,0>));

$$Int1 := 0$$

The contribution from the first leg is zero. Similarly, he calculates the line integral contribution from the third and fourth legs.

> `Int3:=LineInt(V2b,Line(<0,1,0>,<0,1,2>));`
$$Int3 := 2$$
> `Int4:=LineInt(V2b,Line(<0,1,2>,<0,0,0>));`
$$Int4 := -2$$

The total line integral is the sum of the four contributions,

> `Answer:=Int1+Int2+Int3+Int4;`
$$Answer := \frac{3\pi}{2}$$

yielding the final value of $3\pi/2$ for the complete loop.

If Stokes's theorem is valid, integrating $Curl2$ over the surface enclosed by the loop should yield exactly the same answer. From Figure 3.12, there are two planar surfaces involved. In terms of spherical coordinates, the unit normal to the surface in the y-z plane points in the direction of $-\hat{\phi}$. The relevant surface integral is of the form $\int Curl2_\phi \, dA_\phi$. But $Curl2_\phi = 0$, so this integral is zero.

On the other hand, the unit normal to the surface in the x-y plane is in the direction of $-\hat{\theta}$. The relevant surface integral is $-\int_0^{\pi/2}\int_0^1 Curl2_\theta \, r \, dr \, d\phi$,

> `surface_integral:=-Int(Int(Curl2[2]*r,r=0..1),phi=0..Pi/2)`
> `=-int(int(Curl2[2]*r,r=0..1),phi=0..Pi/2);`

$$surface\_integral := -\int_0^{\frac{\pi}{2}} \int_0^1 -6r \, dr \, d\phi = \frac{3\pi}{2}$$

which yields the same answer, $3\pi/2$, thus confirming Stokes's theorem. With the assignment complete, Eoj can hardly wait for Enaj to return to show her his instructive, yet powerful, recipe.

3.3 Supplementary Recipes

03-S01: Bertie Bumblebee Leaves His B and B

Bertie bumblebee flies out from his hive, which he runs as a B and B, in search of nectar to feed his guests. His path during the initial stage of his flight is given by $r(t) = a\,e^{bt}$ meters, $\theta(t) = ct$ radians, with a, b, and c positive constants.

(a) Show that the angle between Bertie's velocity and acceleration vectors remains constant during this portion of the flight.

(b) Taking $a = 0.1$, $b = c = 1$, plot his speed and acceleration magnitude over the first five seconds of flight.

(c) Animate his flight for the same parameter values and time interval. How far is Bertie from his hive after five seconds?

03-S02: Alice in Cylinder Land

Abigail ant's Aunt Alice is uncomfortable living on the surface of a beach ball, and prefers to dwell on the surface of a long cylinder. The cylindrical polar coordinates (ρ, ϕ, η) which characterize cylinder land are related to Cartesian coordinates by $x = \rho \cos \phi$, $y = \rho \sin \phi$, $z = \eta$.

(a) Plot the surfaces corresponding to holding ρ, ϕ, and η constant.

(b) Derive the velocity and acceleration in cylindrical coordinates.

(c) Alice moves in cylinder land, her coordinates at time t being $\rho = A$, $\phi(t) = C \omega t$ and $\eta(t) = (\pi/2)(1 + \cos(B \omega t)/4)$. A, B, C, ω are constants. Determine the magnitude of Alice's velocity and acceleration at time t.

(d) If $A = 1$, $B = 4$, $\omega = 2$, and $C = 3$, plot the magnitude of Alice's velocity and acceleration, and her path in cylinder land for $t = 0..10$. Include the cylindrical surface in the last plot.

03-S03: Car Racing in the Great White North

In the great white north, there lives a race-car driver, Mad Max, who is upset that the summer race season is over. Max has a flash of inspiration and decides to continue to race on a circular racetrack of radius a that he will build on the frozen lake outside his home. The coefficient of friction between the car's tires and the ice is μ and the acceleration due to gravity is g. Max's strategy is to accelerate from rest up to maximum speed without the car slipping on the ice and then maintain that maximum speed for the remainder of the circuit.

(a) If Max's car is able to maintain a maximum total acceleration equal to μg, what is his speed at some time t before reaching the maximum speed?

(b) If $a = 500$ m, $\mu = 0.1$, and $g = 10$ m/s², how long does it take his car to reach its maximum speed and what is the value of this speed?

(c) How far does the car move before it reaches the maximum speed?

(d) How long does it take to make one circuit of the track?

03-S04: The Making of Another Conservative Field

A fluid is characterized by a velocity field,
$$\vec{V} = (a\,x\,y^2\,z - b\,x^2\,y^2\,z - 3\,x^2\,z^3 + x + 2\,y + 4\,z)\hat{e}_x + (a\,x^2\,y\,z - a\,y\,z\,x^3 + 2\,x - 3\,y - z)\hat{e}_y$$
$$+ (x^2\,y^2 - y^2\,x^3 - c\,x^3\,z^2 + 4\,x - y + 2\,z)\hat{e}_z.$$

(a) What values must a, b, and c have to make $\vec{V}$ conservative?

(b) With these parameter values, what is the divergence of $\vec{V}$?

(c) What is the corresponding potential ϕ, for which $\vec{V} = -\nabla \phi$?

(d) Produce a 3-d plot, showing the equipotentials $\phi = \pm 15$ and $\vec{V}$.

03-S05: More Divs and Curls

Determine the divergence and curl of the vectors $\vec{A}$, $\vec{B}$, and $\vec{C}$ expressed in cylindrical, spherical, and spherical coordinates, respectively. Are any of these vector fields conservative? Evaluate the divs and curls at the specified points:

3.3 SUPPLEMENTARY RECIPES

(a) $\vec{A} = \rho z \sin\phi\,\hat{\rho} + 3\rho z^2 \cos\phi\,\hat{\phi} + 0\,\hat{z}$, $\quad (\rho = 5,\ \phi = \pi/2,\ z = 1)$;

(b) $\vec{B} = 2r\cos\theta\cos\phi\,\hat{r} + 0\,\hat{\theta} + \sqrt{r}\,\hat{\phi}$, $\quad (r = 1,\ \theta = \pi/6,\ \phi = \pi/3)$;

(c) $\vec{C} = [(\cos\theta\sin\phi)/r + 2r\,\phi]\,\hat{r}$
$\quad - [(\ln(r)\sin\theta\sin\phi)/r]\,\hat{\theta}$
$\quad + [(\ln(r)\cos\theta\cos\phi + r^2)/(r\sin\theta)]\,\hat{\phi}$, $\quad (r = 1,\ \theta = \pi/6,\ \phi = \pi/3)$.

03-S06: These Operators Have Many Identities

Vector operator identities play an important role in the formal mathematical development of mechanics and electrodynamics. Here $\vec{A}$, $\vec{B}$, V, and f are arbitrary real functions, while $\vec{r}$, $\vec{v}$, and $\vec{a}$ are the time-dependent position, velocity, and acceleration, respectively. Using the coordinate systems indicated, prove the following identities. (The Maple command AddCoordinates is useful in part (c).)

(a) $\nabla \times (\vec{A} \times \vec{B}) = (\vec{B}\cdot\nabla)\vec{A} + \vec{A}(\nabla\cdot\vec{B}) - (\vec{A}\cdot\nabla)\vec{B} - \vec{B}(\nabla\cdot\vec{A})$, (Cartesian)

(b) $\nabla\cdot(\vec{A}\times\vec{B}) = (\nabla\times\vec{A})\cdot\vec{B} - (\nabla\times\vec{B})\cdot\vec{A}$, (spherical polar)

(c) $\nabla\times(\nabla V) = 0$, and $\nabla\cdot(\nabla\times\vec{f}) = 0$, ("our" toroidal coordinates)

(d) $\dfrac{d}{dt}[\vec{r}\times(\vec{v}\times\vec{r})] = r^2\vec{a} + (\vec{r}\cdot\vec{v})\vec{v} - (v^2 + \vec{r}\cdot\vec{a})\vec{r}$, (Cartesian).

03-S07: Thoughts on Flux, From Heraclitas to Gauss

Heraclitas, the Greek philosopher (535–475 B.C.), was qualitatively aware of the concept of flux in the natural world when he stated "All is flux, nothing stays still." Over 2000 years later, the German mathematician Karl Friedrich Gauss (1777–1855) made a more precise statement about flux when he introduced the famous theorem which bears his name. Given a vector field $\vec{F}$, Gauss's theorem (also known as the divergence theorem) states that $\oint_S \vec{F}\cdot d\vec{A} = \int_V (\nabla\cdot\vec{F})\,dv$. The "net flux" of $\vec{F}$ out of a closed surface S surrounding a volume V is equal to the volume integral of the divergence of $\vec{F}$. The element of vector area $d\vec{A}$ in the closed surface integral on the LHS points out of the volume. When surface integrals are messy to do, evaluating the RHS of Gauss's theorem may yield the net flux out of a region S more easily.

A fluid is described by the velocity field $\vec{V} = y\,e^{z^2}\,\hat{e}_x + y^2\,\hat{e}_y + e^{xy}\,\hat{e}_z$. Consider a volume V of fluid enclosed by the surface S bounded by the cylinder $x^2 + y^2 = 9$ and the planes $z = 0$ and $z = y - 3$.

(a) Plot the velocity field over the range $x = -3..3$, $y = -3..3$, $z = -5.9..-6$. Choose a suitable grid.

(b) Make a 3-dimensional plot of the surface S.

(c) Calculate the net flux of $\vec{V}$ out of S by making use of Gauss's theorem.

03-S08: Felonious Fly (Jr.) Flees

Since his father met a sticky end in a Venus flytrap, Felonious Fly Jr. has been on the alert for potential dangers. Despite being reasonably cautious, Jr. has flown too close to a small, hot, heater which produces a temperature profile $T = 400/(1 + x^2/3 + 2y^2/3 + z^2)$, where T is in degrees Celsius and x, y, z are in meters.

(a) If Felonious Jr. is located at the point $P = (1, 1, 2)$, what is the temperature at P?

(b) In which direction should Jr. initially fly in order to move from hot to cold in the shortest distance?

(c) What is the maximum rate of decrease of temperature at P?

(d) Plot the constant temperature surface (isotherm) passing through P. In the same picture, plot the direction in which Felonious Jr. should initially fly as well as the tangent plane to the isotherm. Represent Jr. by a red circle at point P.

03-S09: Designing Gabrielle's Toy Box

Gabrielle needs another toy box because her old one is full. You are asked to create a rectangular toy box without a lid which is to be made from $2\frac{1}{2}$ m^2 of heavy cardboard and have as large a volume V as possible.

(a) What is the length x, the width y, and the height z of the toy box which has maximum volume?

(b) What is the maximum volume?

(c) Confirm that the volume is a maximum by (a) making a suitable 3-dimensional contour plot of V (b) applying the second derivatives test.

(d) Make a 3-dimensional colored plot of Gabrielle's new toy box.

Note: Letting subscripts denote derivatives, the second derivatives test is as follows [Ste87]. If the function $V(x, y)$ has an extremum at $x = a$, $y = b$, form the combination $D(a, b) = V_{xx}(a, b) V_{yy}(a, b) - (V_{xy}(a, b))^2$. If $D(a, b) > 0$ and $V_{xx}(a, b) < 0$, then $V(a, b)$ is a local maximum. (If $V_{xx}(a, b) > 0$, then it is a local minimum.)

Chapter 4

Newtonian Dynamics I

In this chapter, we shall present a wide variety of intellectually delectable recipes involving velocity- and position-dependent forces. The resulting Newtonian equations of motion will be either linear or nonlinear ordinary differential equations (ODEs). Nonlinear ODEs contain one or more terms that are not of first order (linear) in the dependent variable(s). Linear ODEs can usually be solved analytically in terms of known functions, but most nonlinear ODEs of physical interest must be solved numerically. Fortunately, Maple is capable of not only generating analytic solutions (when they exist), but also has various built-in algorithms for numerically solving linear and nonlinear ODEs.

4.1 Velocity-dependent forces

4.1.1 Justine Takes a Dive

If toast always lands butter-side down, and cats always land on their feet, what happens if you strap toast on the back of a cat and drop it?
Steven Wright, contemporary American comedian and Oscar winner (1963–)

While on vacation in Fiji, Jennifer's niece Justine (mass $m = 30$ kg) dove into a 6 m deep lagoon from a low cliff, entering the water perpendicular to the surface at a speed of 10 m/s. She made no swimming motion to ascend but instead let the buoyant force of the water bring her back to the surface in 20 s.

Based on this scenario, Jennifer has created the following problem for her mechanics class to attempt. Assuming that the drag force due to the water is given by $F_{\text{drag}} = -kv$, where v is the velocity, and that Justine's specific gravity is 0.95, (a) determine the drag coefficient k, (b) plot Justine's vertical distance $y(t)$ relative to the smooth water surface over the 20 s interval, (c) determine how far above the lagoon bottom Justine reaches and how long this takes, and (d) animate Justine's motion in the water. The students have been instructed to compute the position of Justine's center of mass and to take $g = 9.8$ m/s^2.

In the following recipe, we reproduce the solution to this problem as given in Jennifer's answer key. She has taken the "final" time for the dive to be $tf=20$ seconds, and the lagoon bottom to be at $y=b=-6$ meters.

> `restart: with(plots):`

> `tf:=20: b:=-6:`

Using Newton's second law, Justine's mass m times her acceleration $(d^2 y(t)/dt^2)$ is equal to the resultant force. The first term on the rhs of `ode` is the drag force of the water, the second is Justine's weight, and the third term the buoyant force of the water. Using Archimedes's principle, the buoyant force is equal to Justine's weight divided by the specific gravity[1] (sp).

> `ode:=m*diff(y(t),t,t)=-k*diff(y(t),t)-m*g+m*g/sp;`

$$ode := m\left(\frac{d^2}{dt^2} y(t)\right) = -k\left(\frac{d}{dt} y(t)\right) - mg + \frac{mg}{sp}$$

Taking $y(t=0)=0$ and an initial velocity $dy(0)/dt=v0$, the linear ODE is analytically solved for $y(t)$ using the `dsolve` command and the differential operator `D(y)`, which computes the derivative of y, for entering the initial velocity.

> `dsol:=dsolve({ode,D(y)(0)=v0,y(0)=0},y(t));`

$$dsol := y(t) =$$
$$-\frac{(mg\,sp - mg + v0\,k\,sp)\,m\,e^{(-\frac{kt}{m})}}{k^2\,sp} - \frac{mg(sp-1)t}{k\,sp} + \frac{(mg\,sp - mg + v0\,k\,sp)\,m}{k^2\,sp}$$

Next, Jennifer has entered the various parameter values, viz., the specific gravity, the acceleration due to gravity, Justine's mass, and the initial velocity.

> `parameters:=sp=0.95,g=9.8,m=30,v0=-10;`

$$parameters := sp = .95,\ g = 9.8,\ m = 30,\ v0 = -10$$

Substituting the *parameters* into the right-hand-side (rhs) of the analytic solution (*dsol*), Justine's vertical position at time t is now known. The drag coefficient k remains to be determined.

> `y:=subs(parameters,rhs(dsol));`

$$y := -\frac{31.57894737\,(-14.700 - 9.50\,k)\,e^{(-\frac{kt}{30})}}{k^2} + \frac{15.47368421\,t}{k}$$
$$+ \frac{31.57894737\,(-14.700 - 9.50\,k)}{k^2}$$

At time tf, Justine has returned to the surface at $y=yf=0$. The k value is obtained by applying the `fsolve` command over the range $k=0$ to 100 to yf.

> `yf:=subs(t=tf,y=0);`

$$yf := -\frac{31.57894737\,(-14.700 - 9.50\,k)\,e^{(-\frac{2k}{3})}}{k^2} + \frac{309.4736842}{k}$$
$$+ \frac{31.57894737\,(-14.700 - 9.50\,k)}{k^2} = 0$$

---

[1] The specific gravity of a body is the ratio of the density of that object to that of water.

4.1. VELOCITY-DEPENDENT FORCES

```
> ksol:=fsolve(yf,k,k=0..100);
```
$$ksol := 49.00000013$$

The drag coefficient of the water was about $k = 49$ kg · s^{-1}. Substituting this value (labeled *ksol*) into y yields Justine's position at time t seconds.

```
> y:=subs(k=ksol,y);
```
$$y := 6.315789455\, e^{(-1.633333338\, t)} + 0.3157894728\, t - 6.315789458$$

Although not requested in the problem, Jennifer has also calculated Justine's velocity v after entering the water.

```
> v:=diff(y,t);
```
$$v := -10.31578947\, e^{(-1.633333338\, t)} + 0.3157894728$$

The velocity and depth are plotted together, along with the bottom, as a function of time in Figure 4.1. The three curves have been labeled in the figure by using the textplot command.

```
> gr1:=plot([y,b,v],t=0..tf,color=[blue,brown,red],thickness=2:
> gr2:=textplot({[10,-3.7,"y"],[15,.9,"v"],[10,-5.5,"bottom"]}):
> display({gr1,gr2},tickmarks=[2,2],labels=["time","depth,vel"]);
```

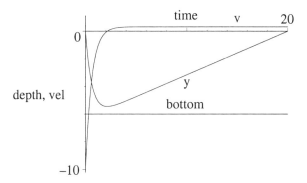

Figure 4.1: Justine's vertical position (y) and velocity (v) for her dive.

When Justine reached maximum depth, her velocity was zero. Setting $v=0$ and solving for t, she reached maximum depth about 2.13 s after entering the water. Evaluating y at time $t=tmd$, her maximum depth was 5.44 m.

```
> tmd:=solve(v=0,t);
```
$$tmd := 2.134503173$$

```
> max_depth:=eval(y,t=tmd);
```
$$max\_depth := -5.448395333$$

Her center of mass was then at a height $h=0.55$ m above the bottom.

```
> h:=max_depth-b;
```
$$h := 0.551604667$$

138 CHAPTER 4. NEWTONIAN DYNAMICS I

Her time (*tup*) to reach the surface then was about 17.9 seconds.

> `tup:=tf-tmd;`

$$tup := 17.86549683$$

To complete the solution to the problem, Jennifer has used the `animate` command with 300 frames to animate Justine's vertical motion over the time interval $t=0$ to $t=tf$. She has represented Justine by a size 16 red circle.

> `animate([0,y],t=0..tf,frames=300,style=point,symbol=circle,`
> `symbolsize=16,color=red);`

By clicking on the resulting computer plot and on the start command, Justine's motion through the water can be observed.

4.1.2 A Rolling Road

The rolling English drunkard made the rolling English road.
A reeling road, a rolling road, that rambles round the shire.
G. K. Chesterton, British author, *The Rolling English Road* (1914)

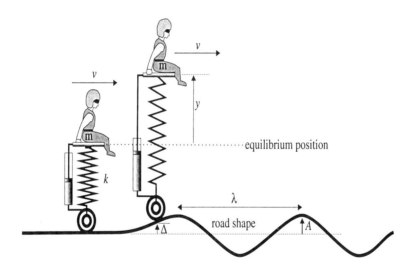

Figure 4.2: Enaj on her unicycle entering the region of rolling road.

Some years ago while driving along a smooth highway in the Mohave desert in southern California, Richard and his family had a stomach-lurching experience when they suddenly encountered a rapidly undulating section of road which persisted for several miles. According to a recent hyperspace e-mail from Erehwon, our friend Enaj (mass m) had a similar bout of queasiness while she was riding her unicycle along a road through the salt flats of Lake Woebegone. Traveling at a fixed horizontal speed v, she suddenly entered a wind-formed

4.1. VELOCITY-DEPENDENT FORCES

sinusoidal region of amplitude A and wavelength λ as shown in Figure 4.2. Somehow, Enaj was able to still maintain her original horizontal speed and vertical orientation, but began to undergo sizeable oscillations in the vertical (y) direction. Fortunately for Enaj, her seat belt was buckled up so she was not thrown off the seat of the unicycle. The spring supporting the seat had a spring constant k, while a built-in shock absorber introduced some damping of the vertical motion. The damping force was proportional to the vertical velocity, c being the damping coefficient. According to Enaj, the parameter values for her bumpy ride were $v = 7$ m/s, $A = 0.05$ m, $\lambda = 5$ m, $\omega \equiv \sqrt{k/m} = 10$ rads/s, and $\gamma \equiv c/(2m) = 1/5$ s^{-1}. Taking $t = 0$ to be the time that she entered the rolling region, the initial conditions for the vertical oscillations were $y(0) = 0$ and $\dot{y}(0) = 0$. Based on this information, Enaj has suggested the following Newtonian mechanics problem.

(a) Derive the general ODE describing Enaj's vertical displacement $y(t)$ from equilibrium in the rolling region. Analytically solve the ODE and identify the transient and steady-state parts of the solution.

(b) Using Enaj's values, plot $y(t)$ for a sufficiently long time for steady-state to be established. Determine the amplitude of the steady-state oscillations.

(c) What speed would yield the maximum amplitude for steady-state oscillations? What is this maximum amplitude? Plot $y(t)$ for this case.

(d) Plot the amplitude of steady-state oscillations vs. horizontal velocity over the range $v = 0$ to 12 m/s and demonstrate that a resonance curve results. Relate this result to some driving experience that you may have had.

After unprotecting γ from its Maple assignment as Euler's constant, the sinusoidal shape $\Delta = A\sin(\Omega t)$ of the rolling road is entered. The frequency Ω is related to the horizontal speed and wavelength by $\Omega = 2\pi v/\lambda$.

```
>  restart: unprotect(gamma):
>  Delta:=A*sin(Omega*t);
```
$$\Delta := A\sin(\Omega t)$$

The damping and restoring forces per unit mass are given by

$$F_{damping} = -\frac{c}{m}\frac{d}{dt}(y(t) - \Delta) = -2\gamma\frac{d}{dt}(y(t) - \Delta),$$

$$F_{restoring} = -\frac{k}{m}(y(t) - \Delta) = -\omega^2(y(t) - \Delta).$$

These forces are now entered and used in Newton's second law (per unit mass).

```
>  F[damping]:=-2*gamma*diff((y(t)-Delta),t);
```
$$F_{damping} := -2\gamma\left(\left(\tfrac{d}{dt}y(t)\right) - A\cos(\Omega t)\Omega\right)$$

```
>  F[restoring]:=-omega^2*(y(t)-Delta);
```
$$F_{restoring} := -\omega^2(y(t) - A\sin(\Omega t))$$

```
> de:=expand(diff(y(t),t,t)-F[damping]-F[restoring])=0;
```

$$de := (\frac{d^2}{dt^2} y(t)) + 2\gamma(\frac{d}{dt} y(t)) - 2\gamma A\cos(\Omega t)\Omega + \omega^2 y(t) - \omega^2 A\sin(\Omega t) = 0$$

The output labeled *de* is the desired equation of motion for the vertical oscillations. The initial conditions (*ic*) at $t = 0$ are specified and the ODE *de* is analytically solved subject to these conditions with the `dsolve` command. The option `method=laplace` produces a solution in terms of trigonometric functions, rather than entirely in terms of exponentials.

```
> ic:=y(0)=0,D(y)(0)=0;
```

$$ic := y(0) = 0, \ D(y)(0) = 0$$

```
> sol:=dsolve({de,ic},y(t),method=laplace);
```

$$sol := y(t) = A\Omega(-\frac{2\gamma\Omega^3 \cos(\Omega t) - \sin(\Omega t)\omega^4 + \sin(\Omega t)\omega^2\Omega^2 - 4\sin(\Omega t)\Omega^2\gamma^2}{\Omega}$$
$$+ e^{(-t\gamma)}(2\gamma\Omega^2 \cosh(t\sqrt{\gamma^2 - \omega^2}) + \frac{(\omega^2\Omega^2 - \omega^4 - 2\Omega^2\gamma^2)\sinh(t\sqrt{\gamma^2 - \omega^2})}{\sqrt{\gamma^2 - \omega^2}}))$$
$$\Big/ (\omega^4 - 2\omega^2\Omega^2 + 4\Omega^2\gamma^2 + \Omega^4)$$

As $t \to \infty$, the term involving $e^{-\gamma t}$ vanishes so must represent the transient part of the solution. The steady-state solution *yss*, which persists for all times, is extracted by neglecting the exponential term (set $e^{-\gamma t}=0$ on the rhs of *sol*).

```
> yss:=subs(exp(-gamma*t)=0,rhs(sol));
```

$$yss := -\frac{A(2\gamma\Omega^3 \cos(\Omega t) - 4\Omega^2 \sin(\Omega t)\gamma^2 + \Omega^2 \sin(\Omega t)\omega^2 - \sin(\Omega t)\omega^4)}{\omega^4 - 2\omega^2\Omega^2 + 4\Omega^2\gamma^2 + \Omega^4}$$

The maximum displacement of the steady-state oscillation must occur when the vertical velocity is zero. The time T at which this occurs is calculated, and the

```
> T:=solve(diff(yss,t)=0,t);
```

$$T := \frac{\arctan(\frac{-4\gamma^2\Omega^2 - \omega^4 + \omega^2\Omega^2}{2\gamma\Omega^3})}{\Omega}$$

result substituted into *yss*. The expression Y for the maximum displacement is then simplified with the `combine` and `simplify` commands.

```
> Y:=combine(subs(t=T,yss),trig);
> Y:=simplify(Y,symbolic);
```

$$Y := -\frac{(4\gamma^2\Omega^2 + \omega^4)A}{\sqrt{16\gamma^4\Omega^4 + 8\gamma^2\Omega^2\omega^4 - 8\gamma^2\Omega^4\omega^2 + \omega^8 - 2\omega^6\Omega^2 + \omega^4\Omega^4 + 4\gamma^2\Omega^6}}$$

A function `Omega` for evaluating the frequency Ω, when the horizontal velocity v is given, is created. Substituting `Omega=Omega(v)` into Y then expresses the maximum vertical displacement in terms of v.

```
> Omega:=v->2*Pi*v/lambda:
> Y1:=subs(Omega=Omega(v),Y);
```

4.1. VELOCITY-DEPENDENT FORCES

$$Y1 := -A\,(16\,\frac{\pi^2 v^2 \gamma^2}{\lambda^2} + \omega^4) \Big/ (256\,\frac{\pi^4 v^4 \gamma^4}{\lambda^4} - \frac{128\,\pi^4 v^4 \gamma^2 \omega^2}{\lambda^4} + \frac{32\,\pi^2 v^2 \gamma^2 \omega^4}{\lambda^2}$$
$$+ \frac{16\,\omega^4 \pi^4 v^4}{\lambda^4} - \frac{8\,\omega^6 \pi^2 v^2}{\lambda^2} + \omega^8 + \frac{256\,\gamma^2 \pi^6 v^6}{\lambda^6})^{(1/2)}$$

To determine the value of v that maximizes $Y1$, we differentiate $Y1$ with respect to v, set the result equal to zero, and solve for v in sol2.

> sol2:=solve(diff(Y1,v)=0,v);

The output *sol2* contains five solutions (not shown here). Four of the solutions must be rejected because one is equal to 0, two are negative, and one is imaginary. The positive real solution is selected. In this particular execution of the recipe, this was the 4th solution in the output, but the ordering can change from one run to the next. So the velocity *vmax* which corresponds to the maximum vertical displacement is known.

> vmax:=sol2[4];

$$vmax := \frac{\sqrt{-\pi^2 \omega^2 + \sqrt{\pi^4 \omega^4 + 8\gamma^2 \pi^4 \omega^2}}\,\omega\,\lambda}{4\,\gamma\,\pi^2}$$

To plot Enaj's vertical motion, the given parameter values are now entered.

> v0:=7: lambda:=5: gamma:=0.2: A:=0.05: omega:=10:

Substituting Omega=Omega(v0) into the rhs of *sol* and evaluating the output to six decimal places yields the expression *y1* for Enaj's vertical displacement.

> y1:=evalf(subs(Omega=Omega(v0),rhs(sol)),6);

$y1 := -0.0259745\cos(8.79645\,t) + 0.217007\sin(8.79645\,t)$
$\quad + 0.000839212\,e^{(-0.2\,t)}(30.9510\cos(9.99800\,t) - 226.888\sin(9.99800\,t))$

The vertical displacement $y1$ is plotted over the interval $t = 0$ to 30 s on the

> plot(y1,t=0..30,numpoints=1000,labels=["t","y1"],
> tickmarks=[2,3]);

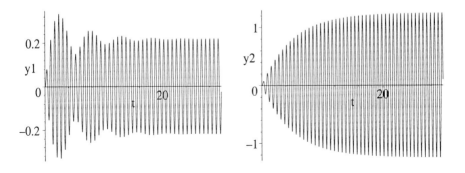

Figure 4.3: Vertical motion for Enaj's ride (lhs); for maximum amplitude (rhs).

lhs of Figure 4.3. The transient part of the solution for early times can be clearly seen. After about 20 seconds, steady-state is essentially achieved, i.e., after a horizontal distance of about 140 meters of rolling road. From the plot, the amplitude of the steady-state oscillations is about 0.2 meters, or 20 cm, which is four times the amplitude A of the road oscillations. A more precise value of 0.219 m for the amplitude can be obtained as follows.

> y1max:=abs(evalf(subs(Omega=Omega(v0),Y)));

$$y1max := 0.2185301242$$

The velocity that corresponds to the largest steady-state amplitude is obtained by numerically evaluating $vmax$ and is found to be about 7.95 m/s.

> vmax:=evalf(vmax);

$$vmax := 7.954568212$$

Making use of $vmax$, the corresponding vertical displacement $y2$ is calculated to six figure accuracy, and then plotted on the rhs of Figure 4.3.

> y2:=evalf(subs(Omega=Omega(vmax),rhs(sol)),6);

$$y2 := -1.25556\cos(9.99597\,t) + 0.0754871\sin(9.99597\,t)$$
$$+ 0.0314141\,e^{(-0.2\,t)}(39.9679\cos(9.99800\,t) - 1.60232\sin(9.99800\,t))$$

> plot(y2,t=0..30,numpoints=1000,labels=["t","y2"],
> tickmarks=[2,3]);

From the figure, the steady-state amplitude in this case is over one meter. A more precise value is now extracted.

> y2max:=abs(evalf(subs(Omega=Omega(vmax),Y)));

$$y2max := 1.251249270$$

The maximum possible steady-state amplitude is 1.25 meters, which is 25 times that of the rolling road amplitude. Enaj was fortunate that she was not traveling at a speed of 7.95 m/s or she might have felt really queasy. Finally, we plot the variation of the steady-state amplitude with the horizontal velocity in Fig. 4.4.

> plot(abs(Y1),v=0..12,labels=["velocity", "amplitude"]);

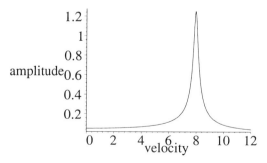

Figure 4.4: Amplitude of steady-state oscillations vs. velocity.

4.1. VELOCITY-DEPENDENT FORCES

A resonance curve results, the amplitude of the steady-state oscillations dropping off again at velocities above *vmax*.

Having solved Enaj's mechanics problem, Richard was reminded of another driving experience that he had in the American Southwest. He had enquired earlier of a friendly old Utah farmer about the condition of a dirt road indicated on the New Mexico map which led to the Chaco Culture National Historic Park. The farmer said that, as long as it didn't rain, the road would be O.K but very "washboardy". His advice was to drive at a reasonably high speed to not only shorten the time but smooth out the ride. And indeed, as Richard increased his speed, he noticed an appreciable amplitude resonance, followed by a much smoother ride as the speed was further increased. Although not scientifically trained, the farmer knew from experience how to drive the washboard road.

4.1.3 Pushing the Envelope

In order to shake a hypothesis, it is sometimes not necessary to do anything more than push it as far as it will go.
Denis Diderot, French philosopher (1713–84)

In supplementary recipe **01-S13**, Sheelo, a part-time *National Geographic* photographer, rented a plane so she could approach as close as possible to an erupting volcanic mountain in the Cascade Range of the Pacific Northwest. The envelope of safety, outside of which the plane could safely fly, was calculated, neglecting the effect of air resistance on the ejected rocks. Qualitatively, the inclusion of air drag will "push" the envelope in and allow the plane to approach more closely. The simplest model of air drag is *Stokes's law of resistance* for which the drag force is given by $\vec{F}_d = -km\vec{v}$ where k is a positive constant, m the mass of the projectile, and $\vec{v}$ the velocity vector. This model has already been used in the previous two recipes of this chapter. Stokes's law is a reasonable approximation [MT95] at very low speeds and also at speeds much greater than the speed of sound (about 340 m/s at 20 °C and 1 atm.). At intermediate speeds, quadratic terms in v should be included in the drag force.

In the following recipe, Sheelo has asked Jennifer to calculate the envelope of safety for the erupting mountain including air resistance. The height of the mountain is neglected and the rocks are ejected in all directions with a maximum speed of 700 m/s. Assuming that Stokes's law prevails, Jennifer enters Newton's equation of motion in the horizontal (x) direction.

```
>   restart: with(plots):
>   xeq:=diff(x(t),t,t)=-k*diff(x(t),t);
```

$$xeq := \frac{d^2}{dt^2} \text{x}(t) = -k \left(\frac{d}{dt} \text{x}(t) \right)$$

Keeping the formulation general for the time being, Jennifer considers a rock to be emitted from the origin with an initial speed *Vo* at an angle θ with respect to the horizontal. The horizontal component of initial velocity is then *Vo* $\cos \theta$, while the vertical component is $Vo \sin\theta$. Then she analytically solves xeq for $x(t)$ subject to the initial conditions and successively collects the coefficients of $\cos\theta$, Vo, and $1/k$ on the rhs of xsol.

> `xsol:=dsolve({xeq,x(0)=0,D(x)(0)=Vo*cos(theta)},x(t));`

$$xsol := \mathrm{x}(t) = \frac{Vo \cos(\theta)}{k} - \frac{Vo \cos(\theta) e^{(-k\,t)}}{k}$$

> `x:=collect(rhs(xsol),[cos(theta),Vo,1/k]);`

$$x := \frac{(1 - e^{(-k\,t)}) Vo \cos(\theta)}{k}$$

The equation of motion in the vertical (y) direction is

> `yeq:=diff(y(t),t,t)=-k*diff(y(t),t)-g;`

$$yeq := \frac{d^2}{dt^2}\mathrm{y}(t) = -k\left(\frac{d}{dt}\mathrm{y}(t)\right) - g$$

which is solved to yield the y-coordinate of the rock as a function of time.

> `ysol:=dsolve({yeq,y(0)=0,D(y)(0)=Vo*sin(theta)},y(t));`

$$ysol := \mathrm{y}(t) = -\frac{(g + Vo \sin(\theta) k) e^{(-k\,t)}}{k^2} - \frac{g\,t}{k} + \frac{g + Vo \sin(\theta) k}{k^2}$$

Here, g is the acceleration due to gravity. The result is made more compact by successively collecting the coefficients of $\sin\theta$, Vo, $1/k$, and g.

> `y:=collect(rhs(ysol),[sin(theta),Vo,1/k,g]);`

$$y := \frac{(1 - e^{(-k\,t)}) Vo \sin(\theta)}{k} - \frac{g\,t}{k} + \frac{(1 - e^{(-k\,t)}) g}{k^2}$$

Jennifer calculates the time T for the rock to reach a horizontal distance X.

> `T:=solve(x=X,t);`

$$T := -\frac{\ln\left(\dfrac{Vo \cos(\theta) - X\,k}{Vo \cos(\theta)}\right)}{k}$$

The y-coordinate of the rock at X is obtained by substituting $t = T$ into y.

> `Y:=simplify(subs(t=T,y));`

$$Y := \frac{Vo \sin(\theta) k^2 X + g \ln\left(\dfrac{Vo \cos(\theta) - X\,k}{Vo \cos(\theta)}\right) Vo \cos(\theta) + g\,X\,k}{Vo \cos(\theta) k^2}$$

To determine the envelope of safety, Y is differentiated with respect to θ and the result set equal to zero. A lengthy equation, relating θ to x, results.

> `eq:=diff(Y,theta)=0;`

Now eq is solved for θ, two lengthy solutions appearing in the Θ output.

> `Theta:=solve(eq,theta);`

The first solution in Θ is selected, substituted into Y, and simplified with the trig option. The formidable appearing equation (Yenv) for the envelope of safety may be viewed on the computer screen.

> `Yenv:=simplify(subs(theta=Theta[1],Y),trig);`

4.1. VELOCITY-DEPENDENT FORCES

As a check on her answer, Jennifer takes the limit as k approaches zero to see if the envelope of safety reduces to that derived in **01-S13** for zero air resistance.

> Yenv0:=expand(limit(Yenv,k=0));

$$Yenv0 := \frac{Vo^2}{2\,g} - \frac{g\,X^2}{2\,Vo^2}$$

The formula given by *Yenv0* is identical to that derived in the earlier recipe. Jennifer now takes the initial speed of the rocks to be $Vo = 700$ m/s, $g \approx 10$ m/s^2, and the coefficient of air drag to be $k = 0.01$ s^{-1}.

> Vo:=700: g:=10: k:=0.01:

The following functions are created for use in the following do loop.

> xx:=(theta,t)->x: yy:=(theta,t)->y:

The do loop calculates the envelope of safety, with and without air resistance, and animates the motion of the rocks over a uniform range of angles from $\theta = 0.1$ radians to 3.1 radians.

> for i from -15 to 15 do
> theta:=1.6+(i*.1):

The time for the ith rock, ejected at an angle $\theta = 1.6 + 0.1\,i$, to hit the ground is numerically determined. The trivial solution $t = 0$ is ruled out by using the **avoid** command.

> T||i:=fsolve(yy(theta,t)=0,t,avoid={t=0},0..1000);

The x and y components of the velocity are calculated at time t.

> Vx:=diff(xx(theta,t),t); Vy:=diff(yy(theta,t),t);

The speed of the ith rock as it hits the ground at time $t = T||i$ is determined.

> speed||i:=evalf(subs(t=T||i,sqrt(Vx^2+Vy^2)));

The motion of the ith rock is animated over the time interval $t = 0$ to $T||i$, 150 frames being used for each rock, and each rock being represented by a randomly colored size 14 circle.

> pl||i:=animate([xx(theta,t),yy(theta,t)],t=0..T||i,
> frames=150,style=point,symbol=circle,symbolsize=14,
> color=COLOR(RGB,rand()/10^12,rand()/10^12,rand()/10^12));

The envelope of safety with air resistance present is plotted as a thick red curve.

> gr1:=plot(Yenv,X=-35000..35000,style=line,color=red,
> thickness=2):

The envelope of safety with no air resistance is plotted as a thick blue curve.

> gr2:=plot(Yenv0,X=-50000..50000,style=line,color=blue,
> thickness=2):
> end do:

On completing the do loop, the graphs **gr1**, **gr2**, and the sequence of **pl||i** are displayed and can be animated by clicking on the computer plot and on the start arrow. The opening frame of the animation is shown on the lhs of Fig. 4.5.

```
> display({gr1,gr2,seq(pl||i,i=-15..15)},view=
> [-50000..50000,0..25000],tickmarks=[3,3],labels=["x","y"]);
```

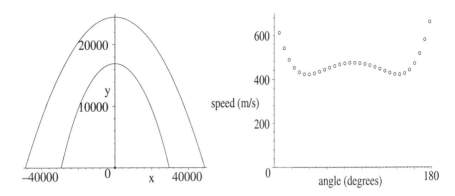

Figure 4.5: Envelopes of safety (lhs) and final speeds of rocks (rhs).

The outer curve is the envelope of safety without air resistance, the inner curve with air resistance. In the animation the rocks are emitted at different angles, none penetrating outside the inner envelope. Sheelo's aircraft is guaranteed to be safe from ejected rocks if it flies outside this envelope.

Jennifer creates a function phi to systematically calculate angles between 0 and 180 degrees for the ejected rocks. She then plots the final speed of the rocks as they hit the ground again for each angle.

```
> phi:=i->evalf((1.6+0.1*i)*180/Pi):
> plot([seq([phi(i),speed||i],i=-15..15)],style=point,
> symbol=circle,symbolsize=14,color=red,view=[0..180,0..700],
> tickmarks=[5,3],labels=["angle (degrees)","speed (m/s)"]);
```

The resulting picture is shown on the rhs of Figure 4.5. Because the rocks lose energy due to air resistance, their final speeds are less than the initial speed of 700 m/s with which they were ejected.

4.1.4 Benny Boffo's Hole in One

If you think it's hard to meet new people,
try picking up the wrong golf ball.
Jack Lemmon, U.S. actor, *Sports Illustrated* (9 Dec. 1985)

Syd and Benny Boffo are again on the links of the Metropolis Country Club for their weekly round of golf. Included in their foursome are Jennifer and Colleen who got paired with Syd and Benny because the course is very busy this afternoon and twosomes are not permitted. After her previous encounter with these two businessmen on these same links, Colleen is pleasantly surprised that Syd

4.1. VELOCITY-DEPENDENT FORCES

and Benny are actually very congenial characters and her initial trepidation of golfing with them is soon overcome. Jennifer, being the analytic mathematician that she is, is more preoccupied with mentally analyzing the shots that are made. On the thirteenth hole, the men's tee is located on a small cliff overlooking the green located on a small island in the middle of a lake below. "This is one scary hole," thinks Jennifer. "I am glad that the women's tee is closer to the green." Syd has already shot and managed to hit the green and Benny is next up. After a few practice swings, Benny hits the ball with authority. "It looks like a good shot," Syd remarks, "Your ball is heading for the hole. I think its going to go in!" Indeed, it does and Benny scores a rare hole in one.

After the golf match is over and Benny has bought celebratory drinks, Jennifer muses about the dynamics of Benny's spectacular golf shot. On being dropped off at her MIT office by Colleen, Jennifer decides to simulate Benny's hole in one as an example for her classical mechanics course. She has already referred her class to a discussion of air resistance in Marion and Thornton [MT95] and has explored some papers on golf ball dynamics in the *American Journal of Physics* [Erl83], [MA88], [MH91]. The simulation will allow her to introduce *Newton's law of resistance* which evidently does a good job of modeling the air resistance experienced by a dimpled golf ball. This resistance law states that the magnitude of the drag force is proportional to the square of the velocity $\vec{V}$. When the ball (mass m) is struck by the grooved golf club, it also acquires lift due to the backspin given it. Assuming that the backspin is about an axis parallel to the ground, the lift force is perpendicular to the plane defined by the backspin axis and $\vec{V}$. The lift force is also proportional to the square of the velocity. The drag ($\vec{F}_d$) and lift ($\vec{F}_L$) forces are shown in the free body diagram, Figure 4.6, along with the gravitational force, $\vec{F}_g$. Jennifer has taken $\vec{V}$ to be in the y-z plane, $\vec{V}$ making an angle θ with the horizontal (y-direction), and the ball's spin axis in the x-direction (out of the page). If the ball were "hooked"

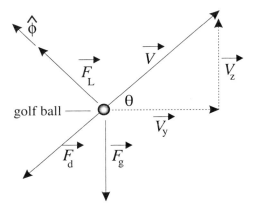

Figure 4.6: Free body diagram for a golf ball, including gravity, drag, and lift.

or "sliced" or a cross-wind were present, $\vec{V}$ would also have a component in the x-direction. From Figure 4.6, Jennifer notes that the unit vectors $\hat{\phi}$ and $\hat{v}$ in the $\vec{F}_L$ and $\vec{V}$ directions, respectively, are related to the unit vectors $\hat{e}_x$ and $\hat{e}_z$ in the y- and z-directions by $\hat{\phi} = -\sin\theta\,\hat{e}_y + \cos\theta\,\hat{e}_z$ and $\hat{v} = \cos\theta\,\hat{e}_y + \sin\theta\,\hat{e}_z$. She will need these results to enter the lift and drag forces.

After loading the plots and VectorCalculus packages, Jennifer enters the velocity $\vec{V}$ of the golf ball and calculates its speed v.

> restart: with(plots): with(VectorCalculus):
> V:=<0,Vy,Vz>;
$$V := Vy\,e_y + Vz\,e_z$$
> v:=sqrt(V.V);
$$v := \sqrt{Vy^2 + Vz^2}$$

From the figure, Jennifer sets $\sin\theta = V_z/v$ and $\cos\theta = V_y/v$ and enters the unit vectors $\hat{\phi}$ and $\hat{v}$ expressed in terms of the Cartesian unit vectors.

> sin(theta):=Vz/v; cos(theta):=Vy/v;
$$\sin(\theta) := \frac{Vz}{\sqrt{Vy^2 + Vz^2}}$$
$$\cos(\theta) := \frac{Vy}{\sqrt{Vy^2 + Vz^2}}$$
> phi_hat:=<0,-sin(theta),cos(theta)>;
$$phi\_hat := -\frac{Vz}{\sqrt{Vy^2 + Vz^2}}\,e_y + \frac{Vy}{\sqrt{Vy^2 + Vz^2}}\,e_z$$
> v_hat:=<0,cos(theta),sin(theta)>;
$$v\_hat := \frac{Vy}{\sqrt{Vy^2 + Vz^2}}\,e_y + \frac{Vz}{\sqrt{Vy^2 + Vz^2}}\,e_z$$

The gravitational force $\vec{F}_g$ on the ball is entered, g being the gravitational acceleration. The drag and lift forces are given by $\vec{F}_d = -C_d\,v^2\,\hat{v}$ and $\vec{F}_L = C_L\,v^2\,\hat{\phi}$, where C_d and C_L are the drag and lift coefficients.

> F[g]:=<0,0,-m*g>;
$$F_g := -m\,g\,e_z$$
> F[d]:=-C[d]*v^2*v_hat;
$$F_d := -C_d\sqrt{Vy^2 + Vz^2}\,Vy\,e_y - C_d\sqrt{Vy^2 + Vz^2}\,Vz\,e_z$$
> F[L]:=C[L]*v^2*phi_hat;
$$F_L := -C_L\sqrt{Vy^2 + Vz^2}\,Vz\,e_y + C_L\sqrt{Vy^2 + Vz^2}\,Vy\,e_z$$

Entering the golf ball's acceleration $\vec{A}$, calculating the total force $\vec{F} = \vec{F}_g + \vec{F}_d + \vec{F}_L$ on the ball, and writing Newton's second law in the y and z directions,

> A:=<0,Ay,Az>;
$$A := Ay\,e_y + Az\,e_z$$

4.1. VELOCITY-DEPENDENT FORCES

> `F:=F[g]+F[d]+F[L];`

$$F := (-C_d \sqrt{Vy^2 + Vz^2}\, Vy - C_L \sqrt{Vy^2 + Vz^2}\, Vz)\, \mathrm{e}_y$$
$$+(-m\,g - C_d \sqrt{Vy^2 + Vz^2}\, Vz + C_L \sqrt{Vy^2 + Vz^2}\, Vy)\, \mathrm{e}_z$$

yields yeq and zeq.

> `yeq:=m*A[2]=F[2]; zeq:=m*A[3]=F[3];`

$$yeq := m\,Ay = -C_d \sqrt{Vy^2 + Vz^2}\, Vy - C_L \sqrt{Vy^2 + Vz^2}\, Vz$$

$$zeq := m\,Az = -m\,g - C_d \sqrt{Vy^2 + Vz^2}\, Vz + C_L \sqrt{Vy^2 + Vz^2}\, Vy$$

The velocity (Vy, Vz) and acceleration (Ay, Az) components are expressed as first and second time derivatives of y and z. As an alternate to entering, e.g., `diff(y(t),t,t)`, Jennifer has used `diff(y(t),t$2)`. Although of no real advantage here, if it was necessary, say, to take the 20th time derivative of some function, the notation `t$20` is simpler to type than entering `t` 20 times.

> `Vy:=diff(y(t),t): Vz:=diff(z(t),t): Ay:=diff(y(t),t$2):`
> `Az:=diff(z(t),t$2):`

To eliminate the mass from the resulting equations, Jennifer writes the coefficients in the form $C_d = K_d\,m$ and $C_L = K_L\,m$, so K_d and K_L are the drag and lift coefficients per unit mass.

> `C[d]:=K[d]*m: C[L]:=K[L]*m:`

Dividing both yeq and zeq by m and expanding yields the equations of motion.

> `yeq2:=expand(yeq/m); zeq2:=expand(zeq/m);`

$$yeq2 := \frac{d^2}{dt^2} y(t) = -K_d \sqrt{(\frac{d}{dt} y(t))^2 + (\frac{d}{dt} z(t))^2}\, (\frac{d}{dt} y(t))$$
$$- K_L \sqrt{(\frac{d}{dt} y(t))^2 + (\frac{d}{dt} z(t))^2}\, (\frac{d}{dt} z(t))$$

$$zeq2 := \frac{d^2}{dt^2} z(t) = -g - K_d \sqrt{(\frac{d}{dt} y(t))^2 + (\frac{d}{dt} z(t))^2}\, (\frac{d}{dt} z(t))$$
$$+ K_L \sqrt{(\frac{d}{dt} y(t))^2 + (\frac{d}{dt} z(t))^2}\, (\frac{d}{dt} y(t))$$

Jennifer notes that the above ODEs are coupled to each other and both are nonlinear due to the presence of the square root terms. These ODEs cannot be solved analytically, so Jennifer must tackle them numerically. To evaluate K_d and K_L, she enters the following expressions for C_d and C_L. Here ρ is the air density, r the golf ball radius, and c_d and c_L are numerical coefficients.

> `C[d]:=rho*Pi*r^2*c[d]/2; C[L]:=rho*Pi*r^2*c[L]/2;`

$$C_d := \frac{1}{2} \rho\,\pi\,r^2\,c_d$$

$$C_L := \frac{1}{2} \rho\,\pi\,r^2\,c_L$$

For her simulation, Jennifer takes $\rho = 1.21$ kg/m^3, which is the air density at 1 atm pressure and 20°C, and $g = 9.81$ m/s^2. From the cited references, a (British) golf ball has a mass $m = 0.046$ kg and a radius $r = 0.0207$ m, and if launched with an initial speed of $Vo = 61$ m/s at a launch angle of 16° and with a backspin of about 60 revolutions per second, the coefficients c_d and c_l are approximately equal to each other with a value 0.28. Although the conditions for Benny's amazing shot probably differed from these, the above parameter values will suffice to illustrate the effect of drag and lift on the flight of a golf ball. She refers the reader to the references for more information on how the drag and lift coefficients change as the launch angle and backspin are varied.

```
>  rho:=1.21: g:=9.81: m:=0.046: r:=0.0207: Vo:=61: Angle:=16:
>  c[d]:=0.28: c[L]:=0.28:
```

The coefficients K_d and K_L are then numerically evaluated and assigned the names KD and KL. This will allow her to also set K_d and K_L equal to zero in the ODEs so that the effects of lift and drag can be seen more clearly.

```
>  KD:=evalf(C[d]/m); KL:=evalf(C[L]/m);
```

$$KD := 0.004957310686$$
$$KL := 0.004957310686$$

The launch *Angle* is converted from degrees into radians and the initial velocity components, *Voy* and *Voz*, calculated in m/s.

```
>  Theta:=convert(Angle,units,degree,radian);
```

$$\Theta := \frac{4\pi}{45}$$

```
>  Voy:=evalf(Vo*cos(Theta)); Voz:=evalf(Vo*sin(Theta));
```

$$Voy := 58.63696345$$
$$Voz := 16.81387871$$

For her animated simulation of Benny Boffos's hole in one, Jennifer takes the cliff edge and top to be at $y0 = 20$ m and $z0 = 20$ m, respectively. Jennifer will consider $j = 3$ different trajectories, corresponding to (a) zero drag and lift, (b) non-zero drag but zero lift, (c) both drag and lift non-zero. The total time is taken to be $TT = 8$ seconds, which is divided into $N = 40$ time steps, so the step size is 0.2 seconds.

```
>  y0:=20: z0:=20: j:=3: TT:=8: N:=40: step:=evalf(TT/N);
```

$$step := 0.2000000000$$

The initial position and velocity of the golf ball are given. The men's tee is located 5 meters back from the cliff's edge.

```
>  ic:=y(0)=y0-5,D(y)(0)=Voy,z(0)=z0,D(z)(0)=Voz:
```

To illustrate to her class how to implement conditional branching with Maple, Jennifer creates the following do loop to numerically solve the ODEs, subject to the initial conditions, for the three different scenarios.

```
>  for n from 1 to j do
```

4.1. VELOCITY-DEPENDENT FORCES

In the following conditional statement, if $n = 1$ then both the drag and lift coefficients are zero, else if $n = 2$ the lift coefficient is zero but the drag coefficient is KD (non-zero), else for $n = 3$ both the lift and drag coefficients are non-zero. The third situation is the one that corresponds to Benny Boffo's hole in one.

```
> if n=1 then K[d]:=0;K[L]:=0 elif n=2 then K[L]:=0;K[d]:=KD
> else K[L]:=KL;K[d]:=KD end if :
```

The ODEs *yeq2* and *zeq2* are solved with the numerical option of the dsolve command. The default method is the Runge–Kutta–Fehlberg method, but other algorithms can be selected. The default number of function evaluations is 30,000, so Jennifer includes the option maxfun=0 which allows more evaluations if necessary. The numerical solution is given as a listprocedure, this output form being most useful when the returned procedure is to be used later. The solution is assigned the name *sol*.

```
> sol:=dsolve({yeq2,zeq2,ic},{y(t),z(t)},numeric,maxfun=0,
> output=listprocedure):
```

Using *sol*, the coordinates of the golf ball are evaluated at arbitrary time t for each of the three cases.

```
> Y[n]:= eval(y(t),sol): Z[n]:=eval(z(t),sol):
> end do:
```

Next, Jennifer wants to plot the numerical solutions for each scenario and to animate the motions. She uses a double do loop to accomplish this. A conditional statement is used in the outer loop, which runs over the three scenarios, to color the ball differently in each case.

```
> for i from 1 to j do
```

In the following command, the ball is colored red for $i = 1$ (no lift or drag), blue for $i = 2$ (drag, but no lift), and brown for $i = 3$ (both drag and lift).

```
> if i=1 then c:=red elif i =2 then c:=blue else c:=brown
> end if;
```

For each value of i, the ball's position is plotted at equal time steps. The ball is represented at each time step by a size 14, appropriately colored, circle.

```
> for k from 0 to N do:
> pp||i[k]:=pointplot([Y[i](step*k),Z[i](step*k)],
> symbol=circle,symbolsize=14,color=c):
> end do:
> end do:
```

Jennifer's final do loop is used to plot the sequence of time steps for each case. The graph gr1 creates a static picture, while gr2 produces an animated plot.

```
> for k from 1 to j do
> gr1[k]:=display(seq(pp||k[i],i=0..N)):
> gr2[k]:=display(seq(pp||k[i],i=0..N),insequence=true):
> end do:
```

Feeling in an artistic mood, Jennifer creates a graph function gr for plotting the cliff, the lake, and the green. The relevant coordinates u and the color c must be supplied. The option filled=true fills in regions between the curves and the horizontal axis.

> gr:=(u,c)->plot(u,style=line,color=c,filled=true):

The cliff is colored brown, the lake blue, and the green (naturally) green.

> Cliff:=gr([[0,z0],[y0,z0],[y0,0]],brown):
> Lake:=gr([[[y0,2],[y0+190,2]],[[y0+230,2],[y0+250,2]]],blue):
> Green:=gr([[y0+190,3],[y0+230,3]],green):

The flag is plotted as a thick vertical red line at the hole.

> Flag:=plot([[236,2],[236,6]],style=line,color=red,
> thickness=2):

The resulting picture, showing all three trajectories, is shown in Figure 4.7.

> display({Cliff,Lake,Green,Flag,seq(gr1[i],i=1..j)},
> view=[0..270,0..70],tickmarks=[3,3]);

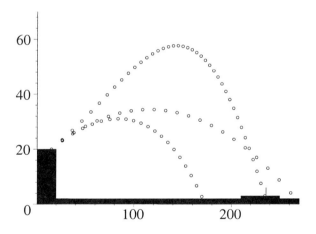

Figure 4.7: The three trajectories. Benny's shot lands in the hole at the flag.

The highest trajectory, which is represented by brown circles on the computer screen, is a simulation of Benny Boffo's hole in one. Both lift and drag have been included, the lift leading to a substantial increase in maximum elevation over the other two trajectories. She will have to reassure her students that the data used in computing the trajectory is realistic, being garnered from the cited references. The longest trajectory (red circles on the screen) corresponds to no lift and no drag. Benny's ball, in this case, would overshoot the green and land in the lake beyond the island. The shortest trajectory (blue circles) includes drag, but no lift. In this case, Benny's shot would have come up short of the green and also landed in the lake. Notice that there are more circles present for the highest trajectory, implying that the ball remains longer in the air when lift and drag are both included. This may be confirmed by executing the following command line which animates all three motions in the same picture.

```
> display({Cliff,Lake,Green,Flag,seq(gr2[i],i=1..j)},
> view=[0..270,0..70],tickmarks=[3,3]);
```

Having completed her simulation, Jennifer is confident that her students will like it. Perhaps, even Benny Boffo might as well. And you never know, in the future Benny and Syd might give an endowment to the university.

4.2 Position-dependent forces

4.2.1 Mike's Mobile Modes

There are two modes of transport in Los Angeles: car and ambulance. Visitors who wish to remain inconspicuous are advised to choose the latter.
Fran Lebowitz, American journalist, *Social Studies, Lesson 1* (1981).

Mike has been asked by his mechanics tutorial supervisor, Jennifer, to develop a computer algebra recipe in linear lattice dynamics. Specifically, he is to animate the normal modes of longitudinal vibration of a one-dimensional chain of $n+2$ identical atoms which interact with their nearest neighbors according to Hooke's law. The gravitational forces on the atoms (mass m) are to be neglected and the two end atoms are pinned down. Jennifer further suggests that Mike take n to be modestly large, e.g., $n = 10$ atoms are allowed to undergo longitudinal vibrations in the animation.

The extension of this problem to very large n and with the inclusion of anharmonic terms in the force law was the subject of a famous numerical investigation in nonlinear lattice dynamics by Fermi, Pasta, and Ulam [FPU55]. For a linear lattice, the energy of each normal mode remains constant. They intended to verify the widely accepted assumption that the introduction of small nonlinearities would lead to an equipartition of energy, i.e., the small nonlinearities would cause energy to flow from one mode to another until all modes in a time-averaged sense, would have the same energy. The existence of such an equilibrium is known as the zeroth law of thermodynamics and is an important assumption of statistical mechanics. What Fermi, Pasta, and Ulam actually found was the that the nonlinear system did not approach equilibrium as expected (exhibiting "non-ergodic" behavior), but instead displayed recurrences of energy in certain modes. This was referred to as the Fermi-Pasta-Ulam anomaly. The resolution of this anomaly is beyond the scope of this book, but a simple version of the nonlinear lattice problem is given in Supplementary Recipe **06-S05: George's Nonlinear Inchworm**. Let's now return to Mike.

Before developing his recipe, Mike establishes the notation that he will be using and the equations to be solved. For the linear chain, the atoms numbered $k = 1, 2, ..., n$ are allowed to vibrate, while the end atoms $k = 0$ and $k = n + 1$ are not. The kth atom, of those free to vibrate, has two nearest neighbors, namely the atoms $k - 1$ and $k + 1$. Mike lets x_k, x_{k-1}, and x_{k+1} be the longitudinal displacements of these atoms from equilibrium. For the end atoms, $x_0 = x_{n+1} = 0$. Hooke's law states that the restoring force is $F = -Kx$, where K is the spring constant and x is the displacement from equilibrium. For the kth atom in the chain, the net force on it due to its two neighbors will be equal to $K(x_{k+1} - x_k) - K(x_k - x_{k-1}) = K(x_{k-1} - 2x_k + x_{k+1})$. Letting $\kappa = K/m$, the equation of motion of the kth vibrating atom is

$$\kappa(x_{k-1} - 2x_k + x_{k+1}) = \ddot{x}_k, \quad k = 1, 2, ..., n. \tag{4.1}$$

To determine the normal modes of vibration, Mike follows standard procedure and substitutes a solution of the form $x_k = a_k \cos(\omega t)$ into Equations (4.1), yielding the following n equations for the amplitudes a_k,

$$-\kappa a_{k-1} + 2\kappa a_k - \kappa a_{k+1} = \omega^2 a_k, \equiv \lambda a_k \quad k = 1, 2, ..., n. \tag{4.2}$$

Solving Equations (4.2) for the frequency ω will yield n eigenfrequencies ω_i, or n eigenvalues $\lambda_i = \sqrt{\omega_i}$. There are n normal mode solutions, one for each ω_i. Normal modes are important because any possible longitudinal motion of the chain will be a linear superposition of the longitudinal normal modes.

Because he intends to use a matrix approach to solving the normal mode problem, Mike begins by loading the LinearAlgebra and LREtools (linear recurrence equation tools) packages. The latter package contains the recurrence equation solve (rsolve) command which he will be employing. The plots and stats[statplots] packages are needed for the display (for superimposing plots) and scatterplot (for plotting atomic positions) commands, respectively. As suggested by Jennifer, he considers the situation when $n = 10$ atoms are allowed to vibrate. This number can be increased, but because of their size the matrices in the recipe will not then be displayed inline. To learn how large matrices can be viewed, the reader should consult the topic "structuredview"

> restart: with(LinearAlgebra): with(LREtools):

> with(plots): with(stats[statplots]): n:=10:

Mike now forms a column vector V (output not shown) of the n amplitudes a_k.

> V:=<seq(a[k],k=1..n)>;

Using the BandMatrix command found in the LinearAlgebra package, he creates an n by n, tridiagonal, matrix A. This matrix has all main diagonal entries equal to 2κ and $-\kappa$ is placed at each entry along the first (1 refers to the number of subdiagonals) subdiagonal just above and below the main diagonal. All remaining entries are by default taken to be equal to zero.

> A:=BandMatrix([-kappa,2*kappa,-kappa],1,n);

4.2. POSITION-DEPENDENT FORCES

$$A := \begin{bmatrix} 2\kappa & -\kappa & 0 & 0 & 0 & 0 & 0 & 0 & 0 & 0 \\ -\kappa & 2\kappa & -\kappa & 0 & 0 & 0 & 0 & 0 & 0 & 0 \\ 0 & -\kappa & 2\kappa & -\kappa & 0 & 0 & 0 & 0 & 0 & 0 \\ 0 & 0 & -\kappa & 2\kappa & -\kappa & 0 & 0 & 0 & 0 & 0 \\ 0 & 0 & 0 & -\kappa & 2\kappa & -\kappa & 0 & 0 & 0 & 0 \\ 0 & 0 & 0 & 0 & -\kappa & 2\kappa & -\kappa & 0 & 0 & 0 \\ 0 & 0 & 0 & 0 & 0 & -\kappa & 2\kappa & -\kappa & 0 & 0 \\ 0 & 0 & 0 & 0 & 0 & 0 & -\kappa & 2\kappa & -\kappa & 0 \\ 0 & 0 & 0 & 0 & 0 & 0 & 0 & -\kappa & 2\kappa & -\kappa \\ 0 & 0 & 0 & 0 & 0 & 0 & 0 & 0 & -\kappa & 2\kappa \end{bmatrix}$$

He similarly forms an n by n diagonal matrix L with all diagonal entries equal to λ and zero subdiagonals. All off-diagonal entries are zero.

> L:=BandMatrix([lambda],0,n);

Using the Multiply command, Mike calculates the matrix equation $AV=LV$ in eq. The resulting set of equations (displayed in matrix form) are the explicit forms of the $n=10$ equations corresponding to Equations (4.2).

> eq:=Multiply(A,V)=Multiply(L,V);

$$eq := \begin{bmatrix} 2\kappa a_1 - \kappa a_2 \\ -\kappa a_1 + 2\kappa a_2 - \kappa a_3 \\ -\kappa a_2 + 2\kappa a_3 - \kappa a_4 \\ -\kappa a_3 + 2\kappa a_4 - \kappa a_5 \\ -\kappa a_4 + 2\kappa a_5 - \kappa a_6 \\ -\kappa a_5 + 2\kappa a_6 - \kappa a_7 \\ -\kappa a_6 + 2\kappa a_7 - \kappa a_8 \\ -\kappa a_7 + 2\kappa a_8 - \kappa a_9 \\ -\kappa a_8 + 2\kappa a_9 - \kappa a_{10} \\ -\kappa a_9 + 2\kappa a_{10} \end{bmatrix} = \begin{bmatrix} \lambda a_1 \\ \lambda a_2 \\ \lambda a_3 \\ \lambda a_4 \\ \lambda a_5 \\ \lambda a_6 \\ \lambda a_7 \\ \lambda a_8 \\ \lambda a_9 \\ \lambda a_{10} \end{bmatrix}$$

Mike now wants to determine the allowed eigenvalues λ_i. To accomplish this, he forms the characteristic matrix of A using the CharacteristicMatrix command, assigning the name cm to the result.

> cm:=CharacteristicMatrix(A,lambda);

$$cm := \begin{bmatrix} -\lambda+2\kappa, & -\kappa, & 0, & 0, & 0, & 0, & 0, & 0, & 0, & 0 \\ -\kappa, & -\lambda+2\kappa, & -\kappa, & 0, & 0, & 0, & 0, & 0, & 0, & 0 \\ 0, & -\kappa, & -\lambda+2\kappa, & -\kappa, & 0, & 0, & 0, & 0, & 0, & 0 \\ 0, & 0, & -\kappa, & -\lambda+2\kappa, & -\kappa, & 0, & 0, & 0, & 0, & 0 \\ 0, & 0, & 0, & -\kappa, & -\lambda+2\kappa, & -\kappa, & 0, & 0, & 0, & 0 \\ 0, & 0, & 0, & 0, & -\kappa, & -\lambda+2\kappa, & -\kappa, & 0, & 0, & 0 \\ 0, & 0, & 0, & 0, & 0, & -\kappa, & -\lambda+2\kappa, & -\kappa, & 0, & 0 \\ 0, & 0, & 0, & 0, & 0, & 0, & -\kappa, & -\lambda+2\kappa, & -\kappa, & 0 \\ 0, & 0, & 0, & 0, & 0, & 0, & 0, & -\kappa, & -\lambda+2\kappa, & -\kappa \\ 0, & 0, & 0, & 0, & 0, & 0, & 0, & 0, & -\kappa, & -\lambda+2\kappa \end{bmatrix}$$

The λ_i can be determined in two ways. The first is to mimic a hand calculation by setting the determinant of cm equal to zero and solving for λ. Alternately, one can directly obtain the eigenvalues. To obtain numerical values for the λ_i and the eigenfrequencies ω_i, Mike takes $\kappa = 1.0$. He increases the number of digits from the default number of 10 to 15 in order to obtain answers using the first approach which are in good agreement with those obtained by the second.

> kappa:=1.0: Digits:=15:

The λi are first obtained by applying the Determinant command to cm, setting the result to 0, and numerically solving for λ. The eigenvalues are ordered from the lowest value to the highest.

> lambda1:=[fsolve(Determinant(cm)=0,lambda)];

$\lambda 1 := [0.0810140527723137, 0.317492934333230, 0.690278532115713,$
$1.16916997399651, 1.71537032203980, 2.28462970269891, 2.83082984723579,$
$3.30972202506184, 3.68250627021611, 3.91898633992483]$

A more direct approach is to apply the Eigenvalues command to A, asking for the output to be given in the form of a list. Unlike the numbers in $\lambda 1$, the eigenvalues in $\lambda 2$ are not systematically ordered. They can be ordered from the lowest value to the highest by using the sort command. The numbers in $\lambda 2$ are in good agreement with those in $\lambda 1$.

> lambda2:=Eigenvalues(A,output='list');

> lambda2:=sort(lambda2);

$\lambda 2 := [0.0810140527710052, 0.317492934337638, 0.690278532109430,$
$1.16916997399623, 1.71537032345343, 2.28462967654657, 2.83083002600377,$
$3.30972146789057, 3.68250706566236, 3.91898594722899]$

The ω_i follow on applying the sequence command to the square root of the λ_i.

> omega:=[seq(sqrt(lambda2[i]),i=1..n)];

$\omega := [0.284629676546570, 0.563465113682860, 0.830830026003773,$
$1.08128163491120, 1.30972146789057, 1.51149914870852, 1.68250706566236,$
$1.81926399070904, 1.91898594722899, 1.97964288376186]$

A normal mode solution will be associated with each ω_i or eigenvalue. To find the solution for each λ_i, Mike creates the following do loop which runs from $i = 1$ to $n = 10$. He enters Equation (4.2) with $\lambda = \lambda 2_i$.

> for i from 1 to n do

> eq[i]:=-kappa*a(k-1)+2*kappa*a(k)-kappa*a(k+1)=lambda2[i]*a(k):

Since $eq[i]$ is a recurrence equation in the amplitudes, Mike solves it with the recurrence equation solve (rsolve) command, setting the amplitudes $a(0)$ and $a(n+1)$ of the fixed end atoms equal to zero.

> sol[i]:=rsolve({eq[i],a(0)=0,a(n+1)=0},a(k));

Each solution $sol[i]$ involves the amplitude $a(n)$ as an overall multiplying factor, so Mike divides $sol[i]$ by $a(n)$ and further simplifies the ratio with the radnormal command. This command normalizes an expression involving radicals. The ith

4.2. POSITION-DEPENDENT FORCES

normal mode, $nm[i]$, is then formed by multiplying the numerically evaluated list of n amplitudes by $\cos(\omega_i t)$.

```
>   nm[i]:=evalf([seq(radnormal(sol[i]/a(n)),k=1..n)]
>   *cos(omega[i]*t));
>   end do:
```

Mike has put a colon on the output so that the 10 normal mode solutions are not displayed. He selects a particular mode, e.g., the 10th one, for animation purposes. By changing the argument, other normal modes may be studied.

```
>   mode:=nm[10]; #select mode to be animated
```

$mode := [-0.999999999999682, 1.91898594722871, -2.68250706566213,$
$3.22870741511938, -3.51333709166601, 3.51333709166606, -3.22870741511952,$
$2.68250706566234, -1.91898594722899, 1.] \cos(1.97964288376186\, t)$

Mike wants to choose an equilibrium spacing for the atoms that is larger than the biggest amplitude in the above output. He does this as follows. The command op(1,mode) selects the first operand (the list of amplitudes) from $mode$. Applying op to this result, removes the amplitude numbers from the list format. The max command then selects the largest amplitude (≈ 3.51). The value of this amplitude is rounded off to the nearest integer (4) with the round command. Finally, Mike chooses a spacing which is four times this value, i.e., an equilibrium spacing of 16 spatial units.

```
>   spacing:=4*round(max(op(op(1,mode))));
```

$$spacing := 16$$

The equilibrium positions of the n atoms are calculated and put into a list.

```
>   equilpos:=[seq(k*spacing,k=1..n)]:
```

A graph function gr is formed using the scatterplot command. The atomic positions will be plotted when the list u of horizontal coordinates and the list v of vertical coordinates are supplied as arguments. The atoms are represented by size 14 red circles.

```
>   gr:=(u,v)->scatterplot(u,v,symbol=circle,symbolsize=14,
>   color=red):
```

In gr1, the (fixed) end atoms are graphed, while the atomic displacements at $t = 0$ are plotted in gr2.

```
>   gr1:=gr([0,(n+1)*spacing],[0,0]):
>   gr2:=gr(equilpos,op(1,mode)):
```

The horizontal displacements of all $n+2$ atoms at time $t = 0$ are now displayed,

```
>   display({gr1,gr2},labels=["equil. position","displacement"]);
```

the resulting picture being shown in Fig. 4.8.

To animate the selected normal mode, Mike takes a total time $T = 50$ and will create $N = 200$ frames. The time step size is then $T/N = 1/4$. He also creates a sequence of n zeros for plotting the longitudinal mode of vibration.

```
>   T:=50: N:=200: step:=T/N; h:=[seq(k*0,k=1..n)]:
```

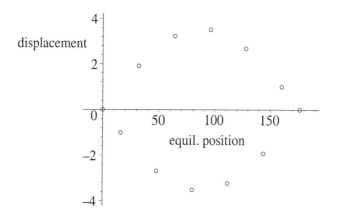

Figure 4.8: Displacements of atoms at $t=0$ for the 10th normal mode.

In the animation, the equilibrium positions are represented by size 24 blue crosses which are created with the scatterplot command in gr3.

> gr3:=scatterplot(equilpos,h,symbol=cross,symbolsize=24,
> color=blue):

Mike uses a do loop to plot the atomic positions at each time step.

> for i from 0 to N do
> t:=step*i:
> gr4[i]:=gr(equilpos+mode,h);

The graphs gr1, gr3 and gr4[i] are superimposed and the do loop ended.

> pl[i]:=display({gr1,gr3,gr4[i]});
> end do:

Mike now unassigns the time. This enables one to quickly look at another normal mode after executing the subsequent display command line by changing the number 10 in mode to some other number between 1 and 10 and re-executing the subsequent command lines. It's not necessary to go back through restart.

> unassign('t'):
> display(seq(pl[i],i=0..N),insequence=true);

On executing the above display command line and clicking on the plot and the start arrow, the longitudinal vibrations of the atoms about their equilibrium positions may be seen for the normal mode that Mike has selected.

Mike is happy with his computer algebra recipe, but intends to check with Jennifer before the tutorial to see if there are any other points that he should address. For example, he could have his students modify the recipe to deal with

4.2. POSITION-DEPENDENT FORCES

the longitudinal or transverse vibrations of molecules which are not undergoing translational or rotational motion. The linear triatomic CO_2 molecule is discussed in all standard mechanics texts, e.g., in Marion and Thornton [MT95] and in Goldstein, Poole, and Safko [GPS02]. In this case, the end (oxygen) atoms are not fixed and the center of mass remains constant during the motion.

4.2.2 A "Hard" Spring's Journey

To work hard, to live hard, to die hard, and then to
go to hell after all would be too damned hard.
Carl Sandburg, American poet, *The People, Yes* (1936)

When lift and drag were included, we saw that the ODEs describing the flight of a golf ball are nonlinear and no analytical solution exists. This is generally the case for most nonlinear ODEs of physical interest. However, the motion of a " hard" spring can be solved analytically in terms of elliptic functions.

George and Richard's research interests lie in the field of nonlinear phenomena, their involvement being from an experimental and theoretical perspective, respectively. In this recipe, they combine forces to create and solve a nonlinear "hard" spring example. The spring configuration in Figure 4.9 lies in the horizontal plane. An airtrack glider (mass m) is free to move horizontally on a linear airtrack. The weight of the glider is balanced by the upward normal

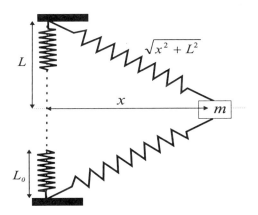

Figure 4.9: Spring configuration for air glider motion.

force of the airtrack and the frictional force is assumed to be negligible. For simplicity in the analysis, air resistance is also ignored. The glider is initially connected to two stretched horizontal linear springs (spring constant k) of equal length L, their unstretched lengths being L_0. The glider is then pulled aside from the equilibrium position a distance x horizontally and allowed to oscillate. The period of the oscillations can be timed with a stop watch and compared with theory, which is now developed.

The DEtools library package is loaded, because the phaseportrait command will be used. We take the nominal values $m=1$ kg, $k=1$ N/m, $L=0.5$ m, and $L_0=0.4$ m.

> `restart: with(plots): with(DEtools):`

> `m:=1: k:=1: L:=0.5: L0:=0.4:`

With the glider pulled aside a distance x, each spring is stretched by an amount $d = \sqrt{x^2 + L^2} - L_0$. Assuming the springs to be linear, the potential energy of the glider is $V = 2 \times (k\,d^2/2)$.

> `d:=sqrt(x^2+L^2)-L0;`

$$d := \sqrt{x^2 + 0.25} - 0.4$$

> `V:=2*(k*d^2/2);`

$$V := (\sqrt{x^2 + 0.25} - 0.4)^2$$

For small x, V can be expanded in a Taylor series in x, say out to 5th order.

> `Vexp:=taylor(V,x,5);`

$Vexp := 0.01000000000 + 0.2000000000\,x^2 + 0.8000000000\,x^4 + \mathrm{O}(x^6)$

The "order of" term ($\mathrm{O}(x^6)$) is removed with the convert(polynom) command, which converts the series to a polynomial.

> `Vexp:=convert(Vexp,polynom);`

$Vexp := 0.01000000000 + 0.2000000000\,x^2 + 0.8000000000\,x^4$

The exact potential V and the expansion $Vexp$ are then plotted together in Figure 4.10 as solid thick green and dashed thick blue curves, respectively.

> `plot([V,Vexp],x=-1..1,color=[green,blue],linestyle=[1,3],`
> `thickness=2,tickmarks=[4,3]);`

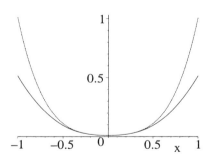

Figure 4.10: Lower (solid) curve: V. Upper (dashed) curve: $Vexp$.

The two curves are quite close over the range $x = -0.5$ to $x = 0.5$, but $Vexp$ rises more rapidly for larger x. The forces $F = -dV/dx$ and $Fexp = -d\,Vexp/dx$ are calculated and x replaced with $x(t)$.

4.2. POSITION-DEPENDENT FORCES

> `F:=-diff(V,x); Fexp:=-diff(Vexp,x);`

$$F := -2\,\frac{(\sqrt{x^2+0.25}-0.4)\,x}{\sqrt{x^2+0.25}}$$

$$Fexp := -0.4000000000\,x - 3.200000000\,x^3$$

> `Fx:=subs(x=x(t),F): Fexpx:=subs(x=x(t),Fexp):`

Newton's second law is written down for F and $Fexp$ in *de1a* and *de1b*, and the velocity v expressed as the time derivative of x in *de2*.

> `de1a:=m*diff(v(t),t)=Fx;`

$$de1a := \tfrac{d}{dt}\mathrm{v}(t) = -\frac{2\,(\sqrt{\mathrm{x}(t)^2+0.25}-0.4)\,\mathrm{x}(t)}{\sqrt{\mathrm{x}(t)^2+0.25}}$$

> `de1b:=m*diff(v(t),t)=Fexpx;`

$$de1b := \tfrac{d}{dt}\mathrm{v}(t) = -0.4000000000\,\mathrm{x}(t) - 3.200000000\,\mathrm{x}(t)^3$$

> `de2:=v(t)=diff(x(t),t);`

$$de2 := \mathrm{v}(t) = \tfrac{d}{dt}\mathrm{x}(t)$$

As initial conditions, let's take $x(0) = 0.5$ m and $v(0) = 0$.

> `ic:=(x(0)=0.5,v(0)=0);`

The `phaseportrait` command can be used to create a "phaseplane portrait", $v(t)$ vs. $x(t)$, as well as to plot $x(t)$ vs. t. A functional operator to accomplish this task is now introduced in `gr`. The relevant pair of first order ODEs (*de1a* and *de2*, or *de1b* and *de2*) must be supplied as the arguments *eq1*, *eq2*, to `gr`. The equation pair (*eq1*, *eq2*) is numerically solved for $x(t)$ and $v(t)$ over the time interval $t = 0...20$, subject to the initial condition (*ic*) with a time step size of 0.05. In the phaseplane portrait, a `tangent field` can be produced indicating the direction of the *trajectory* in the phase plane. Arrows are drawn on a regular grid, the arrow heads indicating the direction of increasing time. The density of arrows is controlled with the `dirgrid` option. The minimum is [2,2] (two arrows horizontally and two vertically) and the default [20,20]. Various types of arrows can be selected in the `arrow` option, including NONE. Two-headed arrows are produced if one uses `arrows=MEDIUM`. The option `scene` controls what type of picture is to be drawn. For example, `scene=[t,x]` produces a plot of x vs. t. If no scene is specified, then v vs. x is plotted by default. The color of the solution curve can also be controlled with the `linecolor` option.

> `gr:=(eq1,eq2,a,lc,b,c)->phaseportrait({eq1,eq2},[x(t),v(t)],`
> `t=0..20,[[ic]],stepsize=0.05,dirgrid=[25,25],arrows=a,`
> `linecolor=lc,scene=[b,c]):`

In the following two `display` commands, the phaseplane portrait ($v(t)$ vs. $x(t)$) and $x(t)$ vs. t are produced for V and $Vexp$. The results are shown in Fig. 4.11.

> `display({gr(de1a,de2,NONE,green,x,v),`
> `gr(de1b,de2,MEDIUM,blue,x,v)});`

```
> display({gr(de1a,de2,NONE,green,t,x),
> gr(de1b,de2,NONE,blue,t,x)});
```

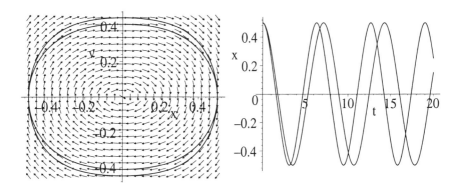

Figure 4.11: Left: phaseplane portrait, v vs. x. Right: x vs. t.

In the phaseplane portrait on the left, the outer curve corresponds to $Vexp$, the inner one to V. Since $Vexp < V$ at $x(0) = 0.5$, the glider does not attain quite as high a velocity for V on passing through $x = 0$. Correspondingly, the glider takes slightly longer to complete one oscillation than it would if $Vexp$ prevailed. This is clearly seen on the right of Figure 4.11, the curve with the longer period being the solution for V. Placing the cursor on top of the maxima, and clicking, one finds that the period for $Vexp$ is about 6.3 s, and about 7.1 s for V.

The equation of motion corresponding to $Vexp$ can be solved analytically, whereas the exact equation cannot. For the former, we write Newton's second law as a second order ODE. This ODE is traditionally referred to as the hard spring equation.

```
> de3:=m*diff(x(t),t,t)=Fexpx;
```

$$de3 := \frac{d^2}{dt^2} x(t) = -0.4000000000\, x(t) - 3.200000000\, x(t)^3$$

An analytic solution of *de3* is sought, by applying the `dsolve` command.

```
> sol:=dsolve(de3,x(t));
```

$$sol := \int^{x(t)} \frac{5}{\sqrt{-10\,\_a^2 - 40\,\_a^4 + 25\,\_C1}}\, d\_a - t - \_C2 = 0,$$

$$\int^{x(t)} -\frac{5}{\sqrt{-10\,\_a^2 - 40\,\_a^4 + 25\,\_C1}}\, d\_a - t - \_C2 = 0$$

Two implicit solutions are generated with two arbitrary coefficients (*_C1* and *_C2*), one solution involving a positive square root, the other a negative root. To obtain a positive period, the positive root (first solution here) must be selected.

```
> ans:=sol[1];
```

4.2. POSITION-DEPENDENT FORCES

Differentiating *ans* with respect to t, the output of *eq* is solved for $dx(t)/dt$.

> `eq:=diff(ans,t);`

$$eq := \frac{5\frac{d}{dt}x(t)}{\sqrt{-10\,x(t)^2 - 40\,x(t)^4 + 25\,\_C1}} - 1 = 0$$

> `eq2:=solve(eq,diff(x(t),t));`

$$eq2 := \frac{1}{5}\sqrt{-10\,x(t)^2 - 40\,x(t)^4 + 25\,\_C1}$$

We replace $x(t)$ in *eq2* with a new dependent variable y so that some ensuing integrals can be carried out.

> `eq3:=subs(x(t)=y,eq2);`

$$eq3 := \frac{\sqrt{-10\,y^2 - 40\,y^4 + 25\,\_C1}}{5}$$

If we substitute $y = A$, where A is the amplitude, the corresponding velocity (*eq3*) must equal zero. Then, *eq4* is solved for the constant *_C1* in terms of A.

> `eq4:=subs(y=A,eq3=0);`

$$eq4 := \frac{\sqrt{-10\,A^2 - 40\,A^4 + 25\,\_C1}}{5} = 0$$

> `_C1:=solve(eq4,_C1);`

$$\_C1 := \frac{2}{5}A^2 + \frac{8}{5}A^4$$

The expression for *_C1* is automatically substituted into *eq3* to yield *eq5*.

> `eq5:=eq3;`

$$eq5 := \frac{\sqrt{-10\,y^2 - 40\,y^4 + 10\,A^2 + 40\,A^4}}{5}$$

So that the integrals can be displayed in a simple form, we assume that the amplitude A is positive and that the solution x lies in the range $-A$ to A.

> `assume(A>0,x>-A,x<A):`

The period T can be obtained by integrating the reciprocal of *eq5* over a half-period from $y = -A$ to $y = A$ and then multiplying by 2.

> `T:=2*(int(1/eq5,y=-A..A));`

$$T := \frac{2\sqrt{10}\,\text{EllipticK}(\frac{2A}{\sqrt{1+8A^2}})}{\sqrt{1+8A^2}}$$

The period is expressed in terms of the complete elliptic integral EllipticK(k), with $k = 2A/\sqrt{1+8A^2}$. In standard math notation, this integral is written as $K(k)$ and defined by

$$K(k) \equiv \int_0^{\pi/2} \frac{d\theta}{\sqrt{1 - k^2\sin^2\theta}}.$$

For the initial condition $x(0) = 0.5$, $v(0) = 0$, the amplitude $A = 0.5$. Substituting this value into T yields a period of 6.33 s, close to the result obtained earlier by clicking on the plot.

> `period:=evalf(subs(A=0.5,T));`

$$period := 6.331369270$$

To obtain an analytic solution for $x(t)$, we integrate $1/eq5$ from x to A and set the result equal to t.

> `eq6:=t=simplify(int(1/eq5,y=x..A));`

$$eq6 := t = \frac{1}{2} \frac{\sqrt{10}\, (2\, \text{EllipticK}(\frac{2A}{\sqrt{1+8A^2}}) - \text{EllipticF}(\frac{\sqrt{A^2-x^2}}{A}, \frac{2A}{\sqrt{1+8A^2}}))}{\sqrt{1+8A^2}}$$

The output is expressed in terms of the complete elliptic integral as well as the incomplete elliptic integral of the first kind, $\text{EllipticF}(\phi, k)$, with $\phi = (\sqrt{A^2 - x^2})/A$ and k already identified. In standard math notation, the incomplete elliptic integral is written as $F(\phi, k)$ and defined by

$$F(\phi, k) \equiv \int_0^\phi \frac{d\theta}{\sqrt{1 - k^2 \sin^2 \theta}}.$$

We have an implicit equation, eq6, with x appearing inside the argument of the elliptic integral. Let's solve eq6 for x.

> `x:=solve(eq6,x); #solve eq6 for x`

$$x := \sqrt{1 - \text{JacobiSN}(\frac{1}{5}(t\sqrt{1+8A^2} - \sqrt{10}\,\text{EllipticK}(k))\sqrt{10},\, k)^2}\, A,$$

$$-\sqrt{1 - \text{JacobiSN}(\frac{1}{5}(t\sqrt{1+8A^2} - \sqrt{10}\,\text{EllipticK}(k))\sqrt{10},\, k)^2}\, A$$

Two solutions are generated in terms of the Jacobian elliptic sine function, $\text{JacobiSN}(z, k)$, with $z = (1/5)(t\sqrt{1+8A^2} - \sqrt{10}\,\text{EllipticK}(k))\sqrt{10},\, k)^2$. This function is defined in Maple's Help and is discussed in standard mathematical physics texts (e.g., [MW70]). It should be regarded as just another "special function", whose properties have been well-studied in the literature (see [AS72]). To plot x, the positive square root solution is required for $x > 0$ and the negative square root for $x < 0$. The solution branches are evaluated at $A = 0.5$.

> `x1:=eval(x[1],A=0.5); x2:=eval(x[2],A=0.5);`

$$x1 := 0.5\sqrt{1 - \text{JacobiSN}(1.095445115\, t,\, 0.5773502690)^2}$$

$$x2 := -0.5\sqrt{1 - \text{JacobiSN}(1.095445115\, t,\, 0.5773502690)^2}$$

The `piecewise` command is used to splice the solution branches $x1$ and $x2$ together over the time interval from $t = 0$ to $9/4$ times the period.

> `p:=piecewise(t<period/4,x1,t<3*period/4,x2,t<=5*period/4,x1,`
> `t<7*period/4,x2,t<9*period/4,x1);`

4.2. POSITION-DEPENDENT FORCES

The piecewise solution p is plotted over the same time interval in Figure 4.12. It may look like a cosine solution, but it is not.

```
> plot(p,t=0..9*period/4,labels=["t","x"],thickness=2,
> color=blue);
```

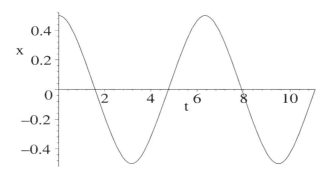

Figure 4.12: Analytic solution of the hard spring equation.

So this completes our intellectual journey exploring some aspects of the hard spring equation. Using computer algebra, solving the hard spring equation wasn't hard at all. For the most part, we simply had to translate what we would have done by hand into the appropriate commands.

4.2.3 Wouldn't Mr. Kepler Be Pleased

Space isn't remote at all. It's only an hour's drive away if your car could go straight upwards.
Fred Hoyle, English astrophysicist, in the *London Observer* (9 Sept. 1979)

Mike is certainly earning his money as a teaching assistant for Jennifer's mechanics course. This week, he has been instructed to create a computer recipe that will:

(a) analytically solve the central force problem for planetary motion;

(b) plot the trajectories of the four inner planets and the asteroid Eros;

(c) animate the motion of the five outer planets and Halley's comet.

Fortunately, Vectoria is free this evening and can help him with the project. She grabs a copy of Marion and Thornton [MT95] off their bookshelf, so that they can look up parameter values for plotting and simulating the planetary motions. Mike turns on the computer, loads the Maple `plots` package, and

```
> restart: with(plots):
```

asks Vectoria about the assumptions that they should feed into the program.

"Mike, we know that a planetary mass (m) moves in a planar orbit about the Sun under the influence of an attractive central force $\vec{F} = -(Km/r^n)\,\hat{e}_r$. Here r is the mass's radial distance from the Sun, $\hat{e}_r$ the radial unit vector, $K \equiv GM_S$ where G is the gravitational constant and M_S the mass of the sun, and $n = 2$. Since the motion is planar, let's use the polar coordinates $(r,\ \theta)$ with θ varying from 0 to 2π. For a closed orbit, there will generally exist two turning points in the radial direction, say $r = A$ and $r = B$, A being the distance of closest approach (perihelion) of m to the Sun and B the furthest distance (aphelion). We can start the mass at $t = 0$ with tangential velocity V at $r = A$, $\theta = 0$. Why don't you enter the assumption that V, A, B, and K are positive."

> assume(V>0,A>0,B>0,K>0):

After following Vectoria's suggestion, Mike enters the radial (*eq1*) and angular (*eq2*) equations[2] of motion for the planetary mass.

> eq1:=diff(r(t),t,t)-r(t)*diff(theta(t),t)^2=-K/r(t)^n;

$$eq1 := (\frac{d^2}{dt^2}\,\mathrm{r}(t)) - \mathrm{r}(t)\,(\frac{d}{dt}\,\theta(t))^2 = -\frac{K}{\mathrm{r}(t)^n}$$

> eq2:=r(t)*diff(theta(t),t,t)+2*diff(r(t),t)*diff(theta(t),t)=0;

$$eq2 := \mathrm{r}(t)\,(\frac{d^2}{dt^2}\,\theta(t)) + 2\,(\frac{d}{dt}\,\mathrm{r}(t))\,(\frac{d}{dt}\,\theta(t)) = 0$$

"Why didn't you set $n = 2$ in the radial equation?" Vectoria asks.

"I thought that I would leave n general for the moment, as it will enable us to deal more easily with other possible central forces. For example, we could look at Einstein's general relativistic force law which accounts for the precession of Mercury's orbit. Let's now solve the angular equation (*eq2*) for $\theta(t)$, subject to the initial conditions $\theta(0) = 0$, $\dot\theta(0) = V/A$."

> eq3:=dsolve({eq2,theta(0)=0,D(theta)(0)=V/A},theta(t));

$$eq3 := \theta(t) = \frac{V\,\mathrm{r}(0)^2}{A} \int_0^t \frac{1}{\mathrm{r}(\_z1)^2}\,d\_z1$$

"The answer is expressed in integral form in terms of the initial radius $r(0)$. Substituting $r(0) = A$, we can eliminate the θ dependence from the radial equation by differentiating *eq3* with respect to t and substituting *eq3b* into *eq1*."

> eq3b:=diff(subs(r(0)=A,eq3),t);

$$eq3b := \frac{d}{dt}\,\theta(t) = \frac{V\,A}{\mathrm{r}(t)^2}$$

> eq1b:=subs(eq3b,eq1);

$$eq1b := (\frac{d^2}{dt^2}\,\mathrm{r}(t)) - \frac{V^2\,A^2}{\mathrm{r}(t)^3} = -\frac{K}{\mathrm{r}(t)^n}$$

"Now, I will specialize to the inverse square law, setting $n = 2$ in *eq1b*."

> eq1c:=subs(n=2,eq1b);

---

[2]The plane polar forms of the acceleration on the lhs of *eq1*, *eq2* were derived in **03-1-1**.

4.2. POSITION-DEPENDENT FORCES

$$eq1c := (\frac{d^2}{dt^2}\,\mathrm{r}(t)) - \frac{V^2\,A^2}{\mathrm{r}(t)^3} = -\frac{K}{\mathrm{r}(t)^2}$$

"To integrate *eq1c*, let's set $p \equiv dr/dt$ so that $d^2 r/dt^2 = p(r)\,(dp(r)/dr)$. Making use of this result and setting $r(t) = r$, *eq1d* results."

> eq1d:=subs({r(t)=r,diff(r(t),t,t)=p(r)*diff(p(r),r)},eq1c);

$$eq1d := \mathrm{p}(r)\,(\frac{d}{dr}\,\mathrm{p}(r)) - \frac{V^2\,A^2}{r^3} = -\frac{K}{r^2}$$

"At $t = 0$, $r = A$ and the radial velocity $dr/dt \equiv p = 0$. So, we can solve the ODE *eq1d* for $p(r)$, subject to the initial condition $p(A) = 0$."

> sol:=dsolve({eq1d,p(A)=0},p(r));

$$sol := \mathrm{p}(r) = \frac{\sqrt{-V^2\,A^2 + 2\,K\,r + \frac{(A\,V^2 - 2\,K)\,r^2}{A}}}{r},$$

$$\mathrm{p}(r) = -\frac{\sqrt{-V^2\,A^2 + 2\,K\,r + \frac{(A\,V^2 - 2\,K)\,r^2}{A}}}{r}$$

"I will select the positive square root solution in *sol* and return to the original variables by substituting $p(r) = dr(t)/dt$ and $r = r(t)$."

> eq1e:=subs({p(r)=diff(r(t),t),r=r(t)},sol[1]);

$$eq1e := \frac{d}{dt}\,\mathrm{r}(t) = \frac{\sqrt{-V^2\,A^2 + 2\,K\,\mathrm{r}(t) + \frac{(A\,V^2 - 2\,K)\,\mathrm{r}(t)^2}{A}}}{\mathrm{r}(t)}$$

"When $r = B$, the radial velocity is again zero. Substituting $r(t) = B$ into the rhs of *eq1e* and setting the result equal to zero yields the following relation (labeled KK) for K."

> KK:=solve(subs(r(t)=B,rhs(eq1e))=0,K);

$$KK := \frac{(A + B)\,A\,V^2}{2\,B}$$

"I have never seen an analytic expression for $r(t)$, Mike, so I am sure that *eq1e* cannot be solved analytically. We could, however, first try to obtain the standard expression for $r(\theta)$ by dividing the rhs of *eq1e* by the rhs of *eq3b* to obtain an ODE for $dr/d\theta$. You can then substitute $K = KK$ into *eq4*,"

> eq4:=diff(r(theta),theta)=subs(r(t)=r(theta),
> rhs(eq1e)/rhs(eq3b));

$$eq4 := \frac{d}{d\theta}\,\mathrm{r}(\theta) = \frac{\sqrt{-V^2\,A^2 + 2\,K\,\mathrm{r}(\theta) + \frac{(A\,V^2 - 2\,K)\,\mathrm{r}(\theta)^2}{A}}\,\mathrm{r}(\theta)}{V\,A}$$

> eq4:=simplify(subs(K=KK,eq4));

$$eq4 := \frac{d}{d\theta}\,\mathrm{r}(\theta) = \frac{\sqrt{-B\,A + \mathrm{r}(\theta)\,A + \mathrm{r}(\theta)\,B - \mathrm{r}(\theta)^2}\,\mathrm{r}(\theta)}{\sqrt{B}\,\sqrt{A}}$$

"and analytically solve the resulting ODE for $r(\theta)$."

```
>  sol2:=dsolve(eq4,r(theta));
```

$$sol2 := \theta - \arctan\left(\frac{-BA + \frac{1}{2}(A+B)\,\mathrm{r}(\theta)}{\sqrt{(-BA + \mathrm{r}(\theta)A + \mathrm{r}(\theta)B - \mathrm{r}(\theta)^2)BA}}\right) + \_C1 = 0$$

Having done as instructed, Mike obtains an implicit solution *sol2* with one arbitrary constant $\_C1$. Using the `isolate` command, which isolates a subexpression to the lhs of an equation, he extracts a lengthy expression in R for $r(\theta)$ in terms of tangent functions.

```
>  R:=isolate(sol2,r(theta));
```

Since the standard formula for $r(\theta)$ is in terms of a cosine function, Mike converts the rhs of R into sines and cosines and simplifies with the symbolic option.

```
>  R:=convert(rhs(R),sincos);
>  R:=simplify(R,symbolic);
```

$$R := \frac{2\left(A + B + \sqrt{1 - \cos(\theta + \_C1)^2}\,A - \sqrt{1 - \cos(\theta + \_C1)^2}\,B\right)AB}{4BA - 2BA\cos(\theta + \_C1)^2 + A^2\cos(\theta + \_C1)^2 + B^2\cos(\theta + \_C1)^2}$$

Examining the form of R, Mike notes that if $\_C1 = \pi/2$, then $\cos(\theta + \pi/2) = \sin\theta$ and $\sqrt{1 - \sin^2\theta} = \cos\theta$. This looks promising, so he evaluates R at $\_C1 = \pi/2$ and simplifies the result.

```
>  R:=simplify(eval(R,_C1=Pi/2),symbolic);
```

$$R := -\frac{2BA}{\cos(\theta)A - \cos(\theta)B - A - B}$$

Opening up [MT95], Vectoria notes that for an ellipse $A = a(1 - \epsilon)$ and $B = a(1 + \epsilon)$, where a is the semimajor axis and ϵ the eccentricity. The standard form[3] of $r(\theta)$ follows on substituting the expressions for A and B.

```
>  R:=simplify(subs({A=a*(1-epsilon),B=a*(1+epsilon)},R));
```

$$R := -\frac{a(-1 + \varepsilon^2)}{\cos(\theta)\varepsilon + 1}$$

"To plot R and to animate the celestial motions, we are going to need the values of a, ϵ, and the period T for the various planets, as well as for the asteroid Eros and Halley's comet. So, Vectoria, why don't you look them up, while I create a functional operator `s` for substituting a and ϵ values into R and another operator, `pl`, to evaluate `s` for different i values, i labeling the celestial object."

```
>  s:=(a0,e0)->subs({a=a0,epsilon=e0},R):
>  pl:=i->s(a||i,e||i):
```

As Vectoria reads off the values for a and ϵ from Table 8-1 of Marion and Thornton, Mike enters them using e for the eccentricity and the following subscripts:

---

[3]Sometimes the solution is given in the form $r = a(-1 + \epsilon^2)/(\cos\theta - 1)$. The consequence is to rotate the orbits in Figure 4.13 by $180°$.

4.2. POSITION-DEPENDENT FORCES

0: the asteroid Eros; 1: Mercury; 2: Venus; 3: Earth; 4: Mars; 5: Jupiter; 6: Saturn; 7: Uranus; 8: Neptune; 9: Pluto; 10: Halley's comet. The units of a are in astronomical units (A.U.) and the earth's semimajor axis is taken to be 1 A.U. in length. One A.U.=1.495×10^{11} m.

```
>   a||0:=1.4583: a||1:=0.3871: a||2:=0.7233: a||3:=1.00:
>   a||4:=1.5237: a||5:=5.2028: a||6:=9.5388: a||7:=19.191:
>   a||8:=30.061: a||9:=39.529: a||10:=18:
>   e||0:=0.2230: e||1:=0.2056: e||2:=0.0068: e||3:=0.0167:
>   e||4:=0.0934: e||5:=0.0483: e||6:=0.0560: e||7:=0.0461:
>   e||8:=0.0100: e||9:=0.2484: e||10:=0.967:
```

Mike then uses the `polarplot` command to plot the orbits of Eros, Mercury, Venus, Earth and Mars. He colors them brown, cyan, green, blue, and red, respectively. A black and white rendition of this colorful picture is shown on the lhs of Figure 4.13. The asteroid Eros's orbit is the one that juts out furthest to the left in the picture.

```
>   polarplot([seq(pl(i),i=0..4)],theta=0..2*Pi,scaling=
>   constrained,thickness=3,color=[brown,cyan,green,blue,red],
>   title="cyan=Mercury, green=Venus, blue=Earth, red=Mars,
>   brown=Eros");
```

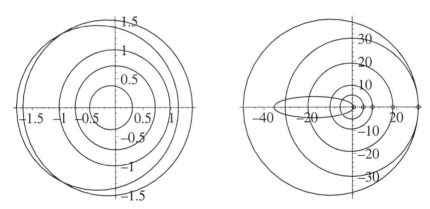

Figure 4.13: LHS: orbits of the four inner planets and the asteroid Eros. RHS: initial frame for the animation of the five outer planets and Halley's comet.

"O.K, Vectoria, let's now animate the five outer planets as well as Halley's comet. I will create a functional operator for calculating A for different i values."

```
>   AA:=i->a||i*(1-e||i):
```

"From Marion, the periods (in years) of Jupiter (subscript 5), Saturn (6), Uranus (7), Neptune (8), Pluto (9), and Halley's comet (10) are as follows."

```
>   T||5:=11.9: T||6:=29.5: T||7:=84.1: T||8:=164.8: T||9:=248.5:
```

> T||10:=76.1:

"The initial tangential velocities V can be determined as follows. From Kepler's second law, a line from the Sun to a planet sweeps out equal areas in equal times. The rate of sweeping out area is the areal velocity v. In one period, the line sweeps out the area S of the planetary ellipse, so $v = S/T$. But for an ellipse, $S = \pi a b$, where $b = a\sqrt{1 - \epsilon^2}$ is the semiminor axis. So, $v = \pi a^2 \sqrt{1 - \epsilon^2}/T$. But, by geometrical considerations, the areal velocity is also given by $v = v_\perp r/2$ (the area of a triangle), where r is the radial distance of the planet from the Sun and $v_\perp$ is the velocity perpendicular to the radius line. At the perihelion $r = a(1 - \epsilon)$ and $v_\perp = V$. So $a(1 - \epsilon) V/2 = \pi a^2 \sqrt{1 - \epsilon^2}/T$ or, solving for V, $V = (2\pi a/T)\sqrt{(1 + \epsilon)/(1 - \epsilon)}$. I will create a function f to calculate V."

> f:=i->evalf((2*Pi*a||i/T||i)*sqrt((1+e||i)/(1-e||i)),4):

"The initial tangential velocities are now calculated. Each initial velocity will allow the celestial object to complete its orbit in a time equal to its period."

> V||5:=f(5); V||6:=f(6); V||7:=f(7); V||8:=f(8); V||9:=f(9);
> V||10:=f(10);

$$V5 := 2.882$$
$$V6 := 2.150$$
$$V7 := 1.501$$
$$V8 := 1.158$$
$$V9 := 1.288$$
$$V10 := 11.41$$

"To animate the masses, we must numerically calculate $r(t)$ and $\theta(t)$. I will create a do loop to do this for the masses labeled 5 to 10."

> for i from 5 to 10 do

"The following line inserts a time dependence into $r(\theta)$ for each mass i."

> r||i:=subs(theta=theta(t),pl(i));

"We can substitute $r(t) = r||i$, $A = AA(i)$, and $V = V||i$ into the ODE $eq3b$ for $\theta(t)$. This yields a time-dependent angular equation $eq||i$ for each mass."

> eq||i:=subs({r(t)=r||i,A=AA(i),V=V||i},eq3b);

"Each angular equation $eq||i$ is then numerically solved for $\theta(t)$, subject to the initial condition $\theta(0) = 0$. I will express the output as a listprocedure and then evaluate θ at any instant t for each mass."

> sol||i:=dsolve({eq||i,theta(0)=0},theta(t),numeric,
> output=listprocedure):
> Theta||i:=eval(theta(t),sol||i):
> end do:

"With $\theta(t)$ determined, we can introduce a functional operator rr for evaluating $r(t)$ for each mass."

> rr:=(i,t)->eval(pl(i),theta=Theta||i(t)):

4.2. POSITION-DEPENDENT FORCES

"What total time should we take for the animation and how many frames should we use?"

"I would suggest that you take the period $(T = T||10)$ of Halley's comet as the duration of the simulation and generate $N = 100$ frames. The time step size then is $T/N = 0.761$ years."

```
> T:=T||10: N:=100: step:=T/N:
```

Having followed Vectoria's advice, Mike again uses the `polarplot` command to plot the orbits of the objects labeled 5 to 10. He colors Jupiter's (J) orbit brown, Saturn's (S) orbit red, Uranus's (U) orbit blue, Neptune's (N) orbit green, Pluto's (P) orbit cyan, and Halley's orbit black.

```
> pp:=polarplot([seq(pl(i),i=5..10)],theta=0..2*Pi,scaling=
> constrained,thickness=3,color=[brown,red,blue,green,cyan,
> black],title="brown=J, red=S, blue=U, green=N, cyan=P,
> black=Halley"):
```

Mike then creates the following double do loop to animate the motion of each of the outer planets and Halley's comet.

```
> for i from 5 to 10 do
```

The inner loop runs over the N time steps for each orbiting mass.

```
> for j from 0 to N do
```

The time t at the jth time step is equal to the step size times j.

```
> t:=step*j:
```

Converting back to Cartesian coordinates, each object's position is represented by a blue circle and plotted at every time step.

```
> gr||i[j]:=pointplot([rr(i,t)*cos(Theta||i(t)),
> rr(i,t)*sin(Theta||i(t))],symbol=circle,symbolsize=16,
> color=blue):
> end do:
```

With the completion of the inner loop, the sequence of time frames for each object is displayed with the `insequence=true` option.

```
> gr2[i]:=display(seq(gr||i[j],j=0..N),insequence=true):
> end do:
```

With the completion of the do loop, Mike executes the following display command, clicks on the plot and on the start arrow. The initial frame in the animation is shown on the rhs of Figure 4.13, the tiny circles indicating the starting positions of the masses. The elongated ellipse is Halley's orbit.

```
> display({pp,seq(gr2[i],i=5..10)});
```

Viewing the animation of the outer planets and Halley's comet on the computer screen, Vectoria remarks, "That's really quite nice. You won't find this in any textbook that I know about. Why don't we wrap up this recipe by calculating the minimum and maximum distances of Halley's comet from the Sun. I notice that the values are given here in Marion and Thornton and this will give us a partial check on our results."

So, Mike enters the value of 1 A.U. in meters and calculates the minimum and maximum distances in meters in $R1$ and $R2$.

> AU:=1.495*10^(11): #1 Astronomical Unit in meters

> R1:=evalf(subs(theta=Pi,pl(10))*AU);

> R2:=evalf(subs(theta=0,pl(10))*AU);

$$R1 := 0.5263790351\,10^{13}$$
$$R2 := 0.8830965002\,10^{11}$$

"Great Mike. The numbers are in complete agreement with those quoted in the text. This has been fun, but I am getting tired. Let's call it a day and modify the recipe later to deal with orbital precession."

4.2.4 The Not-so-Simple Pendulum

Here's a new day. O Pendulum move slowly!
Harold Munro, British poet and critic (1879–1932)

Almost every beginning college engineering or physics student encounters the "simple" plane pendulum problem and is often left with the impression that not only is the motion easy to understand, but the solution is simple as well. This latter belief comes from restricting the pendulum oscillations to small angles θ of the pendulum arm with the vertical. This then leads to simple harmonic motion with $\theta(t)$ being either a sine or cosine function of the time t.

In this recipe, we shall show that the analytic solution of the simple plane pendulum problem for arbitrary θ is not so simple. After deriving the form of $\theta(t)$, we shall animate the motion of the pendulum for the situation where the angular amplitude reaches nearly 180° with the equilibrium position. Referring

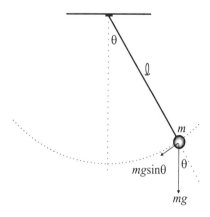

Figure 4.14: Force diagram for the simple plane pendulum.

4.2. POSITION-DEPENDENT FORCES

to Figure 4.14, let us remind the reader how the equation of motion for the plane pendulum is derived. A small pendulum bob (mass m) on the end of a light rigid rod of length ℓ is allowed to swing along a circular arc in a vertical plane. Resolving the gravitational force $m\vec{g}$ into components parallel and perpendicular to the arc, and neglecting air resistance, friction at the pivot point, etc., Newton's second law for motion along the arc yields

$$m(\ell\ddot{\theta}) = -mg\sin\theta, \quad \text{or} \quad \ddot{\theta} + \omega_0^2 \sin\theta, \qquad (4.3)$$

where $\ddot{\theta}$ is the angular acceleration and $\omega_0 = \sqrt{g/\ell}$ is the frequency. The minus sign appearing in the restoring force term is because this force component is in the opposite direction to increasing θ. For small θ, $\sin\theta \approx \theta$, and the equation of motion reduces to that of the simple harmonic oscillator with sine or cosine solutions. The period for small angles is given by $T = 2\pi/\omega_0$.

Let us now tackle the exact problem for arbitrary θ. The angle θ and the angular amplitude Θ are assumed to lie between 0 and π here. A restriction on $\cos(\theta/2)$ is also imposed to simplify the analytic results.

```
> restart: with(plots):
> assume(theta>0,Theta>0); additionally(Theta<Pi,theta<Pi,
> cos(theta/2)<cos(Theta/2));
```

Setting $d\theta/dt \equiv \omega$, where ω is the angular velocity, we can write $d^2\theta/dt^2 = \omega\,(d\omega/d\theta)$. With this change of variables, Equation (4.3) is entered.

```
> de:=omega(theta)*diff(omega(theta),theta)
> +omega||0^2*sin(theta)=0;
```

$$de := \omega(\theta)\,(\tfrac{d}{d\theta}\,\omega(\theta)) + \omega 0^2 \sin(\theta) = 0$$

Imposing the condition $\omega(\Theta) = 0$, the nonlinear ODE de is solved for $\omega(\theta)$.

```
> sol:=dsolve({de,omega(Theta)=0},omega(theta));
```

$$sol := \omega(\theta) = \sqrt{2\,\omega 0^2 \cos(\theta) - 2\,\omega 0^2 \cos(\Theta)},$$
$$\omega(\theta) = -\sqrt{2\,\omega 0^2 \cos(\theta) - 2\,\omega 0^2 \cos(\Theta)}$$

To obtain a positive period, the positive square root solution must be selected in sol. In this run, it was the first answer, which is then simplified.

```
> sol2:=rhs(sol[1]);
> sol2:=simplify(sol2,symbolic);
```

$$sol2 := \omega 0\,\sqrt{2\cos(\theta) - 2\cos(\Theta)}$$

The identities $\cos(\theta) = 1 - 2\sin^2(\theta/2)$ and $\cos(\Theta) = 1 - 2\sin^2(\Theta/2)$ are substituted into $sol2$ to yield $sol3$.

```
> identities:=cos(theta)=1-2*sin(theta/2)^2,
> cos(Theta)=1-2*sin(Theta/2)^2:
> sol3:=subs(identities,sol2);
```

$$sol3 := \omega 0\sqrt{-4\sin(\frac{\theta}{2})^2 + 4\sin(\frac{\Theta}{2})^2}$$

Since $dt = d\theta/\omega$, the reciprocal of *sol3* is integrated with respect to θ to yield an elliptic integral expression for the time t as a function of θ.

>   `ans:=t=simplify(int(1/sol3,theta));`

$$ans := t = \frac{\text{EllipticF}\left(\sin(\frac{\theta}{2}), \frac{1}{\sin(\frac{\Theta}{2})}\right)}{\omega 0 \sin(\frac{\Theta}{2})}$$

The period of the pendulum oscillations is obtained by substituting $\theta = \Theta$ into the rhs of *ans* and multiplying the result (which is for 1/4 of a period) by 4.

>   `Period:=4*subs(theta=Theta,rhs(ans));`

The normalized period follows on dividing *Period* by the small angle period.

>   `Norm_period:=Period/(2*Pi/omega||0);`

$$Norm\_period := 2\,\frac{\text{EllipticF}\left(\sin(\frac{\Theta}{2}), \frac{1}{\sin(\frac{\Theta}{2})}\right)}{\sin(\frac{\Theta}{2})\pi}$$

In `gr1`, a thick, blue, dashed, vertical line is plotted between $(\pi, 0)$ and $(\pi, 4)$.

>   `gr1:=plot([[Pi,0],[Pi,4]],style=line,linestyle=3,`
>   `color=blue,thickness=2):`

The normalized period is plotted in `gr2`, the option `adaptive=false` helping here to speed up the plotting routine. The graphs `gr1` and `gr2` are superimposed in the plot shown on the lhs of Figure 4.15.

>   `gr2:=plot(Norm_period,Theta=0..3.12,numpoints=300,style=`
>   `point,symbol=circle,symbolsize=10,adaptive=false):`
>   `display({gr1,gr2},view=[0..3.2,0..4],labels=["radians",`
>   `"period"],title="normalized period vs. angle");`

An analytic solution $\theta(t)$ is produced in *Angle* by solving *ans* for θ. The answer involves the arcsine of the Jacobian elliptic sine function, a non-trivial result.

>   `Angle:=solve(ans,theta);`

$$Angle := 2\arcsin\left(\text{JacobiSN}\left(t\,\omega 0\sin(\frac{\Theta}{2}), \frac{1}{\sin(\frac{\Theta}{2})}\right)\right)$$

For plotting and animation purposes, *Angle* and *Period* are turned into functional operators by using the `unapply` command.

>   `Angle2:=unapply(Angle,Theta,omega||0):`

4.2. POSITION-DEPENDENT FORCES

```
> Period2:=unapply(Period,Theta,omega||0):
```
We take $g = 9.81$ m/s^2, the pendulum arm length $r = 1.0$ m, angular amplitudes $A1 = 1.0$ and $A2 = 3.11$ radians, and calculate the frequency $w = \sqrt{g/r}$ and the small angle period.

```
> g:=9.81: r:=1.0: A1:=1: A2:=3.11: w:=sqrt(g/r):
> T_small:=evalf(2*Pi/w);
```
$$T\_small := 2.006066681$$

The period of small angle oscillation is about two seconds. The two solutions corresponding to $A1$ and $A2$ are calculated as well as the period for $A2$. The number of digits used is increased to 15 to eliminate a small imaginary part which occurs with the default ten digits.

```
> Angle3:=evalf(Angle2(A1,w)); Angle4:=evalf(Angle2(A2,w));
```
$Angle3 := 2.\arcsin(\text{JacobiSN}(1.50160487136822\,t,\ 2.08582964293349))$

$Angle4 := 2.\arcsin(\text{JacobiSN}(3.13170119483599\,t,\ 1.00012477494271))$

```
> Digits:=15: Period4:=Period2(A2,w);
```
$$Period4 := 7.06823664714696$$

The period corresponding to $A2 = 3.11$ radians is about 7.1 seconds which, as expected, is considerably larger than given by the small angle formula. The analytic results, $Angle3$ and $Angle4$, are now plotted over a time interval twice this period, the resulting graph being shown on the rhs of Figure 4.15. Note that because we let Maple pick the form of the solution, the result for $A1 = 1$ radians corresponds to negative initial angular velocity. If $A1 = -1$, the initial angular velocity is positive.

```
> plot([Angle3,Angle4],t=0..2*Period4,title="theta vs.
> time(T)",labelfont=["SYMBOL",16],labels=["T","q"],color=
> [blue,red],thickness=3);
```

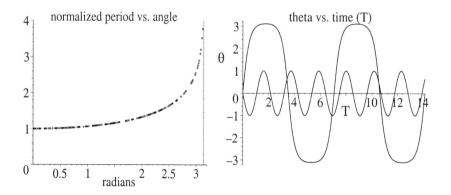

Figure 4.15: LHS: Normalized period. RHS: Solutions for different amplitudes.

We will now animate the solution *Angle4* in the following do loop, taking $N = 300$ frames and a time step size given by *Period4* divided by N.

```
> N:=300: step:=Period4/N;
```
$$step := .0235607888262817$$
```
> for i from 0 to N do
> t:=step*i:
```
The following two command lines plot the pendulum arm and the bob at each time step, coloring the arm green and representing the bob by a red circle.
```
> gr3||i:=plot([[0,0],[r*sin(Angle4),-r*cos(Angle4)]],
> style=line,thickness=3,color=green):
> gr4||i:=pointplot([[r*sin(Angle4),-r*cos(Angle4)]],
> style=point,symbol=circle,symbolsize=16,color=red):
```
The two graphs are superimposed at each time step.
```
> plot||i:=display({gr3||i,gr4||i}):
> end do:
```
Executing the following command line, clicking on the plot, and on the start arrow, animates the motion of the simple pendulum.
```
> display(seq(plot||i,i=0..N),scaling=constrained,
> insequence=true);
```

4.3 Supplementary Recipes

04-S01: Nonlinear Drag on Trout Lake

Richard and his family spent several summer vacations at Trout Lake, high in the mountains above Telluride, Colorado. In addition to hiking in the alpine meadows, a fun time was had trout fishing on the lake from Grandpa's boat. The following problem is inspired by this latter activity.

A boat is traveling on Trout Lake with initial speed v_0 when its electric motor conks out. Suppose that the water exerts a drag force (in newtons per kilogram) given approximately by $F(v) = -a\, e^{b\,v}$, with $a > 0$, $b > 0$. Find an analytic expression for the speed $v(t)$, and determine the time T it takes for the boat to come to rest and the distance d that the boat travels. If $a = 0.2$ N/kg, $b = 0.1$ s/m, and $v_0 = 5$ m/s, evaluate T and d. Animate the boat's motion over the time interval $t = 0$ to T.

04-S02: More Golf, Anyone?

Taking all other parameter values as in *Benny Boffo's Hole in One*, and dividing the angular range into 25 equal steps, **(a)** numerically determine and plot the horizontal range (in m) versus the launch angle (in degrees); **(b)** plot all the golf ball trajectories corresponding to part **(a)** in the same figure; **(c)** plot the maximum golf ball height versus the launch angle; **(d)** discuss the results. Consider only the case that includes air resistance and spin.

4.3 SUPPLEMENTARY RECIPES

04-S03: Dr. No Lets Go
In Follywood, they are shooting another James Bond movie. In one scene, Dr. No is attempting to escape from his archnemesis by moving hand over hand out along a steel spoke of a large horizontal flywheel. Unfortunately for him, Bond has hit the start button on the motor which powers the flywheel and it has begun to rotate. Two possible versions of the scene are being contemplated. In the first, the spoke is coated with yak grease so that Dr. No begins to slip as soon as the wheel starts to move, while in the second there is some delay before slipping occurs because the spoke is not greased and he is wearing metallic coated gloves to grip the spoke. As consultant to the movie, you are asked to deal with the following possibilities:

(a) In the first version, the frictionless spoke of length $L = 5$ m is rotating with constant angular velocity $\omega_0 = 0.2$ rads/s and Dr. No is initially located a distance of 1 m along the spoke from the axis of rotation. Using polar coordinates, derive $r(t)$, $\theta(t)$ for Dr. No as he slides outwards along the spoke. How long does it take for him to reach the end of the spoke? Plot his position in space as a function of time over this time interval.

(b) In the second version, the wheel starts from rest but has an angular acceleration $\alpha = 0.05$ rads/s^2 and the spoke's coefficient of friction $\mu = 1$. Derive $r(t)$, $\theta(t)$ for this situation. At what time after the wheel starts to rotate does Dr. No begin to slip? How long does it take him to reach the end of the spoke? Through what angle has he rotated when slipping begins? when he reaches the end of the spoke? Animate Dr. No's motion along the spoke, being sure to include the rotating spoke.

04-S04: George's Linear Inchworm
George has suggested the following laboratory example of a "linear inchworm". Three airtrack gliders (masses m_1, m_2, m_3) connected by two linear springs (each with spring constant k and unextended length $s0$) are placed on an airtrack of length L. Initially m_1 is at $x10 = 0$ with velocity $v10 = 0$, mass m_2 at $x20 = s0$ with velocity $v20 = 0$, and m_3 at $x30 = 2\,s0$ with non-zero initial velocity $v30$. Given $m_1 = m_2 = m_3 = 1$ kg, $s0 = 0.4$ m, $L = 3$ m, and $v_{30} = 0.5$ m/s, analytically determine the positions $x1(t)$, $x2(t)$, and $x3(t)$ of the gliders at time $t > 0$. Calculate the time that it takes m_3 to reach the other end of the airtrack. Animate the motion of the gliders over this time interval, representing each glider by a differently colored box and the two connecting springs by heavy black lines. The 3-glider motion should resemble that of an inchworm.

04-S05: Period of an Anharmonic Oscillator
This problem is one that Richard and George often assign when they teach their nonlinear physics course. A particle of mass m and total energy E oscillates in the "anharmonic" potential $V = A \mid x \mid^n$ with $A > 0$, $n > 0$. Show that the period T of the anharmonic oscillator is given in term of the Gamma (Γ) function by the following relation,

$$T = (2/n)\sqrt{2\pi m/E}(E/A)^{1/n}\Gamma(1/n)/\Gamma(1/2+1/n).$$

Confirm that T reduces to the standard expression for $n = 2$. For $n = 1, 2, .., 5$, plot V/A, T vs. E with $A = 1$, and T vs. A with $E = 1$. Discuss the plots.

04-S06: In Search of the Central Force Law

A standard task in the study of central forces is to determine the force law given the equation for the orbit. Here's a typical example which we are asking you to push further than is done in any of the standard mechanics texts. Find the force law for a central force field that allows a particle to move in an orbit given by $r = 1 - \cos\theta$, where r is the radial distance from the origin and θ is the angular coordinate. Numerically solve for $r(t)$ and $\theta(t)$ and animate the motion of the particle superimposed on a plot of the orbit. Note that you will have to avoid coming too close to the origin because there is a singularity there, i.e., you will not be able to animate the motion over a complete orbit.

04-S07: Orbital Precession

Einstein's general theory of relativity was able to successfully account for the precession of Mercury's orbit by adding an inverse fourth power of the radial distance to the usual inverse square law. According to Fowles and Cassiday [FC99], precession also occurs for the motion of a mass m moving in the gravitational field of an oblate spheroid, such as the Earth. In this case, the attractive central force is given by

$$\vec{F} = -\frac{Km}{r^2}\left(1 - \frac{\alpha K}{r}\right)\hat{e}_r.$$

Derive the analytic solution $r(\theta)$ for the orbit. Taking the (artificial) parameter values $K = 4\pi^2$, $\alpha = 10^{-6}$, and an initial tangential velocity $V = 1/\sqrt{10}$ at the minimum distance $A = 1$, plot the orbit and show that it is a precessing ellipse.

04-S08: A Perturbing Solution

Perturbation methods are useful for generating approximate analytic solutions to mechanical equations when small nonlinear terms are present. For example, the vibrations of the eardrum can be phenomenologically modeled [EM00] by a simple harmonic oscillator equation to which a quadratic term has been added,

$$\ddot{x}(t) + \omega^2 x(t) + \epsilon x(t)^2 = 0,$$

where $x(t)$ is the displacement of the eardrum from equilibrium at time t and ω and ϵ are positive parameters. If ϵ is small, analytically solve the eardrum equation subject to the initial condition $x(0) = A$, $\dot{x}(0) = 0$ using the *Poisson perturbation method*:

(a) Rewrite the equation in terms of a new time variable $\tau = \Omega t$.

(b) Substitute $x = x_0 + \epsilon x_1 + \epsilon^2 x_2 + \cdots$ and $\Omega = \omega + \epsilon \omega_1 + \epsilon^2 \omega_2 + \cdots$, into the equation and equate equal powers of ϵ.

(c) Solve the relevant ODE in each order, taking care to remove terms involving powers of τ. These "secular" terms would otherwise destroy the periodicity. Also note that since the initial conditions must be satisfied in lowest order, for higher orders $x_i(0) = 0$, $\dot{x}_i(0) = 0$, $i = 1, 2, ...$

4.3 SUPPLEMENTARY RECIPES

Taking $\omega = 1$, $A = 1$, $\epsilon = 1/4$, compare the perturbation result to the numerical solution by plotting the results in the same graph. Explore the accuracy of the perturbation solution as the parameter values are changed.

04-S09: The Force of Love

The only force in this kinematic problem is the attractive force of animal love. Patches, a beagle, is romping in a field when she spots her beloved mistress Heather walking along a straight road. Patches runs towards Heather in such a way as to always aim at her. With distances in km, Patches is initially at $(x = 1, y = 0)$, Heather at $(0,0)$, the road is described by the straight line equation $x = 0$, and the ratio of Heather's speed to the dog's speed is r.

(a) Show that Patches' path is described by the nonlinear ODE

$$x\,(d^2y/dx^2) = r\sqrt{1 + (dy/dx)^2}.$$

(b) Show that the solution of the ODE, subject to the initial conditions, is

$$y = \frac{1}{2}\left(\frac{x^{1+r}}{1+r} - \frac{x^{1-r}}{1-r}\right) + \frac{r}{1-r^2}.$$

(c) If Heather is walking at 3 km/h and Patches runs at 8 km/h, how long does it take the dog to reach her mistress and how far does she run?

(d) Animate Patches' and Heather's motion in the same figure and include the tangent to Patches' instantaneous position to graphically demonstrate that she always aims at her mistress.

04-S10: Lord of the Rings?

In the *Lord of the Rings*, Frodo hobbit is nearly consumed by the power of the ring. In this example, a space-traveling Frodo is attracted to another kind of ring, a thin planar ring of expanding "dust" from a solar explosion which has left no remnant core. In his spacecraft (mass m), Frodo is initially at an initial position $(r(0), \theta(0))$ inside the ring and has initial velocity components $(\dot{r}(0), \dot{\theta}(0))$. The circular ring (mass M) has a radius given by $R = R_0 + vt - (1/2)a\,t^2$ with $R_0 > r(0)$. Determine the gravitational force that the ring exerts on Frodo's spacecraft at some arbitrary point (r, θ) inside the ring. If the spacecraft has lost all manoeuvring ability, determine the trajectory that the spacecraft follows, due to gravitational attraction, from its initial position to the expanding ring. You may use the nominal values $r(0) = 0.5$, $\theta(0) = 0$, $\dot{r}(0) = -0.45$, $\dot{\theta}(0) = 0.1$, gravitational constant $G = 1$, $m = 0.001$, $M = 1000\,m$, $R_0 = 1$, $v = 0.1$, $a = 0$. Animate the motion of the spacecraft along this trajectory. Experiment with other parameter values. Note: The gravitational force becomes singular when the spacecraft reaches the ring.

04-S11: The Growing Raindrop

A raindrop, falling freely downwards from rest under the influence of gravity, grows by condensation on it of vapor previously at rest. After falling for a time t seconds, its mass is $M e^{kt}$ kilograms where M is the initial mass and k is a positive constant. If T is the time required for the mass to increase to $2M$, prove that the speed of the drop at this time is then $gT/(2\ln(2))$ meters per second, and determine how far it has fallen. Here g is the acceleration of gravity in meters per second squared. Calculate the ratio of the distance fallen by the accretive raindrop in time T to the distance traveled by a non-accretive ($k=0$) raindrop in the same time interval. Which drop falls through the larger distance? Discuss this result. Note: Since the mass of the raindrop varies, the more general form of Newton's second law (i.e., external force equals the rate of change of linear momentum) must be used.

04-S12: The Hanging Chain

A flexible chain of uniform density λ and length L lies on a horizontal table top with initially a portion $z0 < L$ of the chain hanging over the edge of the table. The chain is initially at rest and the table top has a coefficient of friction μ less than the critical value, μ_{cr}, needed to prevent sliding of the chain.

(a) Determine the critical value μ_{cr} in terms of L, $z0$, and λ.

(b) For $\mu < \mu_{cr}$, derive the equation of motion for the length $z(t)$ of chain hanging over the edge of the table at time t.

(c) Solve the equation of motion for $z(t)$, subject to the initial conditions, expressing your answer in terms of trig functions.

(d) Determine the time T for the entire length of chain to just leave the table.

(e) Taking $L = 1.6$ m, $z0 = 0.4$ m, $\mu = 0.1$, and $g = 9.8$ m/s^2, evaluate μ_{cr} and T, and plot $z(t)$ from $t = 0$ to $t = T$.

04-S13: Nonlinear Drag on Trout Lake Revisited

A boat of mass m is traveling on Trout Lake with initial speed $v(t=0) = V0$ at $x = 0$ when its motor quits. The boat then slows down under the influence of a drag force (in newtons) given approximately by $F = -k\,m\,x/v$, with $k > 0$.

(a) Determine $v(x)$ for the boat as it slows down.

(b) Determine the distance d that the boat travels before coming to rest and the time T that it takes.

(c) Let $V0 = 5$ m/s and $k = 1/25$ m/s^3 for the remainder of this problem. Evaluate d and T and plot $v(x)$ over the range $x = 0...d$.

(d) Analytically determine $x(t)$ in two different ways: (a) seek an exact solution, (b) derive a series solution in powers of t, dropping terms of $O(t^{12})$.

(e) By plotting the two solutions together, show that the series solution is a very good representation of $x(t)$.

(f) Animate the series solution over the time interval $t = 0$ to $t = T$.

Chapter 5

Newtonian Dynamics II

In this continuing saga of Newtonian dynamics, we shall first look at various examples of time-dependent forces. The first recipe involves an oscillating Lorentz force exerted on a charged object, the second an illustration of how complicated functions can occur when a simple time-dependence is introduced, the third an illustration of the "route to chaos" when a nonlinear oscillator is driven periodically, and the fourth a Monte Carlo simulation.

To this point in the text, only inertial (non-accelerated) frames of reference have been considered. Since we live on a rotating Earth, the subject of non-inertial (accelerated) frames of reference is of some importance. We finish the chapter with two examples of taking the Earth as our frame of reference.

5.1 Time-dependent forces

5.1.1 Mr. Q Feels the Lorentz Force

Magnetism is one of the Six Fundamental Forces of the Universe, with the other five being Gravity, Duct Tape, Whining, Remote Control, and The Force That Pulls Dogs Toward The Groins Of Strangers.
Dave Barry, American Pulitzer prize winner (1988)

A charge q moving with velocity $\vec{v}$ in an external electric field $\vec{E}$ and magnetic field $\vec{B}$ will experience the Lorentz force $\vec{F} = q\vec{E} + q(\vec{v} \times \vec{B})$. Mr. Q is a tiny creature who lives in the ionosphere of Erehwon and spends his time riding his "horse" (a charged metallic sphere of mass m) in the wildly fluctuating fields found there. In this recipe, we will give Mr. Q a gentle ride in the rotating (frequency ω) electric field $E_x = E_0 \cos(\omega t)$, $E_y = 0$, $E_z = E_0 \sin(\omega t)$ and static magnetic field $B_x = B_0$, $B_y = 0$, $B_z = 0$. Our goal is to analytically determine the trajectory that Mr. Q follows as well as his speed at arbitrary time t. Then, we will determine the distance that he travels in 30 seconds and animate his motion over this time interval. Mr. Q starts at the origin with zero speed and $m = 1$ kg, $q = 1$ C, $\omega = 5.1$ rads/s, $E_0 = 2$ V/m, $B_0 = 2.3$ T.

The velocity (v), acceleration (a), and general electric (E) and magnetic (B) field vectors are entered in Cartesian coordinates.

```
> restart: with(plots): with(VectorCalculus):
> v:=<diff(x(t),t),diff(y(t),t),diff(z(t),t)>;
```

$$v := (\frac{d}{dt} x(t))\, e_x + (\frac{d}{dt} y(t))\, e_y + (\frac{d}{dt} z(t))\, e_z$$

```
> a:=diff(v,t);
```

$$a := (\frac{d^2}{dt^2} x(t))\, e_x + (\frac{d^2}{dt^2} y(t))\, e_y + (\frac{d^2}{dt^2} z(t))\, e_z$$

```
> E:=<Ex,Ey,Ez>; B:=<Bx,By,Bz>;
```

$$E := Ex\, e_x + Ey\, e_y + Ez\, e_z$$
$$B := Bx\, e_x + By\, e_y + Bz\, e_z$$

Then, the Lorentz force $\vec{F} = q\vec{E} + q(\vec{v} \times \vec{B})$ is calculated.

```
> F:=q*E+q*(v &x B);
```

$$F := (q\, Ex + q\, ((\frac{d}{dt} y(t))\, Bz - (\frac{d}{dt} z(t))\, By))\, e_x$$
$$+ (q\, Ey + q\, ((\frac{d}{dt} z(t))\, Bx - (\frac{d}{dt} x(t))\, Bz))\, e_y$$
$$+ (q\, Ez + q\, ((\frac{d}{dt} x(t))\, By - (\frac{d}{dt} y(t))\, Bx))\, e_z$$

Applying Newton's second law to each force component yields the general system (sys) of coupled ODEs to be solved.

```
> sys:=m*a[1]=F[1],m*a[2]=F[2],m*a[3]=F[3];
```

$$sys := m\,(\frac{d^2}{dt^2} x(t)) = q\, Ex + q\, ((\frac{d}{dt} y(t))\, Bz - (\frac{d}{dt} z(t))\, By),$$
$$m\,(\frac{d^2}{dt^2} y(t)) = q\, Ey + q\, ((\frac{d}{dt} z(t))\, Bx - (\frac{d}{dt} x(t))\, Bz),$$
$$m\,(\frac{d^2}{dt^2} z(t)) = q\, Ez + q\, ((\frac{d}{dt} x(t))\, By - (\frac{d}{dt} y(t))\, Bx)$$

The specific forms of the electric and magnetic fields are now entered and the resultant system of linear ODEs displayed.

```
> Ex:=Eo*cos(omega*t); Ey:=0; Ez:=Eo*sin(omega*t);
> Bx:=Bo; By:=0; Bz:=0;
> sys;
```

$$m\,(\frac{d^2}{dt^2} x(t)) = q\, Eo \cos(\omega\, t),\ m\,(\frac{d^2}{dt^2} y(t)) = q\,(\frac{d}{dt} z(t))\, Bo,$$
$$m\,(\frac{d^2}{dt^2} z(t)) = q\, Eo \sin(\omega\, t) - q\,(\frac{d}{dt} y(t))\, Bo$$

The ODE system is to be solved for the functions $x(t)$, $y(t)$, and $z(t)$,

5.1. TIME-DEPENDENT FORCES

```
> fcns:={x(t),y(t),z(t)};
```
subject to the given initial conditions.
```
> initcond:=(x(0)=0,y(0)=0,z(0)=0,D(x)(0)=0,
> D(y)(0)=0,D(z)(0)=0);
```
The system is analytically solved and the solution is assigned.
```
> sol:=dsolve({sys, initcond }, fcns); assign(sol):
```
An operator f for simplifying any function u is created. On applying f to $x(t)$, $y(t)$, $z(t)$, Mr. Q's spatial coordinates (X, Y, Z) at time t are determined.
```
> f:=u->simplify(u):
> X:=f(x(t)); Y:=f(y(t)); Z:=f(z(t));
```

$$X := -\frac{q\,Eo\,(\cos(\omega t) - 1)}{m\,\omega^2}$$

$$Y := -\frac{(m^2\,\omega^2\,\cos(\frac{q\,Bo\,t}{m}) - \omega^2\,m^2 + q^2\,Bo^2 - q^2\,Bo^2\,\cos(\omega t))\,Eo}{\omega\,Bo\,(-q^2\,Bo^2 + \omega^2\,m^2)}$$

$$Z := -\frac{(-\sin(\frac{q\,Bo\,t}{m})\,\omega\,m + q\,Bo\,\sin(\omega t))\,Eo\,m}{Bo\,(-q^2\,Bo^2 + \omega^2\,m^2)}$$

His velocity components follow on differentiating X, Y, Z with respect to t and simplifying with f. His speed is determined by calculating $\sqrt{Vx^2 + Vy^2 + Vz^2}$.
```
> Vx:=f(diff(X,t)); Vy:=f(diff(Y,t)); Vz:=f(diff(Z,t));
```

$$Vx := \frac{q\,Eo\,\sin(\omega t)}{m\,\omega}$$

$$Vy := -\frac{q\,(-\sin(\frac{q\,Bo\,t}{m})\,\omega\,m + q\,Bo\,\sin(\omega t))\,Eo}{-q^2\,Bo^2 + \omega^2\,m^2}$$

$$Vz := \frac{q\,\omega\,(\cos(\frac{q\,Bo\,t}{m}) - \cos(\omega t))\,Eo\,m}{-q^2\,Bo^2 + \omega^2\,m^2}$$

```
> Speed:=f(sqrt(Vx^2+Vy^2+Vz^2));
```
The values of the mass m, charge q, frequency w, electric field amplitude E_0, magnetic field amplitude B_0, and time interval T are specified. The number of frames N to be used in the animation is also given.
```
> m:=1: q:=1: omega:=5.1: Eo:=2: Bo:=2.3: T:=30: N:=500:
```
Mr. Q's speed (*Speed2*) is calculated at time t,
```
> Speed2:=Speed;
```

$Speed2 := (0.4454120554 + 0.03926318806\cos(5.1\,t)^2$
$- 0.2185790314\sin(2.3\,t)\sin(5.1\,t) - 0.4846752434\cos(2.3\,t)\cos(5.1\,t))^{(1/2)}$

and plotted over the interval $t = 0$ to $T = 30$ s on the lhs of Figure 5.1. To obtain a smooth curve, 1000 plotting points are used.

> plot(Speed2,t=0..T,numpoints=1000,labels=["t","speed"]);

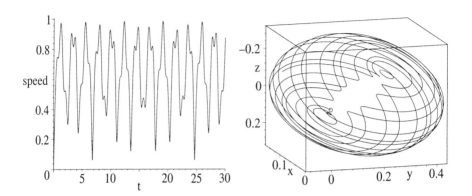

Figure 5.1: LHS: Mr. Q's speed as a function of time. RHS: Mr. Q's trajectory.

The distance (in m) that he travels in the time interval is obtained by integrating *Speed2* with respect to time from $t = 0$ to T.

> Distance:=evalf(int(Speed2,t=0..T));

$$Distance := 19.28747157$$

Q's complete trajectory over the interval is plotted in three dimensions with the spacecurve command, the curve being colored with the shading=zhue option.

> sc:=spacecurve({[X,Y,Z]},shading=zhue,t=0..T,numpoints=1000):

In the following do loop, Mr. Q's position is plotted at N time intervals and superimposed at each time step on the complete trajectory. Mr. Q's position is represented by a red circle. On executing the display line, Mr. Q's motion along the trajectory can be viewed.

> for i from 0 to N do

> t:=i*T/N:

> gr||i:=pointplot3d([[X,Y,Z]],style=POINT,symbol=circle,

> symbolsize=20,color=red);

> pl||i:=display({sc,gr||i}):

> end do:

> display([seq(pl||i,i=0..N)],insequence=true,orientation=

> [16,-103],axes=box,labels=["x","y","z"],tickmarks=[2,3,3]);

The opening frame of the animation is shown on the rhs of Figure 5.1. The reader can experiment with other mathematical forms for the fields and other parameter values. Perhaps you can come up with a more interesting ride.

5.1.2 Jane Rescues Tarzan

There is hardly an American male of my generation who has not... tried to master the victory cry of the great ape as it issued from the... chest of Johnny Weissmuller, to the accompaniment of thousands of arms and legs snapping during attempts to swing from tree to tree....
Gore Vidal, American novelist and critic, in *Esquire* (Dec. 1963)

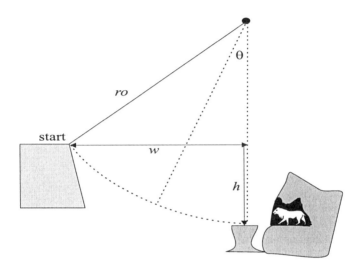

Figure 5.2: Scenario for the new Tarzan movie.

In a new Tarzan movie, it is proposed that the stunt woman who plays Jane will swing across a 10 meter wide canyon to save her beloved Tarzan who is lying injured on a cliff ledge. The difficulty is that the ledge is 15.0 meters lower than Jane's starting position. To overcome this difficulty, Jane will use a light, but strong, vine that will lengthen at a constant speed as she swings across the canyon. To make the scene more exciting, the potential movie viewer will be led to believe that if the vine doesn't lengthen just the right amount, disaster will occur. If the vine lengthens too quickly, Jane will hit the opposite canyon wall below the ledge and go "kersplat". On the other hand, if the vine lengthens too slowly, she will swing above the ledge into a ferocious lions' den. As the technical consultant to this movie, you are asked the following questions.

(a) If the vine is initially 30 meters long, with what speed should the vine lengthen if Jane is to rescue Tarzan? How long will she take to reach him?

(b) What is her horizontal speed at the moment she reaches Tarzan?

(c) Can you animate the motion of the vine for the desired situation?

To solve this problem you might initially proceed as follows. First make some reasonable approximations. The model can be improved afterwards if necessary.

Since the vine is light, completely neglect its weight. Also, the stunt woman is small, so treat Jane like a point mass. Finally, the maximum angle $\theta = \Theta$ from the vertical through which Jane will swing is not too large, so use the small angle approximation $\sin\theta \approx \theta$.

The angular velocity, $\omega = d\theta/dt$, and the vine length, $r = ro + v\,t$, are entered. Here ro is the initial length and v the speed with which it is lengthening.

> `restart: with(plots):`

> `omega:=diff(theta(t),t): r:=ro+v*t:`

For rotational motion, Newton's second law takes the form $dL/dt = \Gamma$, where the angular momentum $L = I\omega$ with I the moment of inertia, and Γ is the torque of the applied force about the pivot point. Here, letting Jane's mass be m and neglecting the weight of the vine, $I = m\,r^2$, while for small angles, $\Gamma = -m\,g\,r\,\sin(\theta) \approx -m\,g\,r\,\theta$. The expressions for L and Γ are entered.

> `L:=(m*r^2)*omega; Gamma:=-m*g*r*theta(t);`

$$L := m\,(ro + v\,t)^2\,(\tfrac{d}{dt}\,\theta(t))$$

$$\Gamma := -m\,g\,(ro + v\,t)\,\theta(t)$$

Newton's second law is entered in *de*, divided by $m\,r$ and simplified in *de2*, and $d^2\theta/dt^2$ terms collected in *de3*.

> `de:=diff(L,t)-Gamma=0;`

> `de2:=simplify(de/(m*r));`

> `de3:=collect(de2,diff(theta(t),t,t));`

$$de3 := (ro + v\,t)\,(\tfrac{d^2}{dt^2}\,\theta(t)) + 2\,(\tfrac{d}{dt}\,\theta(t))\,v + g\,\theta(t) = 0$$

The initial condition (*ic*) is given. It is assumed that Jane pushes off at time $t=0$ with $\theta(0) = \Theta$ and a small initial angular velocity, say $\dot\theta(0) = -0.1$ rads/s. The ODE *de3* is solved analytically for the time-dependent angle, $\theta(t)$, subject to the initial condition. The lengthy output (even after the symbolic simplification) involves a combination of Bessel functions of the first and second kind.

> `ic:=theta(0)=Theta,D(theta)(0)=-0.1:`

> `sol:=dsolve({de3,ic},theta(t)):`

> `sol:=simplify(sol,symbolic);`

The initial vine length $ro = 30$ m, the height $h = 15$ m, the acceleration due to gravity $g = 9.8$ m/s^2, and the width $w = 10$ m are entered. If the time it takes Jane to reach Tarzan is T seconds, the constant speed at which the vine lengthens is given by $v = h/T$. The time T is still unknown.

> `ro:=30: h:=15: g:=9.8: w:=10: v:=h/T:`

The initial angle (*ang*) is evaluated and substituted for Θ on the rhs of *sol*.

> `ang:=evalf(arcsin(w/ro));`

$$ang := .3398369094$$

> `theta:=simplify(subs(Theta=ang,rhs(sol)),symbolic);`

5.1. TIME-DEPENDENT FORCES

The time $t = T$ is substituted into θ, and the resulting equation numerically solved for the unknown time T, the numerical algorithm being started at $T=1$.

> `T:=fsolve(subs(t=T,theta),T=1);`

$$T := 2.581132351$$

Jane will reach Tarzan in about 2.6 seconds. The constant speed v at which the vine must lengthen is then determined,

> `v:=v;`

$$v := 5.811402888$$

and found to be about 5.8 m/s. With v known, the angle θ as a function of time is completely determined and can be simplified by expanding it.

> `theta:=expand(theta);`

$$\theta := -\frac{2.364113641\, \text{BesselJ}(1., 2.597182938\sqrt{5.162264702+t})}{\sqrt{5.162264702+t}}$$
$$- \frac{0.5040337228\, \text{BesselY}(1., 2.597182938\sqrt{5.162264702+t})}{\sqrt{5.162264702+t}}$$

The angle that the vine makes with the vertical is expressed in terms of the Bessel functions, BesselJ and BesselY, of the first and second kinds. The first argument in each function indicates that they are of order 1. In the math literature, these answers are more commonly written as J_1 and Y_1. The second argument gives the time dependence of each Bessel function. With $\theta(t)$ known, the angular velocity at time t is obtained by differentiating $\theta(t)$ with respect to time and applying the `radsimp` command.

> `angvel:=radsimp(diff(theta,t));`

Jane's horizontal velocity when she reaches Tarzan is calculated,

> `vel:=eval(r*angvel,t=T);`

$$vel := -5.395141428$$

so her horizontal speed is about 5.4 m/s. To animate the motion of the vine, the time interval T is divided into $N = 100$ equal time increments. The horizontal location of the cliff edge on which Tarzan is lying is taken to be $a = 20$ m. In gr0, the terrain is plotted as a thick brown line.

> `N:=100: t:=i*T/N: a:=20:`
> `gr0:=plot([[0,-ro-h],[a,-ro-h],[a,-ro-h-a],[a+w,-ro-h-a],`
> `[a+w,-ro],[2*a+w,-ro]],style=line,color=brown,thickness=3):`

Graphs of Jane's position at each time step are created in the following do loop.

> `for i from 0 to N do`

The following line plots the vine at each time step, coloring it green.

> `gr1||i:=plot([[a,0],[a+r*sin(theta),-r*cos(theta)]],`
> `style=line,thickness=3,color=green):`

The next line plots Jane's position as a black circle at each time step.

> `gr2||i:=pointplot([[a+r*sin(theta),-r*cos(theta)]],`

```
> style=point,symbol=circle,symbolsize=16,color=black):
```
The three graphs are combined on each time step into a single figure.
```
> plot||i:=display({gr0,gr1||i,gr2||i}):
```
```
> end do:
```
On executing the following command line, clicking on the resulting plot, and then on the start arrow in the animation tool bar, Jane's motion on the end of the vine may be viewed on the computer screen.
```
> display(seq(plot||i,i=0..N),scaling=constrained,
> insequence=true,tickmarks=[3,6]);
```

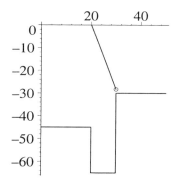

Figure 5.3: Initial frame for the animation of Jane's swing.

So, all the questions have been answered, but you should feel free to refine the model calculation. After all you are consultant to the Tarzan movie, not us!

5.1.3 The Route to Chaos

In all chaos there is a cosmos, in all disorder a secret order.
Carl Jung, Swiss psychiatrist (1875–1961)

In Jennifer's course in classical mechanics, her keener students have come across the phrase "route to chaos" and have requested her to give a lecture on this topic. Let's see what this phrase means and what recipe Jennifer has created to illustrate the concept. She considers a damped ($F_{\text{drag}} = -k\,v$, with $k > 0$ and v the velocity) anharmonic oscillator of mass m which is driven periodically by a force $F_0 \sin(\omega\, t)$, where F_0 is the amplitude and ω the frequency. For the sake of definiteness, she chooses the anharmonic potential energy to be of the form $U = a\,x^4 + b\,x^2$, where a and b are real constants and x is the displacement of the oscillator from equilibrium. For $a = 0$ and $b > 0$, U has

5.1. TIME-DEPENDENT FORCES

a parabolic shape characteristic of the familiar harmonic oscillator. The corresponding force $F_{\text{harmonic}} = -dU/dx = -2bx$ is the usual Hooke's force law, the linear spring constant being equal to $2b$ here. Her students have already learned that if a harmonic oscillator is periodically driven at some frequency, after a transient time interval, the harmonic oscillator will vibrate at the driving frequency. As she will demonstrate, such is not the case for an anharmonic oscillator like the one described by the full form of U. In this case, after a transient interval, the anharmonic oscillator can vibrate at other frequencies than the driving frequency or even display highly irregular ("chaotic") oscillations. The phrase "route to chaos" refers to the change in periodicity of the oscillator response and the onset of chaos as some parameter (e.g., the amplitude F_0) is systematically changed. If the oscillator period is given by nT, with $n = 1, 2, ...$ and $T = 2\pi/\omega$ the driving force period, it is said to have a period-n solution.

In her recipe, Jennifer begins by loading the DEtools package, because it contains the phaseportrait command which she will be using. She takes the parameter values $a = 1/4$, $b = -1/2$, $m = 1$, $k = 0.5$, and $\omega = 1$. Her students can change these numbers, if they so desire, and explore other possible routes to chaos. The anharmonic potential U is entered, the values of a and b being automatically substituted.

>   restart: with(plots): with(DEtools):

>   a:=1/4: b:=-1/2: m:=1: k:=0.5: omega:=1:

>   U:=a*x^4+b*x^2;

$$U := \frac{1}{4}x^4 - \frac{1}{2}x^2$$

So that her students will have a good understanding of what is going on, she plots U over the range $x = -2$ to $x = 2$, coloring the curve blue. The resulting graph is displayed in Figure 5.4. It is referred to as a double-well potential.

>   p[0]:=plot(U,x=-2..2,thickness=2,color=blue):

>   display(p[0],view=[-2..2,-1/2..1]);

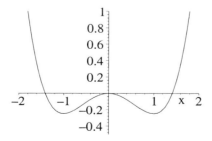

Figure 5.4: A double-well potential.

The time-dependent restoring force corresponding to U is obtained by substituting $x = x(t)$ into $-dU/dx$.

> `Fx:=subs(x=x(t),-diff(U,x));`

$$Fx := -\mathrm{x}(t)^3 + \mathrm{x}(t)$$

Jennifer enters the drag force and writes down Newton's second law in *de1* for the rate of change of the oscillator velocity in the absence of any driving force. In *de2* she relates the velocity to the time derivative of $x(t)$.

> `Fdrag:=-k*v(t);`

$$Fdrag := -0.5\,\mathrm{v}(t)$$

> `de1:=m*diff(v(t),t)=Fx+Fdrag;`

$$de1 := \tfrac{d}{dt}\mathrm{v}(t) = -\mathrm{x}(t)^3 + \mathrm{x}(t) - 0.5\,\mathrm{v}(t)$$

> `de2:=v(t)=diff(x(t),t);`

$$de2 := \mathrm{v}(t) = \tfrac{d}{dt}\mathrm{x}(t)$$

She chooses two different initial conditions for the oscillator. In *ic1*, the oscillator is started at the origin with initial velocity $v(0) = 1$, while in *ic2* the oscillator is given the same initial speed but in the opposite direction.

> `ic1:=(x(0)=0,v(0)=1):ic2:=(x(0)=0,v(0)=-1):`

A functional operator `pp` is created to apply the `phaseportrait` command for different choices of `scene`. The command `pp(x(t),v(t))` produces a phaseplane portrait of $v(t)$ vs. $x(t)$, while `pp(t,x(t))` generates a plot of $x(t)$ vs. t. The resulting curves are shown on the lhs and rhs, respectively, of Figure 5.5.

> `pp:=(s1,s2)->phaseportrait({de1,de2},[x(t),v(t)],t=0..25,`
> `[[ic1],[ic2]],stepsize=0.05,dirgrid=[25,25],arrows=MEDIUM,`
> `linecolor=[blue,green],scene=[s1,s2]):`
> `pp(x(t),v(t));pp(t,x(t));`

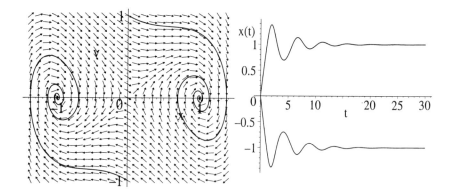

Figure 5.5: LHS: Phaseplane portraits; RHS: Corresponding solutions $x(t)$.

5.1. TIME-DEPENDENT FORCES

In the phaseplane picture on the lhs of the figure, the trajectory with positive initial velocity winds onto the stable equilibrium point $(x = 1, v = 0)$, which corresponds to the bottom of the right potential well in Figure 5.4. The trajectory with negative initial velocity winds onto the other stable equilibrium point $(x = -1, v = 0)$, corresponding to the bottom of the left potential well. Neither initial velocities were large enough here for the oscillator to cross the central hump to the opposite side.

Now, Jennifer will study the effect on the oscillator behavior when a driving force $F_{\text{drive}} = F_0 \sin(\omega t)$ is included, keeping everything else unchanged. She considers four different values of F_0, namely $F_0 = 0.325, 0.35, 0.356, 0.42$. The F_0 are selected to display different periodicities of the oscillator response.

> Fo[1]:=0.325: Fo[2]:=0.35: Fo[3]:=0.356: Fo[4]:=0.42:

A functional operator is created to generate $Fdrive$ for different amplitudes.

> Fdrive:=i->Fo[i]*sin(omega*t);

$$Fdrive := i \rightarrow Fo_i \sin(\omega t)$$

After substituting $de2$ into $Fdrag$, Jennifer enters Newton's second law in $de3$ with the restoring, drag, and driving forces included.

> Fdrag2:=subs(de2,Fdrag);

$$Fdrag2 := -0.5 \left(\frac{d}{dt} x(t)\right)$$

> de3:=diff(x(t),t,t)=Fx+Fdrag2+Fdrive(i);

$$de3 := \frac{d^2}{dt^2} x(t) = -x(t)^3 + x(t) - 0.5 \left(\frac{d}{dt} x(t)\right) + Fo_i \sin(t)$$

For the initial condition, she considers the same situation in $ic3$ as in $ic1$.

> ic3:=x(0)=0,D(x)(0)=1:

Jennifer creates a functional operator odepl for applying the odeplot command to the ith numerical solution $(sol||i)$ of $de3$. To eliminate the transient part of the solution, she starts the plot at $t = 150$ time units and plots up to $t = 200$. The color option is used to color each solution curve differently.

> odepl:=(r1,r2)->odeplot(sol||i,[r1,r2],150..200,
> numpoints=1000,color=c,thickness=1):

In the following do loop, for each of the four force amplitudes,

> for i from 1 to 4 do

the ODE $de3$ is numerically solved for $x(t)$ for the initial condition $ic3$.

> sol||i:=dsolve({de3,ic3},x(t),numeric,maxfun=0,
> output=listprocedure);

If $i = 1$, the curves are colored red, else if $i = 2$ they are colored blue, else if $i = 3$ they are colored green, else they are colored brown.

> if i=1 then c:=red elif i =2 then c:=blue
> elif i=3 then c:=green else c:=brown end if;

The following odeplot line plots the oscillator velocity vs. the displacement, i.e., produces a phaseplane portrait for the ith solution.

```
>  odepl(x(t),diff(x(t),t));
```
The next line plots x vs. t for the ith numerical solution.
```
>  odepl(t,x(t));

>  end do:
```
On completion of the do loop, pictures similar to Figures 5.6, 5.7, 5.8, and 5.9 result. On the lhs of each figure the phaseplane portrait ($v(t)$ vs. $x(t)$) is shown, while on the rhs the solution $x(t)$ is plotted. For $F_0 = 0.325$, a single closed loop occurs in the phaseplane picture. The pattern of the solution on the rhs has a period 2π which is exactly the same as the period $T = 2\pi/\omega = 2\pi$ of the driving force. So a period-1 solution occurs for $F_0 = 0.325$.

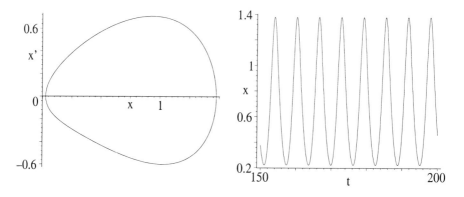

Figure 5.6: Phaseplane portrait and solution for $F_0 = 0.325$.

For $F_0 = 0.35$, the phaseplane picture displays two loops and the $x(t)$ pattern repeats with a period $2T$. This is a period-2 response.

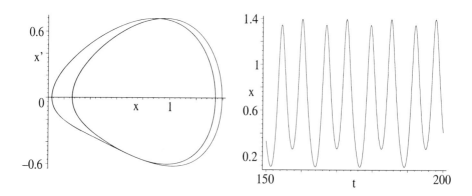

Figure 5.7: Phaseplane portrait and solution for $F_0 = 0.35$.

5.1. TIME-DEPENDENT FORCES

For $F_0 = 0.356$, the phaseplane picture displays four loops and the $x(t)$ pattern repeats with a period $4T$. A period-4 solution has occured.

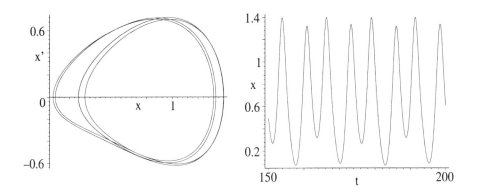

Figure 5.8: Phaseplane portrait and solution for $F_0 = 0.356$.

Finally, for $F_0 = 0.42$, the phaseplane picture is quite complex in appearance and the solution on the rhs of Figure 5.9 displays no discernible repeat pattern, even if a larger time interval is plotted. This is referred to as a chaotic solution.

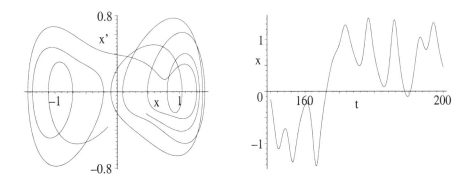

Figure 5.9: Phaseplane portrait and solution for $F_0 = 0.42$.

For increasingly narrower and narrower ranges of F_0 between $F_0 = 0.356$ and 0.42, Jennifer knows that there exist period-8, period-16, period-32, etc., solutions. As the periodicity increases, the transient time also increases, making it difficult to numerically find these higher periodic solutions. The scenario that Jennifer has presented here is part of the "period doubling route" to chaos. This is a common route to chaos in nonlinear dynamics, but not the only one.

For the amusement as well as the education of her students, Jennifer decides to animate the motion of the oscillator for the chaotic case, representing the oscillator as a blue colored circle moving in the double-well potential as well as a red circle moving along the x axis. She consider $N = 2000$ time steps, each step being 0.1 time units.

```
> N:=2000: step:=0.1:
```

The following do loop produces the animation. Jennifer intends to warn those of her students who have slower machines to be patient as this loop takes some time to execute. Or else they can reduce N.

```
> for i from 1 to N do
> X[i]:=rhs(sol||4[2](step*i)):
> V[i]:=subs(x=rhs(sol||4[2](step*i)),U):
> p1[i]:=pointplot([X[i],0],symbol=CIRCLE,symbolsize=16,
> color=red):
> p2[i]:=pointplot([X[i],V[i]],symbol=CIRCLE,symbolsize=16,):
> color=blue):
> gr[i]:=display(p[0],p1[i],p2[i]);
> end do:
```

On executing the following `display` command line, clicking on the plot and on the start arrow, her students (and you the reader) will be treated to a motion that is completely deterministic, but is so complex that it is often impossible to guess how the colored circles representing the oscillator will move in the next instant of time.

```
> display(seq(gr[i],i=1..N),insequence=true,view=[-2..2,-1..1],
> tickmarks=[2,2]);
```

Jennifer's recipe is an extremely powerful one as her students can not only experiment with different parameter values, but can also change the nature of the potential function as well as the driving term.

5.1.4 Blowing in the Wind, Monte Carlo Style

Indoors or out, no one relaxes, In ... that month of wind and taxes, The wind will presently disappear, The taxes last us all the year.
Ogden Nash, American poet, *Thar She Blows* (1949)

In the spirit of Ogden Nash, this recipe uses a Monte Carlo approach to numerically simulate the effect of a gusting wind on a raindrop as it falls from a rain cloud towards the ground in the vicinity of one of Metropolis's skyscrapers. Monte Carlo methods involve the use of a random number generator, and here we shall use the `rand` command to determine what happens to the raindrop on each step that it takes towards the ground. The motion of the raindrop will be animated, with the skyscraper included. For simplicity, the problem is treated as two-dimensional.

5.1. TIME-DEPENDENT FORCES

We take the height of the cloud bottom to be $Hc = 1000$ m above the ground and the height of the building to be $Hb = 450$ m. Taking the initial horizontal coordinate of the raindrop as it leaves the cloud to be $x = 0$, the nearest edge of the building is placed to the right (East) at $x = Xb = 200$ m. The parameter $R = 500$ m controls the plot range in the resulting figure and animation.

```
> restart: with(plots):
> Hc:=1000: Hb:=450: Xb:=200: R:=500:
```

We will let a and b refer to the horizontal and vertical displacements of the raindrop on each time step (1 second). The subscript 1 will indicate up, the subscript 2 down, the subscript 3 to the right (East), and the subscript 4 to the left (West). According to Fowles and Cassiday[FC99], a small raindrop of diameter 0.1mm will attain a terminal speed of 0.33 m/s in about 0.034 seconds. In our simulation, we consider a larger raindrop which quickly achieves a terminal velocity of 2 m/s, thus $a_2 = 0$, $b_2 = -2$. Gusts blowing to the right (East) are at 2.5 m/s (so $a_3 = 2.5$, $b_3 = 0$), while gusts blowing to the left (West) are at 1 m/s ($a_4 = -1$, $b_4 = 0$). Updrafts of 1 m/s ($a_1 = 0$, $b_1 = 1$) can also occur in the simulation. The probability of an updraft is taken to be 0.1, of falling vertically downward is 0.7, of being blown to the right is 0.15, and of being blown to the left is 0.05. The position of the raindrop on the $(n+1)$st step is related to its position on the nth step by the finite difference equations $x_{n+1} = x_n + a_i$, $y_{n+1} = y_n + b_i$, with $i = 1, 2, 3, 4$. On each step, a random number r lying between 0 and 1 will be generated. If $r < p_1 = 0.1$, then $i = 1$ is selected; else if $r < p_2 = 0.8$, then $i = 2$ is chosen; else if $r < p_3 = 0.95$, then $i = 3$ is selected; else $i = 4$ is chosen. The values of the a_i, b_i, and p_i are now entered. Clearly, these values can be altered to create other wind conditions. We have chosen rather gentle gusts here.

```
> a[1]:=0: a[2]:=0: a[3]:=2.5: a[4]:=-1:
> b[1]:=1: b[2]:=-2: b[3]:=0: b[4]:=0:
> p[1]:=0.1: p[2]:=0.8: p[3]:=0.95:
```

If the raindrop were to fall vertically, it would take 1000/2=500 time steps (seconds) to strike the ground. With the wind present it will take longer, so let's consider a maximum of $N = 2000$ steps. The input coordinates ($x_0 = 0$, $y_0 = Hc$) of the raindrop, as it leaves the bottom of the cloud, are entered separately and as a list. The step number n is initialized to zero.

```
> N:=2000: x[0]:=0: y[0]:=Hc: pnt[0]:=[x[0],y[0]]: n:=0:
```

The polygonplot command is used in gr1 to plot the cloud. The proportion of red (R), green (G), and blue (B) is chosen to produce a dark blue cloud.

```
> gr1:=polygonplot([[-R,Hc],[R,Hc],[R,Hc+50],[-R,Hc+50]],
> style=patch,color=COLOR(RGB,0.2,0.3,0.6)):
```

In gr2, the building is plotted and given a brownish color.

```
> gr2:=polygonplot([[Xb,0],[R,0],[R,Hb],[Xb,Hb]],
> style=patch,color=COLOR(RGB,0.5,0.5,0.3)):
```

196   CHAPTER 5. NEWTONIAN DYNAMICS II

The skyscraper has a multistory entrance door which is colored red in gr3.

> gr3:=polygonplot([[330,0],[370,0],[370,40],[330,40]],
> style=patch,color=COLOR(RGB,1,0,0)):

The command randomize() is entered before the main body of the code to set the random number seed to a value based on the computer system clock. Each time the recipe is restarted, a different seed number will be produced. If this command is omitted, the trajectory of the raindrop will not change from one run to the next.

> randomize():

The following repetitive loop begins with a while statement. Two possibilities are considered. In the first round brackets, we allow the possibility that the raindrop could hit the top of the skyscraper roof. With the value that we have chosen for the wind velocity to the right, this is highly improbable, but if we increased the wind speed, this could happen. The loop proceeds as long as $n \leq N$, $x_n \leq R$, and $y_n \geq Hb$. Alternately, the raindrop misses the rooftop and either strikes the side of the building or hits the ground. Thus, in the second round brackets the loop continues provided $n \leq N$ and $x_n \leq Xb$ and $y_n \geq 0$.

> while (n<=N and x[n]<=R and y[n]>=Hb) or
> (n<=N and x[n]<=Xb and y[n]>=0) do

The command rand() generates a random 12 digit, positive integer, number. Dividing by 10^{12} produces a random fractional number between 0 and 1, and applying evalf converts it to a decimal number.

> r:=evalf(rand()/10^12);

The probability condition is entered along with the difference equations.

> if r<p[1] then i:=1 elif r<p[2] then i:=2
> elif r<p[3] then i:=3 else i:=4   end if;
> x[n+1]:=x[n]+a[i];
> y[n+1]:=y[n]+b[i];

The raindrop coordinates on step $(n+1)$ are formed into a list and a plot created for the nth point. The raindrop is represented by a circle and given the same color as the cloud.

> pnt[n+1]:=[x[n+1],y[n+1]];
> Gr[n]:=pointplot(pnt[n],style=point,symbol=circle,
> symbolsize=14,color=COLOR(RGB,0.2,0.3,0.6)):

The graphs gr3, gr1, gr2, and Gr[n] are placed in a list and displayed. See what happens if the order of the graphs in the list is changed.

> pl[n]:=display([gr3,gr1,gr2,Gr[n]]):

The number n is incremented by one and the loop ended.

> n:=n+1;
> end do:

5.1. TIME-DEPENDENT FORCES

We record the total number of steps before the raindrop was stopped by the rooftop, or the side of the building, or the ground. The horizontal displacement and vertical position of the raindrop at this last step are also recorded. The numbers will vary from one execution of the recipe to the next.

```
> Total_steps:=n; horizontal_displacement:=x[n-1];
> vertical_position:=y[n-1];
```

$$Total\_steps := 597$$
$$horizontal\_displacement := 198.5$$
$$vertical\_position := 216$$

In this particular run, the raindrop underwent 597 steps, was displaced 198.5 meters horizontally from its starting position, and had a final vertical height of 216 meters. In this case, the raindrop struck the side of the skyscraper. This may be confirmed by plotting the trajectory in gr4 and superimposing it with the display command on the plots of the cloud, skyscraper, and door. The result is shown in Figure 5.10.

```
> gr4:=pointplot([seq(pnt[j],j=0..Total_steps)],style=line,
> color=COLOR(RGB,0.2,0.3,0.6)):
> display([gr3,gr1,gr2,gr4],axes=boxed,scaling=constrained,
> view=[-R..R,0..Hc+50],labels=["x","y"],tickmarks=[3,3]);
```

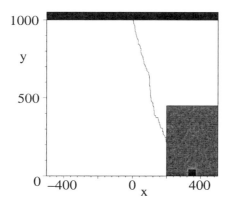

Figure 5.10: A representative trajectory for the falling raindrop.

The motion of the raindrop is animated by displaying the sequence of graphs pl and using the insequence=true option. Click on the computer plot and start arrow to make the raindrop move.

```
> display(seq(pl[j],j=0..Total_steps-1),insequence=true,
> axes=boxed,scaling=constrained,view=[-R..R,0..Hc+50],
> labels=["x","y"],tickmarks=[3,3]);
```

The reader may experiment with different wind speeds, probabilities, sizes of the building, and so on. You could generalize the recipe to have several raindrops falling simultaneously. Go in whichever direction the wind blows you!

5.2 Accelerated Reference Frames

The effective force [MT95] on a mass m in a non-inertial coordinate system that is undergoing translation and rotation relative to a fixed inertial frame is given by

$$\vec{F}_{\text{eff}} \equiv m\,\vec{a}_r = \vec{F} - m\,\ddot{\vec{R}}_f - m\,\dot{\vec{\omega}} \times \vec{r} + \vec{F}_{\text{centrifugal}} + \vec{F}_{\text{Coriolis}} \qquad (5.1)$$

with $\vec{F}_{\text{centrifugal}} = -m\,\vec{\omega} \times (\vec{\omega} \times \vec{r})$ and $\vec{F}_{\text{Coriolis}} = -2\,m\,\vec{\omega} \times \vec{v}_r$. Here $\vec{F}$ is the vector sum of the forces acting on m as measured in the fixed frame. $\vec{R}_f$ is the position vector of the origin of the non-inertial frame relative to the origin of the fixed frame. Thus, the force term $-m\,\ddot{\vec{R}}_f$ results from the translational acceleration of the non-inertial frame relative to the fixed one. $\vec{\omega}$ is the angular velocity of the non-inertial frame and $\vec{r}$ the vector position of m relative to the origin of this frame. The force term $-m\,\dot{\vec{\omega}} \times \vec{r}$ results from the angular acceleration of the non-inertial frame relative to the fixed frame. The remaining two force terms in (5.1) are the centrifugal and Coriolis forces, where $\vec{v}_r$ is the velocity of m relative to the non-inertial frame.

In this section, we look at two examples, the first involving a rotating merry-go-round, the second dealing with the Earth rotating about its axis. For the former, the Earth is regarded as the inertial frame and the origin of the rotating (non-inertial) frame is placed at the center of the merry-go-round, while for the latter the inertial frame is taken at the center of the Earth and the non-inertial frame is attached to the Earth's surface.

5.2.1 Merry-go-round Merriment

Love makes the world go round? Not at all.
Whisky makes it go round twice as fast.
Compton Mackenzie, English novelist, *Whisky Galore* (1967)

Kevin and his sister Ruth are having a merry time riding on the merry-go-round at the local park. The surface of the merry-go-round is quite slippery and they are holding onto the supports for dear life as their friends Justine and Gabrielle try to spin the merry-go-round faster. A hockey puck, which Kevin has inadvertently pushed on the merry-go-round surface, slides on a curved trajectory before flying off the rim and nearly beaning the pushers. This recipe idealizes the motion of the puck as it slides on the surface and looks at the puck's trajectory from the viewpoint of Kevin and Ruth on the merry-go-round.

We mentally attach an inertial frame to the Earth and a rotating coordinate system to a merry-go-round of radius R, with the origin at its center. The x and

5.2. ACCELERATED REFERENCE FRAMES

y axes of the rotating frame are in the plane of the merry-go-round, while the z axis is perpendicular to the surface. For simplicity, let's completely neglect friction, treat the puck as a point particle, and assume that the merry-go-round rotates counterclockwise with a constant angular speed Ω. The hockey puck is initially at $x = aR$, $y = bR$ with $a < 1$, $b < 1$. If the puck has the initial velocity components (in the rotating frame) $\dot{x}(0) = c\Omega R$, $\dot{y}(0) = d\Omega R$, analytically determine $x(t)$ and $y(t)$. Taking $a = -0.5$, $b = 0$, $c = 0.35$, $d = 0.35$, $\Omega = 1$ rad/s, and $R = 1$ m (a small merry-go-round!), plot the trajectory of the puck as seen in the rotating frame and animate its motion.

To calculate the cross products appearing in the Coriolis and centrifugal forces, the VectorCalculus package is loaded. Convenient aliases are introduced to save on the typing. When, e.g., W is entered, Ω appears in the output.

> restart: with(plots): with(VectorCalculus):

> alias(omega=w,Omega=W):

The angular velocity vector $\vec{\omega}$ is entered as well as the position vector $\vec{r}$ of the puck relative to the rotating frame attached to the merry-go-round. The velocity $\vec{v}$ in the rotating frame is calculated.

> w:=<0,0,W>;

$$\omega := \Omega\, e_z$$

> r:=<x(t),y(t),0>; v:=diff(r,t);

$$r := x(t)\, e_x + y(t)\, e_y$$

$$v := (\frac{d}{dt} x(t))\, e_x + (\frac{d}{dt} y(t))\, e_y$$

The gravitational force on the puck is canceled out by the normal force of the surface and there is no friction, so the first term $\vec{F} = 0$ in $\vec{F}_{\text{eff}}$. The merry-go-round is not being translated relative to the Earth and has a constant angular velocity, so the second and third terms in $\vec{F}_{\text{eff}}$ are zero as well. We calculate the Coriolis and centrifugal forces (per unit mass) and form the resultant force.

> F[Coriolis]:=2*v &x w;

$$F_{Coriolis} := 2(\frac{d}{dt} y(t))\, \Omega\, e_x - 2(\frac{d}{dt} x(t))\, \Omega\, e_y$$

> F[centrifugal]:=w &x (r &x w);

$$F_{centrifugal} := \Omega^2\, x(t)\, e_x + \Omega^2\, y(t)\, e_y$$

> F[resultant]:=F[Coriolis]+F[centrifugal];

$$F_{resultant} := (2(\frac{d}{dt} y(t))\, \Omega + \Omega^2\, x(t))\, e_x + (-2(\frac{d}{dt} x(t))\, \Omega + \Omega^2\, y(t))\, e_y$$

Newton's second law is applied in the x and y directions, yielding xeq and yeq.

> xeq:=diff(x(t),t,t)=F[resultant][1];

$$xeq := \frac{d^2}{dt^2} x(t) = 2(\frac{d}{dt} y(t))\, \Omega + \Omega^2\, x(t)$$

> yeq:=diff(y(t),t,t)=F[resultant][2];

$$yeq := \frac{d^2}{dt^2}\,y(t) = -2\,(\frac{d}{dt}\,x(t))\,\Omega + \Omega^2\,y(t)$$

The initial conditions ic are entered and the system (xeq, yeq) of linear ODEs analytically solved for $x(t)$ and $y(t)$, subject to the initial condition. The Laplace transform option is used here in the dsolve command. The solution is then assigned and $x(t)$ and $y(t)$ written out and labeled as X and Y.

> `ic:=x(0)=a*R,y(0)=b*R,D(x)(0)=c*W*R,D(y)(0)=d*W*R;`

$$ic := \mathrm{x}(0) = a\,R,\ \mathrm{y}(0) = b\,R,\ \mathrm{D}(x)(0) = c\,\Omega\,R,\ \mathrm{D}(y)(0) = d\,\Omega\,R$$

> `sol:=dsolve({xeq,yeq,ic},{x(t),y(t)},method=laplace);`
> `assign(sol):`
> `X:=x(t); Y:=y(t);`

$$X := R\,((\Omega\,(c - b)\,t + a)\cos(\Omega\,t) + (\Omega\,(a + d)\,t + b)\sin(\Omega\,t))$$
$$Y := R\,((\Omega\,(a + d)\,t + b)\cos(\Omega\,t) + (\Omega\,(-c + b)\,t - a)\sin(\Omega\,t))$$

X and Y are the coordinates of the puck at time t, as seen in the rotating frame. To animate the puck, let's now enter the coordinate values a, b, c, and d that were given, as well as the radius ($R0$) and angular speed ($W0$). The parameter values are substituted into X and Y.

> `P:=a=-0.5,b=0,c=0.35,d=.35: R0:=1: W0:=1:`
> `X:=subs({P,R=R0,W=W0},X);`

$$X := (0.35\,t - 0.5)\cos(t) - 0.15\,t\sin(t)$$

> `Y:=subs({P,R=R0,W=W0},Y);`

$$Y := -0.15\,t\cos(t) + (-0.35\,t + 0.5)\sin(t)$$

When the puck reaches the rim of the merry-go-round, we have $X^2 + Y^2 = R0^2$. The time T for the puck to reach the rim is now numerically calculated.

> `T:=fsolve(X^2+Y^2=R0^2,t=0..100);`

$$T := 3.781583538$$

The trajectory of the puck, as observed in the rotating frame, is now plotted over the time interval $t = 0$ to $t = T$, the curve being a thick red line.

> `gr1:=plot([X,Y,t=0..T],style=line,color=red,thickness=2):`

The circular rim of the merry-go-round is graphed with the polarplot command, and is represented by a thick green line.

> `gr2:=polarplot(R0,color=green,thickness=2):`

We take $N = 100$ frames in our animation, so the time step size is T/N.

> `N:=100: step:=T/N:`

In the following do loop, we animate the motion of the puck, superimposing the graph gr3||i on each time step on the plots gr1 and gr2.

> `for i from 1 to N do`
> `t:=step*i:`
> `gr3||i:=pointplot([X,Y],symbol=circle,symbolsize=14,`

5.2. ACCELERATED REFERENCE FRAMES

```
>     color=blue):
>   pl||i:=display({gr1,gr2,gr3||i}):
>   end do:
```
Executing the `display` line, clicking on the computer plot and on the start arrow, the reader will see the animation of the hockey puck, the opening frame being shown in Figure 5.11.

```
>   display(seq(pl||i,i=1..N),insequence=true,
>   scaling=constrained);
```

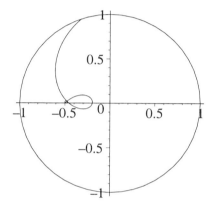

Figure 5.11: Looped trajectory of the puck as observed on the merry-go-round.

The puck executes a looped trajectory before reaching the rim. You can experiment with different initial conditions and see if any other interesting trajectories can be observed. Remember, these trajectories are those observed in the rotating frame. In the inertial frame attached to the Earth, the puck moves in a straight line because the net force in that frame $\vec{F} = 0$.

5.2.2 Falkland Fiasco

If we were doing this in the Falklands they would love it. It's part of our heritage. The British have always been fighting wars.
Soccer fan on hooliganism charge, quoted in the *London Independent* (1988)

According to Marion and Thornton [MT95], in World War I, British naval gunners fighting near the Falkland Islands took the Coriolis effect into account in sighting their guns. Therefore, they were probably quite stunned to see their shots falling 100 yards to the left of the enemy ships. Evidently the sighting mechanism on the British guns had been corrected assuming that sea battles would be fought at about 50 degrees North latitude. The only problem was that this naval engagement took place at around 50 degrees South latitude. In the Southern Hemisphere, all deflections due to the Coriolis force are in the opposite direction to those in the Northern Hemisphere. Instead of hitting the enemy ships, the British shells missed by a distance equal to twice the Coriolis deflection.

In this recipe, we consider a shell fired due East from a point on the surface of the rotating Earth at a northern latitude α. The muzzle velocity is V_0 at an angle θ to the horizontal. Neglecting air resistance, we shall show that the lateral deflection of the shell when it strikes the Earth is to the right (South) and given by
$$d = 4\,V_0^3\sin^2\theta\,\cos\theta\,\Omega\,\sin\alpha/g^2,$$
where Ω is the rotation frequency of the Earth. We shall also determine the distance the shell travels in the easterly direction. Then we will evaluate our analytical formulas for $\alpha = 45°$, $V_0 = 500$ m/s, $\theta = 60°$, $g = 9.8$ m/s^2, and (for the Earth) $\Omega = 7.27 \times 10^{-5}$ rad/s. For those readers who object to leaving out air resistance, we shall revisit this problem in Supplementary Recipe **05-S06**.

For the non-inertial frame attached to the Earth's surface, we shall take x to be to the South, y to the East, and z perpendicular to the Earth's surface. The VectorCalculus package is needed for calculating the cross product. We assume that $g > 0$, $\Omega > 0$, $V_0 > 0$, $\theta > 0$, and α is between 0 and $\pi/2$. Convenient aliases are introduced to help with the typing.

> `restart: with(VectorCalculus):`

> `assume(g>0,Omega>0,V0>0,theta>0,alpha>0,alpha<=Pi/2):`

> `alias(omega=w,Omega=W,alpha=a):`

The angular velocity of the Earth at latitude α is $\vec{\omega} = -\Omega\cos(\alpha)\,\hat{e}_x + \Omega\sin(\alpha)\,\hat{e}_z$.

> `w:=<-W*cos(a),0,W*sin(a)>;`

$$\omega := -\Omega\cos(\alpha)\,e_x + \Omega\sin(\alpha)\,e_z$$

Relative to the origin of the non-inertial frame, the shell is located at time t at $\vec{r} = x(t)\,\hat{e}_x + y(t)\,\hat{e}_y + z(t)\,\hat{e}_z$ and its velocity $\vec{v}$ is calculated.

> `r:=<x(t),y(t),z(t)>; v:=diff(r,t);`

With all forces per unit mass, the gravitational, Coriolis, and centrifugal forces acting on the shell are entered and the resultant force formed.

> `F[gravity]:=<0,0,-g>;`

$$F_{gravity} := -g\,e_z$$

> `F[Coriolis]:=-2*w &x v;`

$$F_{Coriolis} := 2\,\Omega\sin(\alpha)\,(\frac{d}{dt}\,y(t))\,e_x - (2\,\Omega\sin(\alpha)\,(\frac{d}{dt}\,x(t)) + 2\,\Omega\cos(\alpha)\,(\frac{d}{dt}\,z(t)))\,e_y$$
$$+ 2\,\Omega\cos(\alpha)\,(\frac{d}{dt}\,y(t))\,e_z$$

> `F[centrifugal]:=-w &x (w &x r);`

5.2. ACCELERATED REFERENCE FRAMES

$F_{centrifugal} := \Omega \sin(\alpha) \left(\Omega \sin(\alpha) \operatorname{x}(t) + \Omega \cos(\alpha) \operatorname{z}(t) \right) e_x +$
$(\Omega^2 \sin(\alpha)^2 \operatorname{y}(t) + \Omega^2 \cos(\alpha)^2 \operatorname{y}(t)) e_y + \Omega \cos(\alpha) \left(\Omega \sin(\alpha) \operatorname{x}(t) + \Omega \cos(\alpha) \operatorname{z}(t) \right) e_z$

```
> F[resultant]:=F[gravity]+F[Coriolis]+F[centrifugal];
```

A functional operator eq is formed for calculating Newton's second law in the x, y, and z directions. Then, calculating eq(x,1), the x equation is determined, the y and z equations being similarly calculated.

```
> eq:=(u,i)->simplify(diff(u(t),t,t)=F[resultant][i]):
> xeq:=eq(x,1); yeq:=eq(y,2); zeq:=eq(z,3);
```

$xeq := \dfrac{d^2}{dt^2} \operatorname{x}(t) = \Omega \sin(\alpha) \left(2 \left(\dfrac{d}{dt} \operatorname{y}(t) \right) + \Omega \sin(\alpha) \operatorname{x}(t) + \Omega \cos(\alpha) \operatorname{z}(t) \right)$

$yeq := \dfrac{d^2}{dt^2} \operatorname{y}(t) = \Omega \left(-2 \sin(\alpha) \left(\dfrac{d}{dt} \operatorname{x}(t) \right) - 2 \cos(\alpha) \left(\dfrac{d}{dt} \operatorname{z}(t) \right) + \Omega \operatorname{y}(t) \right)$

$zeq := \dfrac{d^2}{dt^2} \operatorname{z}(t) = -g + 2\Omega \cos(\alpha) \left(\dfrac{d}{dt} \operatorname{y}(t) \right) + \Omega^2 \cos(\alpha) \sin(\alpha) \operatorname{x}(t)$
$+ \Omega^2 \cos(\alpha)^2 \operatorname{z}(t)$

To solve the coupled linear ODEs given in xeq, yeq, and zeq, the initial conditions are specified. The shell is fired from the origin of the non-inertial frame at time $t=0$ and has an initial velocity $\vec{v}(0) = V_0 \cos\theta\, \hat{e}_y + V_0 \sin\theta\, \hat{e}_z$.

```
> ic:=x(0)=0,y(0)=0,z(0)=0,D(x)(0)=0,D(y)(0)=V0*cos(theta),
> D(z)(0)=V0*sin(theta);
```

The system of equations (xeq, yeq, zeq) is analytically solved for $x(t)$, $y(t)$, $z(t)$, using the Laplace transform method subject to the initial condition. The solution is then assigned.

```
> sol:=dsolve({xeq,yeq,zeq,ic},{x(t),y(t),z(t)},method=laplace):
> assign(sol):
```

A functional operator f is created to expand and simplify a function $u(t)$. The solutions (labeled X, Y, Z) for each direction follow on entering f(x), etc. Only X is shown here because the expressions, which are exact, are quite lengthy.

```
> f:=u->simplify(expand(u(t))): X:=f(x); Y:=f(y); Z:=f(z);
```

$X := \dfrac{1}{2} \sin(\alpha)(\cos(\alpha)\, g\, t^2\, \Omega^2 - 2\cos(\alpha)\, t\, V0 \sin(\theta)\, \Omega^2 + 2\cos(\alpha)\, g$
$- 2\cos(\alpha)\, g \cos(\Omega t) + 2t\cos(\alpha)\, V0 \sin(\theta) \cos(\Omega t)\, \Omega^2$
$+ 2t\sin(\Omega t)\, V0 \cos(\theta)\, \Omega^2 - 2t\sin(\Omega t)\cos(\alpha)\, g\, \Omega) \Big/ \Omega^2$

We shall now take advantage of the fact that Ω is very small and can be used as an expansion parameter. We create a functional operator P which Taylor expands a function u in powers of Ω out to order n. Applying convert(polynom) to the Taylor expansion removes order of terms.

```
> P:=(u,n)->convert(taylor(u,W=0,n),polynom):
```

Because Ω^2 occurs in the denominators of X, Y, and Z, these solutions must be Taylor expanded to fourth order in Ω to obtain results which are linear in Ω. The results are assigned the names $Xexp$, $Yexp$, and $Zexp$. The error is of second order in Ω, which is very small

> `Xexp:=P(X,4); Yexp:=P(Y,4); Zexp:=P(Z,4);`

$$Xexp := \sin(\alpha)\, t^2\, V0\, \cos(\theta)\, \Omega$$

$$Yexp := t\, V0\, \cos(\theta) + (\frac{1}{3} t^3 \cos(\alpha)\, g - t^2 \cos(\alpha)\, V0\, \sin(\theta))\, \Omega$$

$$Zexp := -\frac{t^2\, g}{2} + t\, V0\, \sin(\theta) + t^2 \cos(\alpha)\, V0\, \cos(\theta)\, \Omega$$

The time for the shell to strike the ground again is obtained in tt by setting $Zexp = 0$ and solving for the time t.

> `tt:=solve(Zexp=0,t);`

$$tt := 0,\ -\frac{2\, V0\, \sin(\theta)}{-g + 2 \cos(\alpha)\, \Omega\, V0\, \cos(\theta)}$$

The first solution in tt is the time at which the shell was fired, so the second solution is selected and Taylor expanded. Since $Xexp$ is already of first order in Ω, we retain only the zeroth order in $T1$. Evaluating $Xexp$ at $t = T1$ yields the desired analytical result d_x for the deflection of the shell in the South direction.

> `T1:=P(tt[2],1); d[x]:=eval(Xexp,t=T1);`

$$T1 := \frac{2\, V0\, \sin(\theta)}{g}$$

$$d_x := \frac{4 \sin(\alpha)\, V0^3\, \sin(\theta)^2\, \cos(\theta)\, \Omega}{g^2}$$

For the y direction, the first term in $Yexp$ is of zeroth order in Ω, the second term of first order. So in $T2$, terms to order Ω are retained. Then $Yexp$ is evaluated at $t = T2$ and the result Taylor expanded one more time, so no terms higher than first order in Ω appear. This yields the distance d_y traveled in the East direction. The result is simplified by collecting $\cos\alpha$, $1/g^2$, $V0^3$, and Ω.

> `T2:=P(tt[2],2); d[y]:=P(eval(Yexp,t=T2),2);`

$$T2 := \frac{2\, V0\, \sin(\theta)}{g} + \frac{4\, V0^2\, \sin(\theta)\, \cos(\alpha)\, \cos(\theta)\, \Omega}{g^2}$$

> `d[y]:=collect(d[y],[cos(a),1/g^2,V0^3,W]);`

$$d_y := \frac{(4 \sin(\theta) \cos(\theta)^2 - \frac{4}{3} \sin(\theta)^3)\, \Omega\, V0^3\, \cos(\alpha)}{g^2} + \frac{2\, V0^2\, \sin(\theta)\, \cos(\theta)}{g}$$

The second term in d_y is the standard expression for the horizontal range when $\Omega = 0$, the first term being the very small correction. The given parameters are now entered as a set and d_x and d_y numerically evaluated.

> `parameters:={a=Pi/4,theta=Pi/3,V0=500,W=7.27*10^(-5),g=9.8}:`
> `d[x]:=eval(d[x],evalf(parameters));`

$$d_x := 100.3618213$$
> d[y]:=eval(d[y],evalf(parameters));
$$d_y := 22092.48480$$

The shell is deflected about 100 meters in the positive x (South) direction, while traveling a distance of about 22,000 m or 22 km in the y (East) direction.

5.3 Supplementary Recipes

05-S01: Will Ellen be Yellin'?
In an effort to attract more customers to his Metropolis amusement park, Mr. X is considering a ride where a typical teenager (say, Ellen of mass m) is attached to the bottom of a light spring arrangement and given a large initial velocity directed toward the ground far below. The spring is to have a built-in fatigue factor which causes the spring to execute a small number of oscillations before undergoing total fatigue, causing the spring to extend indefinitely and plunging Ellen to the ground below. Ellen is supposed to be scared, not injured or killed, so the velocity with which she hits the ground must be kept close to zero. Mr. X carries out a computer simulation, considering a time-dependent spring "constant" ke^{ct} for the spring and initially ignoring any damping.

(a) If Ellen has a mass $m = 57$ kg, is given an initial velocity $\dot{x}(0) = -30$ m/s, and the spring parameters are $k = 59$ N/kg and $c = -m/(10k)$, analytically determine Ellen's vertical displacement $x(t)$ at time t, taking $x(0) = 0$. Identify the functions involved in the solution and their order. Calculate Ellen's velocity $v(t)$.

(b) Plot $x(t)$ and $v(t)$ over the time interval $t = 0...150$ s. How many oscillations does Ellen undergo before total fatigue of the spring sets in? What is her asymptotic speed? How high above the ground should she be initially to just attain this asymptotic speed? Ellen's size may be ignored.

(c) In an effort to improve his model, Mr. X adds a linear damping term $F_{\text{drag}} = -b\dot{x}$, with $b = m/k$ to Ellen's equation of motion. By calculating $x(t)$ and $v(t)$ and plotting the results over a suitable time interval, discuss the behavior of the solution of Mr. X's new model equation. Animate her motion for this new model.

05-S02: Poincare Sections and Strange Attractors
An important approach to studying the forced motion of an anharmonic oscillator is to form a *Poincaré section*. If the driving frequency is ω, we can take a "snapshot" of the phaseplane after each period $T_0 = 2\pi/\omega$ of the driving force. After an initial transient time, the steady-state motion of the oscillator will be either periodic or chaotic. For a period-1 solution (oscillator period $T = T_0$), the Poincaré section will consist of a single point which is reproduced at each multiple of the driving period. For a period-2 solution ($T = 2T_0$), the Poincaré

section will consist of two points between which the oscillator "jumps" as multiples of T_0 elapse. And so on. For chaotic motion, on the other hand, a point is produced at a different location at each multiple of T_0 and the "sum" of the individual snapshots can produce strange, localized, patterns (referred to as *strange attractors*) of points with complex boundaries in the phase plane.

Consider the forced anharmonic Duffing oscillator described by

$$\ddot{x} + k\dot{x} + \alpha x + \beta x^3 = F\cos(\omega t),$$

with $k = 0.5$, $\alpha = -1$, $\beta = 1$, $\omega = 1$, and $F = 0.325, 0.35, 0.356, 0.42$. The initial conditions are $x(0) = 0$ and $\dot{x}(0) = 1$.

(a) Setting $\dot{x} = y$ and $\dot{z} = \omega$, write the above ODE as a system of three first order ODEs for $x(t)$, $y(t)$, $z(t)$. What are the initial conditions for this system?

(b) Numerically solve the system of three ODEs for each value of F, expressing the output as a listprocedure.

(c) For each F value, plot the phaseplane points (x, y) for the times $t = iT0$, where i runs from $i = 25$ to $i = 250$ and $T0$ is the period of the driving force. For clarity, represent the points by crosses. Identify the periodicity of the solution for each F value. Are there any strange attractors?

05-S03: Tug of War: Math vs. Physics

At the annual mathematics-physics summer picnic, a friendly tug of war takes place between the two departments. Mike is on the math team, his wife Vectoria on the physics team. Each team consists of five persons, the average mass per person being 60 kg. Initially, each individual on the physics team exerts an average force $F_P = 900\,e^{-0.1\,t}$ N as a function of time t in seconds, while each person on the math team exerts an average force $F_M = 950\,e^{-0.13\,t}$ N. The negative exponents indicate a tiring effect on the part of both teams. Assuming that the mass of the connecting rope can be neglected and that the two teams are initially at rest, which team is the first to pull the other team a distance of 0.75 m? At what time does this occur? Which team is the first to pull the other team a distance of 1.0 m? At what time does this occur? If at this time, the average force per person on the physics team changes suddenly to $F2_P = 600\,e^{-0.15\,t}$ and the winning team is the one which first pulls the other through a distance 2m, which team wins, Vectoria's or Mike's? At what time does the win take place? Hint: Plotting the displacement versus time is useful in the analysis.

05-S04: Statistical Approach to Blowing in the Wind

Using relevant statistical library packages, modify the recipe **05-1-4 (Blowing in the Wind, Monte Carlo Style)** to determine the average final height of 200 raindrops, all starting from the same position, and the percentage of raindrops which hit the ground. All parameter values are the same as in the earlier recipe. If a_3 is increased to 3.0, what percentage of raindrops hit the ground? What percentage hit the roof?

05-S05: Kids Will Be Kids
While hiking in the North Cascades range of Washington and waiting impatiently for her grandparents to catch up, Justine wiles away the time by dropping small rocks from a cliff edge into the creek far below. Neglecting air resistance and the variation of g with altitude, show that if a rock is dropped through a vertical distance h, the lateral deflections d_x and d_y of the rock to the South and to the East are given by

$$d_x = \frac{3}{2} \frac{h^2 \Omega^2 \sin\alpha \cos\alpha}{g}, \quad d_y = \frac{2\sqrt{2}}{3} \frac{h^{3/2} \Omega \cos\alpha}{\sqrt{g}},$$

where α is the latitude, Ω the Earth's rotation frequency, and g the gravitational acceleration. Taking $\alpha = 49°$, $\Omega = 7.27 \times 10^{-5}$ rad/s, $h = 100$ m, and $g = 9.8$ m/s^2, determine d_x and d_y in cm for the rock when it hits the creek.

05-S06: Falkland Fiasco, Revisited
In the recipe **05-2-2** (**Falkland Fiasco**), we considered a shell fired due East from a point on the surface of the Earth at a northern latitude α with a muzzle speed V_0 at an angle θ to the horizontal. Air resistance was neglected. Modifying the recipe to include a drag force $F_{\text{drag}} = -k v$, analytically determine the trajectory of the shell to first order in Ω until it hits the ground again. Taking $k = 0.01$ and all other parameter values as in the text recipe, plot the solution derived above in a 3-dimensional picture. For comparison sake, also plot the curve which results if $\Omega = 0$. How does the deflection of the shell towards the South compare to the value obtained with no air resistance?

05-S07: Power Spectrum: The Idea
Still another approach to studying the response of an anharmonic oscillator to a time-dependent driving force is to calculate its *power spectrum*. The basic idea underlying the power spectrum is as follows [PFTV90]:

Consider a time-dependent solution, $x(t)$, which has been evaluated (e.g., numerically or experimentally) at N evenly spaced time intervals T_s, i.e., the values $x_n = x(t = nT_s)$ with $n = 0, 1, ..., N-1$ are known. With the kth frequency component (in Hz) given by $f_k = k/(NT_s)$, $k = 0, 1, 2, ..., N-1$, one can form the *discrete Fourier transform* X_k and its inverse,

$$X_k \equiv \sum_{n=0}^{N-1} x_n e^{-(2\pi i f_k T_s)n} = \sum_{n=0}^{N-1} x_n e^{-2\pi i k n/N}, \quad x_n = \frac{1}{N} \sum_{k=0}^{N-1} X_k e^{2\pi i k n/N}.$$

Parseval's theorem states that $\sum_{n=0}^{N-1} |x_n|^2 = \sum_{k=0}^{N-1} |X_k|^2 / N$. If $x(t)$ is a mechanical displacement, this theorem is a statement of the total mechanical energy. The power spectrum, defined as $S_k = |X_k|^2/N$, is a measure of the energy content in the kth frequency component f_k. To calculate the discrete Fourier transform X_k rapidly, one makes use of the "fast" Fourier transform

(FFT) (see [PFTV90]). The Maple FFT requires that N must be expressible as $N = 2^m$, where m is a positive integer. The sampling time T_s should be sufficiently small to capture all the distinctive features of the solution in the time or frequency domains. A more precise statement on this aspect, known as the sampling theorem, may be found in [PFTV90] and [EM00]. Now try the following illustrative problem, where the answer is known.

Consider the displacement $x(t) = A_1 \sin(2\pi f_1 t) + A_2 \sin(2\pi f_2 t)$, with $A_1 = 1$, $A_2 = 0.5$, $f_1 = 1$, and $f_2 = 1.5$. Taking $m = 9$, so $N = 2^m = 512$, and $T_s = 1/(10 f_2)$, plot $x(t)$ over the range $t = 0...NT_s/2$ to see what the wave profile looks like. Then confirm that there are two frequencies, 1 Hz and 1.5 Hz, in the power spectrum by performing the following steps:

(a) Evaluate the $x_n = x(t = nT_s)$ and place the values in a list, labeled x. Create a second list, y, of zeros which is of the same length as the x list. The y list is needed because Maple's FFT command assumes that generally the input data may have real (x) and imaginary (y) parts.

(b) Calculate the FFT of the x, y data, using the command FFT(m,x,y).

(c) Calculate S_k for each frequency f_k and plot the results over the range $f = 0...1/(2T_s)$. Sharp spikes should occur at 1 and 1.5 Hz.

05-S08: Power Spectrum: Driving Miss Duffing

The power spectrum approach of **05-S07** may be applied to the driven Duffing oscillator example of **05-1-3** (**The Route to Chaos**). Calculate the power spectrum for each of the four force amplitudes F_0 in that recipe. Take $m = 10$, use six digits accuracy, solve the oscillator ODE numerically, take the sampling frequency (the reciprocal of the sampling time) to be four times the driving frequency, and start the sampling at $t0 = 50\pi$ to eliminate the transient. In the power spectrum, the spike corresponding to the driving frequency is quite tall, so use $|X_k|$ as a measure of the power spectrum and suitably limit the vertical scale so smaller spikes may be observed. Interpret the power spectrum for each F_0, making use of the conclusions in recipe **05-1-3**.

05-S09: Rocket Sled.

A rocket sled accelerated from rest experiences (per unit mass) a thrust force $F = F0\, e^{-ct}$ and a drag force $Fdrag = -a\, v(t) - b\, v(t)^2$. Taking $F0 = 250$, $a = 1/10$, $b = 1/100$, $c = 1/2$ (in MKS units), analytically determine $v(t)$ and plot v vs. t. Determine the maximum speed in m/s and km/h and the time at which it occurs. Plot the acceleration in Gs, taking $g = 9.81$ m/s^2. Plot the distance $x(t)$ and find the maximum distance the sled travels.

Part III

THE DESSERTS

Mechanics is the paradise of the mathematical sciences because by means of it one comes to the fruits of mathematics.
Leonardo da Vinci, *The Notebooks* (1508–1518)

The Answer to the Great Question Of... Life, the Universe, and Everything...[is] Forty-two.
Douglas Adams, *The Hitch Hiker's Guide to the Galaxy* (1979)

It is often said that there is no such thing as a free lunch.
The Universe, however, is a free lunch.
Alan Guth, American physicist, in *Harpers* (November, 1994)

Chapter 6

Lagrangian & Hamiltonian Dynamics

In the Desserts, we look at a wide variety of interesting mechanics examples for which we use either the Lagrangian or Hamiltonian approach.

6.1 Some Lagrangian Examples

God used beautiful mathematics in creating the world.
Paul Dirac, Nobel laureate in physics (1933)

When the forces and geometry are more complicated, the Lagrangian approach can prove to be easier to work with than Newton's second law as it is often easier to determine the kinetic energy T and the potential energy V rather than the forces. Introducing the Lagrangian function $L = T - V$, Lagrange's equation of motion [FC99] for a conservative system is

$$\frac{d}{dt}\left(\frac{\partial L}{\partial \dot{q}_k}\right) - \frac{\partial L}{\partial q_k} = 0, \tag{6.1}$$

where q_k is the kth generalized coordinate and $\dot{q}_k$ is the generalized velocity.

In applying Equation (6.1), we shall demonstrate two different approaches. The first is to mimic a "hand" calculation and carry out each of the mathematical operations on the left-hand side of Eq. (6.1) and set the resulting expression equal to zero. In earlier versions of Maple, this was the only approach possible. In Maple 8, a `VariationalCalculus` library package has been added, this package containing the `EulerLagrange` command, which generates expressions corresponding to (minus) the lhs of Equation (6.1). In some instances, when there are ignorable coordinates, the first integrals of these expressions are also given, the integration constants being labeled K_1, K_2, etc.

6.1.1 Rock in the Rim

Diplomacy is the art of saying "Nice doggie"
until you can find a rock.
Will Rogers, American humorist (1879–1935)

On the planet Erehwon, a circular wheel of mass M and radius a is "lop-sided" because it has a small rock of mass m stuck in its rim as shown in Figure 6.1. The wheel is allowed to roll in a straight line along a level road. At time t, the rock makes an angle $\theta(t)$ with the vertical from the wheel's center to the instantaneous point (P) of contact with the ground. At $t = 0$, the rock is in contact with the ground ($\theta(0) = 0$) and is given some initial angular velocity ($\dot\theta(0) \neq 0$) so the wheel begins to roll. Enaj's goal is to determine the ODE that $\theta(t)$ satisfies and solve it for varying initial velocities.

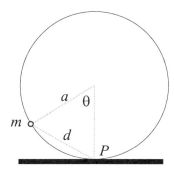

Figure 6.1: Geometry for the rolling wheel with a rock in its rim.

Enaj enters an interface command to free I for use as the moment of inertia.
> `restart: with(plots): interface(imaginaryunit=j):`

Treating the wheel as a solid disk, its moment of inertia I_{cm} about its center of mass is entered. Using the parallel-axis theorem, its moment of inertia about the contact point P with the road then is $I_P = I_{cm} + M a^2$.
> `I[cm]:=(1/2)*M*a^2;`

$$I_{cm} := \frac{M a^2}{2}$$

> `I[P]:=I[cm]+M*a^2;`

$$I_P := \frac{3 M a^2}{2}$$

When the rock is located at an angle θ, its distance from the contact point P is $d = 2a\sin(\theta/2)$. The moment of inertia of the rock about the point P is $I_{rock} = m d^2$. The total moment of inertia of the wheel and the rock about the point P then is $I_{total} = I_P + I_{rock}$.
> `d:=2*a*sin(theta(t)/2);`

6.1. SOME LAGRANGIAN EXAMPLES

$$d := 2a\sin(\frac{1}{2}\theta(t))$$

> `I[rock]:=m*d^2;`

$$I_{rock} := 4ma^2\sin(\frac{1}{2}\theta(t))^2$$

> `I[total]:=I[P]+I[rock];`

$$I_{total} := \frac{3Ma^2}{2} + 4ma^2\sin(\frac{1}{2}\theta(t))^2$$

The kinetic energy is $KE = (1/2)I_{total}\,\dot{\theta}^2$ and the potential energy (measured from the road) is $PE = mga(1-\cos\theta)$, where g is the acceleration due to gravity. Enaj only considers the potential energy of the rock, because the potential energy of the wheel doesn't change.

> `KE:=(1/2)*I[total]*(diff(theta(t),t))^2;`

$$KE := \frac{1}{2}(\frac{3Ma^2}{2} + 4ma^2\sin(\frac{1}{2}\theta(t))^2)(\frac{d}{dt}\theta(t))^2$$

> `PE:=m*g*a*(1-cos(theta(t)));`

$$PE := mga(1-\cos(\theta(t)))$$

The Lagrangian function $L = KE - PE$ is formed.

> `L:=KE-PE;`

$$L := \frac{1}{2}(\frac{3Ma^2}{2} + 4ma^2\sin(\frac{1}{2}\theta(t))^2)(\frac{d}{dt}\theta(t))^2 - mga(1-\cos(\theta(t)))$$

Lagrange's equation of motion takes the following form here,

$$\frac{d}{dt}\left(\frac{\partial L}{\partial \dot{\theta}}\right) - \frac{\partial L}{\partial \theta} = 0. \tag{6.2}$$

To carry out the partial derivatives in Equation (6.2), Enaj temporarily sets $\dot{\theta}(t) = \omega$ and $\theta(t) = \alpha$ in L, labeling the resulting expression LL. The inverse *relations* are also given so that she can convert back to the original variables.

> `LL:=subs(diff(theta(t),t)=omega,theta(t)=alpha,L);`

$$LL := \frac{1}{2}(\frac{3Ma^2}{2} + 4ma^2\sin(\frac{\alpha}{2})^2)\omega^2 - mga(1-\cos(\alpha))$$

> `relations:=omega=diff(theta(t),t),alpha=theta(t):`

Enaj tackles the first term in Equation (6.2) first. The partial derivative of LL with respect to ω is performed in *eq1* and then the *relations* are substituted.

> `eq1:=diff(LL,omega);`

$$eq1 := (\frac{3Ma^2}{2} + 4ma^2\sin(\frac{\alpha}{2})^2)\omega$$

> `eq1:=subs(relations,eq1);`

$$eq1 := (\frac{3Ma^2}{2} + 4ma^2\sin(\frac{1}{2}\theta(t))^2)(\frac{d}{dt}\theta(t))$$

She finishes evaluating the first term by taking the time derivative of *eq1*.

```
> term1:=diff(eq1,t);
```

$$term1 := 4\, m\, a^2 \sin(\frac{1}{2}\theta(t))\cos(\frac{1}{2}\theta(t))(\frac{d}{dt}\theta(t))^2$$
$$+(\frac{3\,M\,a^2}{2}+4\,m\,a^2\sin(\frac{1}{2}\theta(t))^2)(\frac{d^2}{dt^2}\theta(t))$$

To evaluate the second term in (6.2), *LL* is differentiated with respect to α in *eq2* and the *relations* are substituted.

```
> eq2:=diff(LL,alpha);
```

$$eq2 := 2\,m\,a^2\sin(\frac{\alpha}{2})\cos(\frac{\alpha}{2})\omega^2 - m\,g\,a\sin(\alpha)$$

```
> term2:=subs(relations,eq2);
```

$$term2 := 2\,m\,a^2\sin(\frac{1}{2}\theta(t))\cos(\frac{1}{2}\theta(t))(\frac{d}{dt}\theta(t))^2 - m\,g\,a\sin(\theta(t))$$

The ODE *de* describing the angular motion of the rock is obtained by subtracting *term2* from *term1* and equating to zero. The equation is simplified in *de2* and *de3* by combining trig terms and collecting $d^2\theta(t)/dt^2$ and a^2.

```
> de:=term1-term2=0;
```

$$de := 2\,m\,a^2\sin(\frac{1}{2}\theta(t))\cos(\frac{1}{2}\theta(t))(\frac{d}{dt}\theta(t))^2$$
$$+(\frac{3\,M\,a^2}{2}+4\,m\,a^2\sin(\frac{1}{2}\theta(t))^2)(\frac{d^2}{dt^2}\theta(t))+m\,g\,a\sin(\theta(t))=0$$

```
> de2:=combine(de,trig);
> de3:=collect(de2,[diff(theta(t),t,t),a^2]);
```

$$de3 := (\frac{3\,M}{2}+2\,m-2\,m\cos(\theta(t)))\,a^2\,(\frac{d^2}{dt^2}\theta(t))$$
$$+m\,a^2\,(\frac{d}{dt}\theta(t))^2\sin(\theta(t))+m\,g\,a\sin(\theta(t))=0$$

Observing that the equation of motion *de3* is nonlinear in the dependent variable θ, Enaj decides to solve it numerically by taking some representative values for the parameters. She takes $M = 100$ kg, $m = 1$ kg, $a = 1$ m, and $g = 10$m/s^2. She considers $N = 6$ different initial angular velocities, $\dot\theta(0) = step \times i$, where $i = 1, 2, ..., N$ and $step = 0.1$ rad/s. The total time is $tt = 100$ seconds.

```
> M:=100: m:=1: a:=1: g:=10:
> N:=6: step:=0.1: tt:=100:
```

Enaj creates a do loop to numerically solve *de3* and plot $\theta(t)$ over the time interval $t = 0$ to $t = tt$ for the N different initial velocities. The curves are given different colors. On completion of the do loop, the `display` command is used to produce Figure 6.2.

```
> for i from 1 to N do
> sol||i:=dsolve({de3,theta(0)=0,D(theta)(0)=step*i},
> theta(t),type=numeric,output=listprocedure);
```

6.1. SOME LAGRANGIAN EXAMPLES

```
>   gr||i:=odeplot(sol||i,[t,theta(t)],0..tt,axes=boxed,
>   numpoints=200,labels=["t","theta"],color=
>   COLOR(RGB,.15*i,0,.4),thickness=2);
>   end do:
>   display(seq(gr||i,i=1..N),view=[0..tt,-3..10]);
```

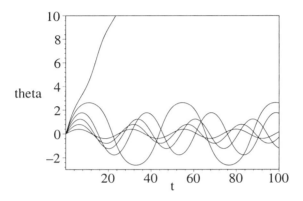

Figure 6.2: Rocking motion for $i = 1, ..., 5$. Rolling motion for $i = 6$.

For $i = 1$ to 5, the rock undergoes a periodic rocking motion as the wheel has insufficient initial angular velocity for the rock to go "over the top". The amplitude of the oscillation increases as i increases from 1 to 5. The maximum angular range for which rocking motion can occur is $\theta = -\pi$ to $+\pi$ radians. Enaj notes that the rocking behavior is similar to that of the simple plane pendulum. When $\dot\theta(0)$ is large enough, as it is for $i = 6$, the rock goes over the top and θ increases indefinitely with time. This is again similar angular behavior to that of the pendulum. In this case, the wheel no longer rocks back and forth about a fixed point on the road, but rolls along the road.

6.1.2 Airtrack Mechanics

We haven't got the money, so we've got to think.
Ernest Rutherford, Nobel laureate in chemistry (1908)

An adherent of Rutherford's philosophy, George is always thinking of possible experiments for the lab which are inexpensive to create, simple to perform, yet display interesting nonlinear behavior. Many of his mechanical experiments (see [EM00]) involve the use of the linear airtrack. This Lagrangian recipe was created to simulate one of George's suggestions. Referring to Figure 6.3, an airtrack glider (mass m_1) is connected by an inextensible string passing over two pulleys, separated by a distance L, to a mass $m_2 < m_1$. In the recipe

216   CHAPTER 6. LAGRANGIAN & HAMILTONIAN DYNAMICS

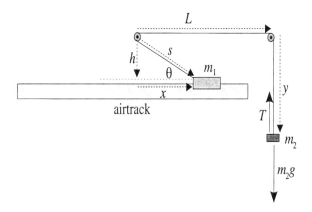

Figure 6.3: The geometry of the airtrack experiment.

we will assume that the airtrack is completely frictionless, that both pulleys are light and frictionless, and that air resistance can be neglected. Our goal is to determine the equation of motion for the glider, numerically solve it, and determine the period. In the lab, the period is an easily measurable quantity.

In the previous recipe, we mimicked a "hand" derivation in obtaining the equation of motion. The various steps in applying Lagrange's equation of motion may, however, be bypassed by loading the VariationalCalculus package which contains the EulerLagrange command. The desired equation of motion will follow on applying this command to the Lagrangian.

> restart: with(plots): with(VariationalCalculus):

Letting x and y be the coordinates of the masses m_1 and m_2 as shown in the figure, their velocities v_1 and v_2 at time t are entered.

> v1:=diff(x(t),t); v2:=diff(y(t),t);

The kinetic energy, KE, of the two masses is formed and the potential energy, PE, of m_2 entered. The potential energy of m_1 doesn't change as long as it stays on the airtrack. The *Lagrangian* is then calculated.

> KE:=(1/2)*m1*v1^2+(1/2)*m2*v2^2;

$$KE := \frac{1}{2} \, m1 \, (\frac{d}{dt} \text{x}(t))^2 + \frac{1}{2} \, m2 \, (\frac{d}{dt} \text{y}(t))^2$$

> PE:=-m2*g*y(t);

$$PE := -m2 \, g \, \text{y}(t)$$

> Lagrangian:=KE-PE;

$$Lagrangian := \frac{1}{2} \, m1 \, (\frac{d}{dt} \text{x}(t))^2 + \frac{1}{2} \, m2 \, (\frac{d}{dt} \text{y}(t))^2 + m2 \, g \, \text{y}(t)$$

The length of the string, $s(t) + L + y(t)$, is a constant C. This *constraint* is entered. From Fig. 6.3, the theorem of Pythagoras yields $s(t) = \sqrt{x(t)^2 + h^2}$.

6.1. SOME LAGRANGIAN EXAMPLES

$s(t)$ is automatically substituted into the *constraint* equation.

```
> constraint:=C=s(t)+L+y(t);
```
$$\mathit{constraint} := C = s(t) + L + y(t)$$

```
> s(t):=sqrt(x(t)^2+h^2);
```
$$s(t) := \sqrt{x(t)^2 + h^2}$$

```
> constraint;
```
$$C = \sqrt{x(t)^2 + h^2} + L + y(t)$$

The *constraint* equation is solved for $y(t)$, which is automatically substituted into the Lagrangian, leaving us with LL, which depends only on $x(t)$.

```
> y(t):=solve(constraint,y(t));
```
$$y(t) := C - \sqrt{x(t)^2 + h^2} - L$$

```
> LL:=Lagrangian;
```

$$LL := \frac{1}{2}\, m1\, (\frac{d}{dt} x(t))^2 + \frac{1}{2}\, \frac{m2\, x(t)^2\, (\frac{d}{dt} x(t))^2}{x(t)^2 + h^2} + m2\, g\, (C - \sqrt{x(t)^2 + h^2} - L)$$

The `EulerLagrange` command is applied to LL, the independent variable being the time t, the dependent variable being the coordinate $x(t)$ of mass m_1.

```
> eq:=EulerLagrange(LL,t,x(t));
```

$$eq := \left\{ \frac{m2\, x(t)^3\, (\frac{d}{dt} x(t))^2}{(x(t)^2 + h^2)^2} - \frac{m2\, x(t)\, (\frac{d}{dt} x(t))^2}{x(t)^2 + h^2} - \frac{m2\, g\, x(t)}{\sqrt{x(t)^2 + h^2}} \right.$$

$$- m1\, (\frac{d^2}{dt^2} x(t)) - \frac{m2\, x(t)^2\, (\frac{d^2}{dt^2} x(t))}{x(t)^2 + h^2},$$

$$\frac{1}{2}\, m1\, (\frac{d}{dt} x(t))^2 + \frac{1}{2}\, \frac{m2\, x(t)^2\, (\frac{d}{dt} x(t))^2}{x(t)^2 + h^2} + m2\, g\, (C - \sqrt{x(t)^2 + h^2} - L)$$

$$\left. - (\frac{d}{dt} x(t)) \left(m1\, (\frac{d}{dt} x(t)) + \frac{m2\, x(t)^2\, (\frac{d}{dt} x(t))}{x(t)^2 + h^2} \right) = K_1 \right\}$$

The output in *eq* displays two entries, the first being the lhs of Lagrange's equation multiplied by -1, the second entry being the first integral with K_1 an arbitrary constant (the "constant of the motion"). Because both entries involve nonlinear functions of the dependent variable, a solution can only be obtained numerically. We shall choose to work directly with Lagange's equation, rather than the first integral result. The first integral entry can be removed from *eq* with the following `remove` command. This will leave only a single entry (not shown here) in *eq2*.

```
> eq2:=remove(has,eq,K[1]);
```

Setting the first, and only, entry in $-eq2$ equal to zero completes the evaluation of Lagrange's equation of motion. We make the resulting second order nonlinear ODE more compact by collecting derivative terms in de.

> `de:=collect(-eq2[1],[diff(x(t),t,t),diff(x(t),t)^2])=0;`

$$de := (m1 + \frac{m2\,\mathrm{x}(t)^2}{\mathrm{x}(t)^2 + h^2})(\frac{d^2}{dt^2}\mathrm{x}(t)) + (-\frac{m2\,\mathrm{x}(t)^3}{(\mathrm{x}(t)^2 + h^2)^2} + \frac{m2\,\mathrm{x}(t)}{\mathrm{x}(t)^2 + h^2})(\frac{d}{dt}\mathrm{x}(t))^2$$
$$+ \frac{m2\,g\,\mathrm{x}(t)}{\sqrt{\mathrm{x}(t)^2 + h^2}} = 0$$

To solve de, we consider some representative parameter values, viz., $m_1 = 1$ kg, $m_2 = 0.9$ kg, $h = 0.1$ m, $g = 9.81$ m/s^2, and the initial condition $x(0) = 1$ m, $\dot{x}(0) = 0$. For $m_2 < m_1$, the glider will remain on the airtrack.

> `m1:=1: m2:=0.9: h:=.1: g:=9.81:`

> `ic:=x(0)=1,D(x)(0)=0:`

Then de is numerically solved for $x(t)$ subject to ic and a 3-dimensional plot of t vs. x vs. $v = dx/dt$ created with the odeplot command.

> `de;`

> `sol:=dsolve({de,ic},{x(t)},type=numeric,output=listprocedure);`

> `odeplot(sol,[t,x(t),diff(x(t),t)],0..10,numpoints=1000,`

> `axes=framed,labels=["t","x","v"],tickmarks=[3,3,3],`

> `orientation=[-60,45],thickness=2);`

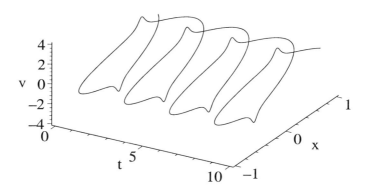

Figure 6.4: 3-dimensional plot of the periodic motion of the mass m_1.

The solution is shown in Figure 6.4. By rotating the computer plot to view x vs. t, the reader can determine the (approximate) period of the cyclic motion.

6.1.3 Can You Top This Nutation?

Happiness makes up in height for what it lacks in length.
Robert Frost, American poet, title of poem (1942)

Gabrielle is very happy with the new top that she received as a birthday present. This symmetric top has its tip held fixed and Gabrielle starts it moving by

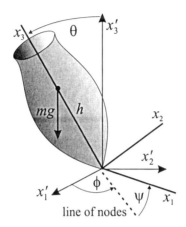

Figure 6.5: Euler angles ϕ, θ, and ψ for the spinning top.

spinning the top around its axis, giving it an initial tilt, and then releasing it. Referring to Fig. 6.5, we shall determine the equations of motion for the Euler angles ϕ, θ, ψ [MT95], [GPS02], and then animate the motion of the top's tilt angle θ for conditions approximating those used by Gabrielle. The fixed ("space") axes are x_1', x_2', x_3', while the "body" axes attached to the top are x_1, x_2, and x_3. The "line of nodes" is the line common to the planes containing the x_1 and x_2 axes and the x_1' and x_2' axes. The top has a mass m and the distance from its fixed tip to its center of mass is h. Because the top is symmetric, its moments of inertia $I1$ and $I2$ about the x_1 and x_2 axes, respectively, are equal.

```
>   restart: with(plots): I2:=I1:
```
Note that we have not loaded the `Variational Calculus` package here, having chosen to mimic a hand calculation. From Marion and Thornton [MT95], the angular velocity components ω_1, ω_2, ω_3 along the body coordinate axes x_1, x_2, and x_3 are as follows:

```
>   omega1:=sin(theta(t))*sin(psi(t))*diff(phi(t),t)
>   +cos(psi(t))*diff(theta(t),t);
```

$$\omega 1 := \sin(\theta(t))\sin(\psi(t))\left(\tfrac{d}{dt}\phi(t)\right) + \cos(\psi(t))\left(\tfrac{d}{dt}\theta(t)\right)$$

```
>   omega2:=sin(theta(t))*cos(psi(t))*diff(phi(t),t)
>   -sin(psi(t))*diff(theta(t),t);
```

$$\omega 2 := \sin(\theta(t))\cos(\psi(t))\,(\tfrac{d}{dt}\phi(t)) - \sin(\psi(t))\,(\tfrac{d}{dt}\theta(t))$$

```
> omega3:=cos(theta(t))*diff(phi(t),t)+diff(psi(t),t);
```

$$\omega 3 := \cos(\theta(t))\,(\tfrac{d}{dt}\phi(t)) + (\tfrac{d}{dt}\psi(t))$$

The top's kinetic energy KE in the body coordinate system is entered along with its potential energy PE and the Lagrangian is formed.

```
> KE:=simplify(I1*omega1^2/2+I2*omega2^2/2+I3*omega3^2/2);
```

$$KE := \tfrac{1}{2} I1\,(\tfrac{d}{dt}\phi(t))^2 - \tfrac{1}{2} I1\,(\tfrac{d}{dt}\phi(t))^2 \cos(\theta(t))^2 + \tfrac{1}{2} I1\,(\tfrac{d}{dt}\theta(t))^2$$
$$+ \tfrac{1}{2} I3 \cos(\theta(t))^2 (\tfrac{d}{dt}\phi(t))^2 + I3 \cos(\theta(t))\,(\tfrac{d}{dt}\phi(t))\,(\tfrac{d}{dt}\psi(t)) + \tfrac{1}{2} I3\,(\tfrac{d}{dt}\psi(t))^2$$

```
> PE:=m*g*h*cos(theta(t));
```

$$PE := m\,g\,h\cos(\theta(t))$$

```
> Lagrangian:=KE-PE;
```

The following two relations will be used for removing and reinserting the time dependence of the variables. This is necessary for performing the differentiations. The first relation is substituted into the Lagrangian, producing L.

```
> rel1:=diff(phi(t),t)=w1,diff(psi(t),t)=w2,diff(theta(t),t)=w3,
> phi(t)=a1,psi(t)=a2,theta(t)=a3:
> rel2:=w1=diff(phi(t),t),w2=diff(psi(t),t),w3=diff(theta(t),t),
> a1=phi(t),a2=psi(t),a3=theta(t):
> L:=subs(rel1,Lagrangian);
```

$$L := \frac{I1\,w1^2}{2} - \tfrac{1}{2} I1\,w1^2 \cos(a3)^2 + \frac{I1\,w3^2}{2} + \tfrac{1}{2} I3 \cos(a3)^2 w1^2$$
$$+ I3 \cos(a3)\,w1\,w2 + \frac{I3\,w2^2}{2} - m\,g\,h\cos(a3)$$

The equation of motion obtained by taking $q_k = \phi(t)$ in Lagrange's equation is now derived. Differentiating L with respect to $a1$ (i.e., $\phi(t)$) yields zero.

```
> diff(L,a1);
```

$$0$$

This means that $d(\partial L/\partial \dot\phi)/dt = 0$ or $\partial L/\partial \dot\phi = P_\phi$, where P_ϕ is a constant (angular momentum) of the motion. Entering this latter equation with the time dependence reinserted yields the equation of motion $eq1$.

```
> eq1:=subs(rel2,diff(L,w1))=P[phi];
```

$$eq1 := I1\,(\tfrac{d}{dt}\phi(t)) - I1\,(\tfrac{d}{dt}\phi(t))\cos(\theta(t))^2 + I3\cos(\theta(t))^2\,(\tfrac{d}{dt}\phi(t))$$
$$+ I3 \cos(\theta(t))\,(\tfrac{d}{dt}\psi(t)) = P_\phi$$

Similarly, taking $q_k = \psi(t)$, we differentiate L with respect to $a2 \equiv \psi(t)$, again obtaining zero. As above, this means that $\partial L/\partial \dot\psi = P_\psi$, where P_ψ is a second

6.1. SOME LAGRANGIAN EXAMPLES

constant of the motion. Entering this latter equation with the time dependence reinserted yields the equation of motion *eq2*.

> `diff(L,a2);`

$$0$$

> `eq2:=subs(rel2,diff(L,w2))=P[psi];`

$$eq2 := I3\cos(\theta(t))\,(\tfrac{d}{dt}\,\phi(t)) + I3\,(\tfrac{d}{dt}\,\psi(t)) = P_\psi$$

Finally we take $q_k = \theta(t)$ and calculate $(d/dt)(\partial L/\partial\dot\theta) - \partial L/\partial\theta = 0$.

> `eq3:=diff(subs(rel2,diff(L,w3)),t)-subs(rel2,diff(L,a3))=0;`

$$eq3 := I1\,(\tfrac{d^2}{dt^2}\,\theta(t)) - I1\,(\tfrac{d}{dt}\,\phi(t))^2 \cos(\theta(t))\sin(\theta(t)) + I3\cos(\theta(t))$$
$$(\tfrac{d}{dt}\,\phi(t))^2\sin(\theta(t)) + I3\sin(\theta(t))\,(\tfrac{d}{dt}\,\phi(t))\,(\tfrac{d}{dt}\,\psi(t)) - m\,g\,h\sin(\theta(t)) = 0$$

Equations *eq1* and *eq2* are linear in $\dot\phi$ and $\dot\psi$, so they may be analytically solved for these variables. Substituting the trig relation $\cos^2\theta(t) = 1 - \sin^2\theta(t)$ into *sol*[1] and *sol*[2] yields the equations *eq1b* and *eq2b* for $\dot\phi$ and $\dot\psi$.

> `sol:=solve({eq1,eq2},{diff(phi(t),t),diff(psi(t),t)});`
> `trigrel:=cos(theta(t))^2=1-sin(theta(t))^2:`
> `eq1b:=algsubs(trigrel,sol[1]); eq2b:=algsubs(trigrel,sol[2]);`

$$eq1b := \tfrac{d}{dt}\,\phi(t) = \frac{P_\phi - \cos(\theta(t))\,P_\psi}{I1\sin(\theta(t))^2}$$

$$eq2b := \tfrac{d}{dt}\,\psi(t) = \frac{-I3\cos(\theta(t))\,P_\phi + I3\,P_\psi - I3\,P_\psi\sin(\theta(t))^2 + P_\psi\,I1\sin(\theta(t))^2}{I3\,I1\sin(\theta(t))^2}$$

We substitute *eq1b* and *eq2b* into *eq3* and simplify the output in *eq3c*. The numerator of the lhs of *eq3c* is set equal to zero in *eq3d*, trig terms combined in *eq3e*, and finally various terms collected in *eq3f*.

> `eq3b:=subs({eq1b,eq2b},eq3);`
> `eq3c:=simplify(eq3b);`
> `eq3d:=numer(lhs(eq3c))=0;`
> `eq3e:=combine(eq3d,trig);`
> `eq3f:=collect(eq3e,[diff(theta(t),t,t),cos(theta(t)),`
> `cos(2*theta(t))]);`

$$eq3f := (\tfrac{1}{4}\,I1^2\sin(3\,\theta(t)) - \tfrac{3}{4}\,I1^2\sin(\theta(t)))\,(\tfrac{d^2}{dt^2}\,\theta(t)) + (P_\psi{}^2 + P_\phi{}^2)\cos(\theta(t))$$
$$+ (-\tfrac{m\,g\,h\,I1}{2} - \tfrac{1}{2}\,P_\phi\,P_\psi)\cos(2\,\theta(t)) - \tfrac{3}{2}\,P_\phi\,P_\psi + \tfrac{3\,m\,g\,h\,I1}{8}$$
$$+ \tfrac{1}{8}\,m\,g\,h\,I1\cos(4\,\theta(t)) = 0$$

The three equations of motion (*eq1b*, *eq2b*, and *eq3f*) are nonlinear and must be solved numerically. For Gabrielle's top, we take $I1 = 1$ kg·m^2, $I3 = 1/2$

kg·m², $m = 0.5$ kg, $g = 10$ m/s², $h = 0.2$ m, $P_\phi = 1/2$ kg·m²/s, $P_\psi = 1/\sqrt{3}$ kg·m²/s, and an initial tilt angle $\theta 0 = \pi/6$ rads (30°). These parameter values will produce a motion similar to that observed by Gabrielle when she spun her top. Other types of motion will occur if these values are changed.

> I1:=1: I3:=1/2: m:=0.5: g:=10: h:=0.2:

> P[phi]:=1/2: P[psi]:=1/sqrt(3): theta0:=Pi/6:

> eq3f:=eq3f; eq1b:=eq1b; eq2b:=eq2b;

As initial conditions, we take $\theta(0) = \theta 0$, $\phi(0) = 0$, $\psi(0) = 0$, $\dot\theta(0) = 0$, and consider a total time of 50 s.

> ic:=theta(0)=theta0,phi(0)=0,psi(0)=0,D(theta)(0)=0:

> tt:=50:

The three equations of motion are numerically solved, and a three-dimensional plot of θ vs. ϕ vs. ψ is produced using the odeplot command.

> sol2:=dsolve({eq3f,eq1b,eq2b,ic},{theta(t),phi(t),psi(t)},

> numeric,output=listprocedure):

> gr1:=odeplot(sol2,[theta(t),phi(t),psi(t)],t=0..tt,

> axes=normal,labels=["theta","phi","psi"],thickness=2,

> numpoints=500):

To see how the tilt angle θ of the top's spin axis varies as the top precesses (ϕ increases) around the vertical space axis, we choose the orientation [0,0] in the following display command. The result is shown in Figure 6.6. The tilt axis displays cusplike "nutation", periodically tilting from the start angle of $\pi/6 \approx 0.5$ rads to about 2.6 rads.

> display(gr1,orientation=[0,0]);

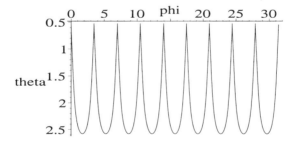

Figure 6.6: Variation of tilt angle θ with ϕ.

The three angles are evaluated at arbitrary time t and labeled Θ, Φ and Ps. We have not used Psi for the latter because it is a protected Maple function. Of course, we could have unprotected Psi if we had wished to use it as a symbol.

> Theta:=eval(theta(t),sol2): Phi:=eval(phi(t),sol2):

6.1. SOME LAGRANGIAN EXAMPLES

```
>   Ps:=eval(psi(t),sol2):
```
The **polarplot** command with constrained scaling is used to plot $\Theta \equiv \theta(t)$ vs. $\Phi \equiv \phi(t)$. In this graph, θ is the radial coordinate and ϕ the angular coordinate.
```
>   gr2:=polarplot([Theta(t),Phi(t),t=0..tt],scaling=constrained,
>   thickness=2,numpoints=1000):
```
To animate the motion of the top's spin axis, we will create $N = 500$ time frames, so the step size is $tt/N = 1/10$ seconds.
```
>   N:=500: step:=tt/N;
```
The motion is now animated with the following do loop.
```
>   for i from 0 to N do
```
The time at the ith step is evaluated,
```
>   t:=step*i;
```
and the upper end of the top's spin axis is plotted in polar coordinates at each time step, being represented by a blue circle.
```
>   gr3||i:=pointplot([Theta(t),Phi(t)],coords=polar,
>   symbol=circle,color=blue,symbolsize=16):
```
At each time step, the graph gr3||i is superimposed on the graph gr2.
```
>   pl||i:=display({gr2,gr3||i}):
>   end do:
```
On executing the following command line, clicking on the plot, and on the start arrow, the nutatational motion of the top's axis can be observed. The opening frame in the animation is shown in Figure 6.7.
```
>   display(seq(pl||i,i=0..N),insequence=true,scaling=constrained);
```

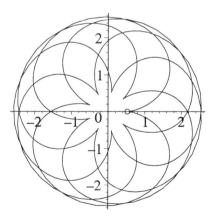

Figure 6.7: Nutation of spinning top.

Feel free to change the parameter values and initial conditions in the recipe and see if you can create another beautiful nutational pattern.

6.1.4 This Double Pendulum is "Dynamite"

If A is a success in life, then A equals x plus y plus z.
Work is x; y is play; and z is keeping your mouth shut.
Albert Einstein, Nobel laureate in physics (1921)

When it comes to extolling the intellectual excitement of exploring nonlinear phenomena, Jennifer just can't keep her mouth shut, sometimes to the annoyance of a few of her "linear minded" colleagues. In particular she has criticized the practice of deriving the equations of motion for such pendula as the double pendulum, the spherical pendulum, the rotating pendulum, etc., and then ending up looking only at the small angle solutions to the "linearized" equations. Jennifer feels that mechanics students would become more excited about the subject if they could explore solutions at any angles and observe the sometimes unexpected or even bizarre behavior which can occur. Therefore, she has decided to create a number of what she considers to be "dynamite" pendulum recipes based on this philosophy. The first one involves using the Lagrangian approach to derive the double pendulum equations of motion and then animating the numerical solution.

Referring to Figure 6.8, a double pendulum consists of two masses m_1 and m_2 suspended by massless rigid rods of lengths r_1 and r_2. The rods make angles

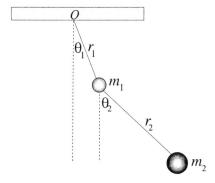

Figure 6.8: The double pendulum.

θ_1 and θ_2 with the vertical and move in the same plane. To make the motion interesting, Jennifer assumes that both rods can execute complete revolutions about their respective pivot points. Frictional effects are neglected.

After loading the VariationalCalculus package, Jennifer introduces aliases for the angles. Entering A1 and A2 generates θ_1 and θ_2 in the output.

>   restart: with(plots): with(VariationalCalculus):
>   alias(theta[1]=A1,theta[2]=A2):

The Cartesian coordinates $(x1, y1)$ of $m1$ and $(x2, y2)$ of $m2$ are entered and the corresponding velocity components $vx1$, $vy1$, and $vx2$, $vy2$ calculated.

6.1. SOME LAGRANGIAN EXAMPLES

```
>  x1:=r1*sin(A1(t)); y1:=r1*cos(A1(t));
```
$$x1 := r1\sin(\theta_1(t))$$
$$y1 := r1\cos(\theta_1(t))$$
```
>  x2:=x1+r2*sin(A2(t)); y2:=y1+r2*cos(A2(t));
```
$$x2 := r1\sin(\theta_1(t)) + r2\sin(\theta_2(t))$$
$$y2 := r1\cos(\theta_1(t)) + r2\cos(\theta_2(t))$$
```
>  vx1:=diff(x1,t); vy1:=diff(y1,t);
>  vx2:=diff(x2,t); vy2:=diff(y2,t);
```
The total kinetic energy T and the potential energy V are entered and the Lagrangian formed and simplified with the combine command.
```
>  T:=1/2*m1*(vx1^2+vy1^2)+1/2*m2*(vx2^2+vy2^2);
>  V:=-m1*g*y1-m2*g*y2;
>  Lagrangian:=combine(T-V);
```

$$Lagrangian := \frac{1}{2}\,m1\,r1^2\,(\tfrac{d}{dt}\theta_1(t))^2 + \frac{1}{2}\,m2\,r1^2\,(\tfrac{d}{dt}\theta_1(t))^2$$
$$+ m2\,r1\,(\tfrac{d}{dt}\theta_1(t))\,r2\,(\tfrac{d}{dt}\theta_2(t))\cos(\theta_1(t)-\theta_2(t)) + \frac{1}{2}\,m2\,r2^2\,(\tfrac{d}{dt}\theta_2(t))^2$$
$$+ m1\,g\,r1\cos(\theta_1(t)) + m2\,g\,r1\cos(\theta_1(t)) + m2\,g\,r2\cos(\theta_2(t))$$

Jennifer applies the EulerLagrange command, with two dependent variables ($A1(t)$, $A2(t)$, or $\theta_1(t)$, $\theta_2(t)$) present, to the Lagrangian. The output, which is not displayed here, evaluates $\partial L/\partial\theta_1 - (d/dt)(\partial L/\partial\dot\theta_1)$, $\partial L/\partial\theta_2 - (d/dt)(\partial L/\partial\dot\theta_2)$, as well as generating a first integral with integration constant K_1.

```
>  eq:=EulerLagrange(Lagrangian,t,[A1(t),A2(t)]);
```
Since it is not of much help here, the first integral expression is removed from eq and the two relevant ODEs extracted and simplified.
```
>  de1:=simplify(remove(has,eq,K[1])[1]/(-r1))=0;
```

$$de1 := m1\,g\sin(\theta_1(t)) + m2\,g\sin(\theta_1(t)) + m1\,r1\,(\tfrac{d^2}{dt^2}\theta_1(t))$$
$$+ m2\,r1\,(\tfrac{d^2}{dt^2}\theta_1(t)) + m2\,r2\,(\tfrac{d^2}{dt^2}\theta_2(t))\cos(\theta_1(t)-\theta_2(t))$$
$$+ m2\,r2\,(\tfrac{d}{dt}\theta_2(t))^2\sin(\theta_1(t)-\theta_2(t)) = 0$$

```
>  de2:=simplify(remove(has,eq,K[1])[2]/(-m2*r2))=0;
```

$$de2 := g\sin(\theta_2(t)) + r1\,(\tfrac{d^2}{dt^2}\theta_1(t))\cos(\theta_1(t)-\theta_2(t))$$
$$- r1\,(\tfrac{d}{dt}\theta_1(t))^2\sin(\theta_1(t)-\theta_2(t)) + r2\,(\tfrac{d^2}{dt^2}\theta_2(t)) = 0$$

The resulting equations $de1$ and $de2$ are complicated nonlinear ODEs in the dependent variables θ_1 and θ_2. Since $de1$ and $de2$ cannot be solved exactly analytically, the standard practice in most mechanics texts is to linearize them by making the small angle approximations $\sin\theta \approx \theta$ and $\cos\theta \approx 1$. To show her students more interesting behavior than that obtained in the linear approximation, Jennifer chooses some parameter values in order to solve the ODEs numerically. She takes $m_1 = 1$ kg, $m_2 = 0.1$ kg, $r_1 = 1$ m, $r_2 = 2$ m, and $g = 9.8$ m/s^2. The total time is $tt = 50$ s. As initial conditions, she chooses $\theta_1(0) = 1.1$ rad, $\theta_2(0) = 1.2$ rad, and $\dot\theta_1(0) = \dot\theta_2(0) = 0$ rad/s.

> m1:=1: m2:=0.1: r1:=1: r2:=2: g:=9.8: tt:=50:

> ic:=A1(0)=1.1,A2(0)=1.2,D(A1)(0)=0,D(A2)(0)=0;

The ODEs *de1* and *de2* are numerically solved for $\theta_1(t)$ and $\theta(t)$ subject to the *ic* and then plotted separately with the odeplot command over the time interval $t = 0$ to $t = tt = 50$ s. The results are shown in Figure 6.9, the behavior of θ_1 and θ_2 displayed in the left and right plots, respectively.

> sol:=dsolve({de1,de2,ic},{A1(t),A2(t)},type=numeric,

> output=listprocedure,maxfun=0):

> odeplot(sol,[t,A1(t)],0..tt,numpoints=500);

> odeplot(sol,[t,A2(t)],0..tt,numpoints=500);

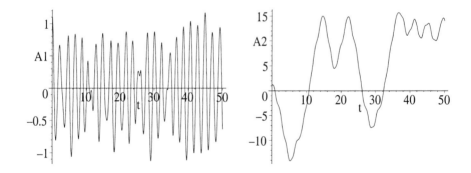

Figure 6.9: Temporal evolution of $\theta_1 \equiv A1$ (left) and $\theta_2 \equiv A2$ (right).

Realizing that the static plots in Fig. 6.9 do not really convey what is happening to the double pendulum, Jennifer animates the motion. She uses *sol* to evaluate θ_1 and θ_2 at time t, labeling the results *angle1* and *angle2*. In the animation, $N = 500$ frames are used, the time step size being $tt/N = 50/500 = 1/10$ s.

> angle1:=eval(A1(t),sol): angle2:=eval(A2(t),sol):

> N:=500: step:=tt/N:

The following do loop produces $N = 500$ graphs, one for each time step. In the various plot commands, Jennifer uses $-y1$ and $-y2$, instead of $y1$ and $y2$, so that the angles are measured from the downward vertical axis instead of the upward axis. The masses are represented by blue circles, the connecting rods by thick green lines.

```
>    for n from 0 to N do
>    t:=step*n:
>    x1:=r1*sin(angle1(t)): y1:=r1*cos(angle1(t)):
>    x2:=x1+r2*sin(angle2(t)): y2:=y1+r2*cos(angle2(t)):
>    gr1[n]:=pointplot([[x1,-y1],[x2,-y2]],symbol=circle,
>    symbolsize=14,color=blue):
>    gr2[n]:=plot([[[0,0],[x1,-y1]],[[x,-y1],[x2,-y2]]],
>    style=line,thickness=2,color=green):
>    pl[n]:=display({gr1[n],gr2[n]}):
>    end do:
>    display(seq(pl[n],n=0..N),insequence=true,
>    view=[-r1-r2..r1+r2,-r1-r2..r1+r2],scaling=constrained);
```

On executing the display command line, clicking on the computer plot and on the start arrow, the exciting animation of the double pendulum will begin. Jennifer hopes that this recipe will prove useful and interesting to her students as they play with the different parameter values and initial conditions.

6.2 Calculus of Variations

The two recipes in this section are based on a mathematical framework developed by the mathematicians Euler and Lagrange. Consider a function $y(x)$ with fixed values $y(x_0) = y_0$ and $y(x_1) = y_1$ at two distinct points $A(x_0, y_0)$ and $B(x_1, y_1)$, respectively. Form a specific function F of x, $y(x)$ and $y' \equiv dy(x)/dx$, i.e., $F = F(x, y, y')$. In our mechanics examples, the form of F will be determined by the specification of the problem. Amongst all functions $y(x)$ connecting A and B, find those which give an extremum (minimum or maximum, usually) to the integral

$$I[y] = \int_{x_0}^{x_1} F(x, y(x), y'(x))\ dx. \qquad (6.3)$$

The value I of the integral depends on what functional form is chosen for $y(x)$. Rather than trying different forms of $y(x)$ ad infinitum, it turns out that the $y(x)$ which yields an extremum value for I must satisfy the Euler–Lagrange differential equation:

$$\frac{\partial F}{\partial y} - \frac{d}{dx}\left(\frac{\partial F}{\partial y'}\right) = 0. \qquad (6.4)$$

The proof may be found in most standard mechanics texts such as Marion and Thornton [MT95] and Goldstein, Poole, and Safko [GPS02]. The resulting ODE is often nonlinear in nature, but may be susceptible to analytic solution. If not, it may be solved numerically.

6.2.1 Of Wine Goblets and Whisky Tumblers

Adulthood is the ever-shrinking period between childhood and old age. It is the apparent aim of modern industrial societies to reduce this period to a minimum.
Thomas Szasz, American psychiatrist, *The Second Sin*, "Social Relations" (1973)

A classic problem in "variational" calculus is to determine the curve $y(x)$ connecting two fixed points, (x_0, y_0) and (x_1, y_1), which generates the minimum surface area when the curve is rotated through one revolution about the x-axis. The element dA of surface area is given by $dA = (2\pi y)\, d\ell$ where $d\ell$ is an element of length along the curve. But $d\ell = \sqrt{(dx)^2 + (dy)^2} = \sqrt{1 + (dy/dx)^2}\, dx$. So the total surface area is given by $A = 2\pi \int_{x_0}^{x_1} y\sqrt{1 + (y')^2}\, dx$, where the prime indicates a derivative with respect to x. To determine the $y(x)$ that minimizes A, one can solve the Euler–Lagrange equation (6.4) with $F = y\sqrt{1 + (y')^2}$.

Jennifer has already discussed this problem in class and has asked Mike to illustrate in his tutorial how it would be tackled using computer algebra. Taking $(x_0 = 0,\ y_0 = 1)$ and $(x_1 = 1,\ y_1 = 2)$, Mike has entitled his recipe "Of Wine Goblets and Whisky Tumblers" for reasons which will become apparent. For convenience, he introduces the alias yp for y' and then enters F.

> `restart: with(plots): alias('y''=yp):`
> `F:=y*sqrt(1+yp^2);`

$$F := y\sqrt{1 + y'^2}$$

The partial derivatives $\partial F/\partial y$ and $\partial F/\partial y'$ are evaluated in *Dy* and *Dyp*. Introducing the relation *rel* and substituting it into *Dy* and *Dyp* restores the x dependence to these quantities.

> `Dy:=diff(F,y); Dyp:=diff(F,yp);`

$$Dy := \sqrt{1 + y'^2}$$

$$Dyp := \frac{y\, y'}{\sqrt{1 + y'^2}}$$

> `rel:=y=y(x),yp=diff(y(x),x):`
> `Dyx:=subs(rel,Dy); Dypx:=subs(rel,Dyp);`

The Euler–Lagrange equation is applied, the result being simplified with the symbolic option.

> `ELeq:=simplify(Dyx-diff(Dypx,x),symbolic)=0;`

$$ELeq := \frac{1 + (\frac{d}{dx}\, y(x))^2 - y(x)\,(\frac{d^2}{dx^2}\, y(x))}{(1 + (\frac{d}{dx}\, y(x))^2)^{(3/2)}} = 0$$

The relevant differential equation, *ode*, to be solved for $y(x)$ is obtained by taking the numerator of the left-hand side of *ELeq* and setting it equal to zero.

> `ode:=numer(lhs(ELeq))=0;`

6.2. CALCULUS OF VARIATIONS

$$ode := 1 + (\tfrac{d}{dx} y(x))^2 - y(x) \, (\tfrac{d^2}{dx^2} y(x)) = 0$$

Although *ode* is a nonlinear ODE, it can be solved analytically for $y(x)$.

> `sol:=dsolve(ode,y(x));`

$$sol := y(x) = \frac{1}{2} \frac{(1 + \frac{1}{(e^{(x-C1)})^2 \, (e^{(-C2 - C1)})^2}) \, e^{(x - C1)} \, e^{(-C2 - C1)}}{\_C1},$$

$$y(x) = \frac{1}{2} \frac{1 + (e^{(x - C1)})^2 \, (e^{(-C2 - C1)})^2}{e^{(x - C1)} \, e^{(-C2 - C1)} \, \_C1}$$

Two solutions are generated, each involving two arbitrary constants $\_C1$ and $\_C2$. Mike selects the first one, substituting $\_C1 = a$, $\_C2 = b$ and simplifying the result. He then converts the exponential solution to a trig form and achieves further simplification with the `combine` command.

> `Y:=simplify(subs(_C1=a,_C2=b,rhs(sol[1])),symbolic);`

$$Y := \frac{1}{2} \frac{(1 + e^{(-2a\,(x+b))}) \, e^{(a\,(x+b))}}{a}$$

> `Y:=combine(convert(Y,trig),symbolic);`

$$Y := \frac{\cosh(x\,a + b\,a)}{a}$$

This form of the solution is similar to that found in standard mechanics texts. To evaluate a and b, Mike substitutes $x = 0$ into Y in *bc1* and sets the result equal to 1. The second boundary condition, *bc2*, makes use of the other end point coordinates.

> `bc1:=subs(x=0,Y)=1; bc2:=subs(x=1,Y)=2;`

Using the floating point solve command, Mike finds that he can obtain two sets of values for a and b, depending on the search range chosen for b.

> `sol2:=fsolve({bc1,bc2},{a,b},{b=0..5});`

$$sol2 := \{b = 0.3069167873,\ a = 1.052643959\}$$

> `sol3:=fsolve({bc1,bc2},{a,b},{b=-5..0});`

$$sol3 := \{a = 5.465940659,\ b = -0.4360106988\}$$

Not yet knowing which a, b values will yield the curve that generates the minimum area of revolution, he decides to examine both possibilities. He labels the two solutions *Y1* and *Y2*, and then uses the evaluation command to substitute *sol2* and *sol3* into Y.

> `Y1:=eval(Y,sol2); Y2:=eval(Y,sol3);`

$$Y1 := 0.9499888271 \cosh(1.052643959\,x + 0.3230741021)$$

$$Y2 := 0.1829511263 \cosh(5.465940659\,x - 2.383208606)$$

Mike plots *Y1* and *Y2*, the resulting curves being shown in Figure 6.10. The top curve is *Y1*, the bottom one *Y2*. Each curve will generate a different surface of revolution when it is rotated around the x-axis, but only one can correspond to the minimum area.

```
> plot({Y1,Y2},x=0..1,view=[0..1,0..2.5],tickmarks=[3,3],
> labels=["x","Y"],color=[red,blue],thickness=2);
```

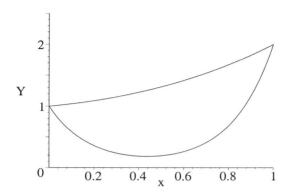

Figure 6.10: The curves $Y1$ (upper) and $Y2$ (lower).

From the figure, Mike can see that $Y1$ must be the one that produces the minimum area. To confirm this and calculate the minimum area, he introduces a functional operator for calculating the area when the form of Y is specified.

```
> A:=Y->2*Pi*int(Y*sqrt(1+diff(Y,x)^2),x=0..1.0):
```

Taking $Y1$ and $Y2$ as arguments in A, Mike finds that $Y1$ generates a surface area of revolution of about 4.16π, while $Y2$ produces a larger surface area 5.15π.

```
> A1:=A(Y1); A2:=A(Y2);
```

$$A1 := 4.157660180\,\pi$$
$$A2 := 5.149302424\,\pi$$

By now, the reader may be wondering (as Mike's students certainly will be), what this straightforward mathematical exercise has to do with wine goblets and tumblers. Replacing x with z in Y, Mike uses the `implicitplot3d` to create pictures of the surfaces of revolution for the two curves $Y1$ and $Y2$. The orientation is chosen so that the resulting pictures have suggestive shapes and various color options are included.

```
> eq:=Y->subs(x=z,Y):
> gr:=eq->implicitplot3d(eq=sqrt(x^2+y^2),x=-2..2,y=-2..2,
> z=0..1,style=patch,shading=zhue,lightmodel=light4,axes=
> none,grid=[15,15,15],numpoints=2000,orientation=[90,50]):
```

On executing the following line, vividly colored pictures will appear on the computer screen. Fig. 6.11 shows a black and white version of what is observed.

```
> gr(eq(Y1)); gr(eq(Y2));
```

6.2. CALCULUS OF VARIATIONS

Figure 6.11: The whisky tumbler (left) and the wine goblet (right).

On the left, Mike has created a flared "whisky tumbler" and on the right a "wine goblet". The top and bottom planar areas of each are the same, but the whisky tumbler has the smaller area of revolution. According to Mike, the glasses resemble a set that Vectoria and he were given as a wedding present. One little problem with Mike's whisky tumbler and wine goblet, however. If you rotate the 3-dimensional pictures on the computer screen by dragging with the mouse, you will discover that they have no bottoms!

Mike notes that the surface of revolution of, e.g., $Y2$ could also be plotted by loading the Student[CalculusI] package and using the SurfaceOfRevolution command with the option output=plot. A 3-dimensional picture of the "wine goblet" is produced, the goblet lying on its side. You can drag on the goblet with the mouse to orient it vertically.

```
> with(Student[Calculus1]):
> SurfaceOfRevolution(Y2,x=0..1,output=plot,style=patch,
> shading=zhue,lightmodel=light4,axes=none);
```

If no option is specified, the numerical value of the surface area of revolution

```
> SurfaceOfRevolution(Y2,x=0..1);
                16.17701067
```

is produced, in agreement with a floating point evaluation of $A2$.

6.2.2 Suzy Spider's Speedy Strand

Little Miss Muffet Sat on a tuffet, Eating some curds and whey;
There came a great spider, And sat down beside her,
And frightened Miss Muffet away.
Mother Goose Nursery Rhymes

Ever since this incident, Miss Muffet, having regained her confidence, has been rather hard on any spider that she encounters. To avoid Miss Muffet's wrath,

Suzy spider (mass m) wishes to create a speedy escape route consisting of a single strand connecting a point P on her web to a hole P_1 in the fence 6 m horizontal distance away and 2 m below P. Assuming that Suzy slides down the strand from rest and frictional effects are neglected, what shape should the strand have for Suzy to reach the hole in the quickest time? How many seconds would it take? What distance would she have traveled? How much faster would this escape route be than for a straight strand from P to the hole, even though the distance for the straight route is shorter? Try some other strand shapes, calculating the distances traveled and the time to reach the hole. Plot the shape of the speediest path as well as the shapes of the other slower routes.

Choosing our origin at P and using energy conservation, Suzy's speed v when she falls through a vertical distance y is given by $(1/2)\,m\,v^2 = m\,g\,y$, so $v = \sqrt{2\,g\,y}$. Since the element of arclength is $d\ell \equiv \sqrt{(dx)^2 + (dy)^2}$, the time for her to slide from P to P_1 ($x_1=6$, $y_1=2$) is

$$T = \int_P^{P_1} d\ell/v = \int_0^{x_1} \sqrt{(1+(dy/dx)^2)/(2\,g\,y)}\,dx.$$

The constant $\sqrt{2g}$ will cancel out of the Euler–Lagrange equation, so we can take $F = \sqrt{1+(dy(x)/dx)^2}/\sqrt{y(x)}$ which is now entered after loading the plots and VariationalCalculus packages.

> restart: with(plots): with(VariationalCalculus):
> F:=sqrt(1+diff(y(x),x)^2)/sqrt(y(x));

$$F := \frac{\sqrt{1+(\frac{d}{dx}\,\mathrm{y}(x))^2}}{\sqrt{\mathrm{y}(x)}}$$

The EulerLagrange command is applied to F and the first integral selected in *eq2*. The integration constant is K_1.

> eq:=EulerLagrange(F,x,y(x));
> eq2:=select(has,eq,K[1])[1];

$$eq2 := \frac{\sqrt{1+(\frac{d}{dx}\,\mathrm{y}(x))^2}}{\sqrt{\mathrm{y}(x)}} - \frac{(\frac{d}{dx}\,\mathrm{y}(x))^2}{\sqrt{1+(\frac{d}{dx}\,\mathrm{y}(x))^2}\,\sqrt{\mathrm{y}(x)}} = K_1$$

Solving *eq2* for $dy(x)/dx$ yields two answers in *eq3*, involving positive and negative square roots. The positive square root is selected in *eq4* and the substitution $y(x) = y$ is made.

> eq3:=solve(eq2,diff(y(x),x));

$$eq3 := \frac{\sqrt{-\mathrm{y}(x)\,(K_1{}^2\,\mathrm{y}(x)-1)}}{\mathrm{y}(x)\,K_1},\ -\frac{\sqrt{-\mathrm{y}(x)\,(K_1{}^2\,\mathrm{y}(x)-1)}}{\mathrm{y}(x)\,K_1}$$

> eq4:=subs(y(x)=y,eq3[1]);

6.2. CALCULUS OF VARIATIONS

$$eq4 := \frac{\sqrt{-y(K_1^2 y - 1)}}{y K_1}$$

To obtain an analytic form for Suzy Spider's path, we assume that y depends on a parameter θ through the relation $y = (\sin(\theta/2)/K_1)^2$. This expression will be automatically substituted into *eq4*.

```
> y:=(sin(theta/2)/K[1])^2;
```

$$y := \frac{\sin(\frac{\theta}{2})^2}{K_1^2}$$

Since $dy/dx = eq4$, then $x = \int (1/eq4)\, dy = \int (1/eq4)\, (dy/d\theta)\, d\theta$. We calculate the latter *integrand* in the following command line and then simplify the result in the subsequent two command lines.

```
> integrand:=(1/eq4)*diff(y,theta);
```

$$integrand := \frac{\sin(\frac{\theta}{2})^3 \cos(\frac{\theta}{2})}{K_1^3 \sqrt{-\frac{\sin(\frac{\theta}{2})^2 (\sin(\frac{\theta}{2})^2 - 1)}{K_1^2}}}$$

```
> integrand:=radsimp(integrand,symbolic);
```

$$integrand := \frac{\sin(\frac{\theta}{2})^2 \cos(\frac{\theta}{2})}{K_1^2 \sqrt{-\sin(\frac{\theta}{2})^2 + 1}}$$

```
> integrand:=combine(simplify(integrand,symbolic));
```

$$integrand := \frac{\frac{1}{2} - \frac{1}{2}\cos(\theta)}{K_1^2}$$

The dependence of x on θ is obtained by integrating *integrand* with respect to θ. The arbitrary constant is set equal to zero so that $x = 0$ when $\theta = 0$.

```
> x:=int(integrand,theta);
```

$$x := \frac{\frac{\theta}{2} - \frac{1}{2}\sin(\theta)}{K_1^2}$$

The coordinates $x_1 = 6$, $y_1 = 2$ of the hole are entered, and the boundary conditions $x = x1$ and $y = y1$ imposed.

```
> x1:=6: y1:=2:
> bc1:=x=x1; bc2:=y=y1;
```

$$bc1 := \frac{\frac{\theta}{2} - \frac{1}{2}\sin(\theta)}{K_1^2} = 6$$

$$bc2 := \frac{\sin(\frac{\theta}{2})^2}{K_1^2} = 2$$

Then $bc1$ and $bc2$ are numerically solved for K_1 and θ and the parameter values assigned. We then explicitly write out K_1 and θ, assigning the name $\theta 1$ to the latter. $\theta 1$ is the value of θ corresponding to $x = x_1$, $y = y_1$.

> `parameters:=fsolve({bc1,bc2},{K[1],theta});`

$$parameters := \{\theta = 4.051628024, K_1 = -0.6351609686\}$$

> `assign(parameters):`

> `K[1]:=K[1]; theta1:=theta;`

$$K_1 := -0.6351609686$$
$$\theta 1 := 4.051628024$$

To free up the parameter θ, it is now unassigned. The equation of the path that minimizes the time of descent is given in parametric form by the output of the subsequent command line.

> `unassign('theta'):`

> `x:=x; y:=y;`

$$x := 1.239374053\,\theta - 1.239374053\sin(\theta)$$

$$y := 2.478748106\sin(\frac{\theta}{2})^2$$

We take $g = 9.8$ m/s^2 which is needed to calculate the time of descent. In terms of θ, this time is given by $T = \int_0^{\theta 1} d\theta (dx/d\theta)\sqrt{1+s^2}/\sqrt{2\,g\,y}$, where the slope s is expressed in the form $s = (dy/d\theta)/(dx/d\theta)$.

> `g:=9.8:`

> `s:=combine(diff(y,theta)/diff(x,theta));`

$$s := -\frac{1.239374053\sin(\theta)}{-1.239374053 + 1.239374053\cos(\theta)}$$

> `T:=evalf(int(sqrt(1+s^2)*diff(x,theta)/sqrt(2*g*y)`
> `theta=0..theta1));`

$$T := 1.440846542$$

Suzy spider would take about 1.44 seconds to slide from her web to the hole. The distance d that she would travel along the minimum time path is given by $d = \int_0^{\theta 1} d\theta (dx/d\theta) \sqrt{1+s^2}$.

> `d:=evalf(int(sqrt(1+s^2)*diff(x,theta),theta=0..theta1));`

$$d := 7.136207722$$

Suzy would travel about 7.14 m. To illustrate that alternate routes will produce a longer time even though the distances are shorter, we consider the straight-line path $Y1 = (y_1/x_1)\,X1$ and the curved path $Y2 = (y_1/\sqrt{x_1})\,\sqrt{X2}$.

> `Y1:=(y1/x1)*X1; Y2:=(y1/sqrt(x1))*sqrt(X2);`

6.2. CALCULUS OF VARIATIONS

Functional expressions TT and dd are created to determine the time of descent and the distance traveled.

```
> TT:=(X,Y)->evalf(int(sqrt(1+diff(Y,X)^2)/sqrt(2*g*Y),X=0..x1)):
> dd:=(X,Y)->evalf(int(sqrt(1+diff(Y,X)^2),X=0..x1)):
```

Substituting the arguments X1, Y1 and X2, Y2 into TT yields a time $T1 = 2.02$ seconds to descend the straight-line path and a time $T2 = 1.55$ seconds to descend the curved path. So both times $T1$ and $T2$ are longer than T.

```
>   T1:=TT(X1,Y1); T2:=TT(X2,Y2);
```
$$T1 := 2.020305088$$
$$T2 := 1.546973630$$

Calculating the distances traveled yields $D1 = 6.32$ m for the straight-line path and $D2 = 6.5$ m for the curved path. Both of these alternate paths are shorter than the minimum time path.

```
>   D1:=dd(X1,Y1); D2:=dd(X2,Y2);
```
$$D1 := 6.324555320$$
$$D2 := 6.498059177$$

Finally we plot the three paths. In gr1, the minimum time path is plotted as a thick blue line. In gr2, the straight line is plotted as a red curve, while in gr3 the curved path is plotted as a green line. The paths are superimposed in the same figure with the display command.

```
>   gr1:=plot([x,-y,theta=0..theta1],thickness=2,color=blue):
>   gr2:=plot([X1,-Y1,X1=0..x1],thickness=2,color=red):
>   gr3:=plot([X2,-Y2,X2=0..x1],thickness=2,color=green):
>   display({gr1,gr2,gr3},scaling=constrained,labels=["x","y"]);
```

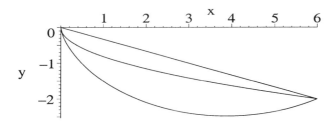

Figure 6.12: The bottom path will produce the fastest time of descent.

The bottom curve corresponds to the path that minimizes the time of descent. The other curved path lies between this path and the straight-line path.

6.3 Some Hamiltonian Examples

Consider a system consisting of a single particle[1] with three degrees of freedom, i.e., three independent coordinates are required to describe the motion. If the Lagrangian $L = L(q_i, \dot{q}_i, t)$, with q_i the coordinates, $\dot{q}_i$ the "generalized" velocities, and t the time, a Hamiltonian function $H(q_i, p_i, t)$ can be defined quite generally through the so-called "Legendre transformation" as

$$H = \sum_{i=1}^{3} p_i \dot{q}_i - L \qquad (6.5)$$

where the p_i are generalized momenta defined by $p_i = \partial L/\partial \dot{q}_i = p_i(q_1, ..., \dot{q}_1, ...)$. If the system is conservative (the potential energy is velocity independent) and does not contain t explicitly in the Hamiltonian or in its coordinates, then [MT95] the Hamiltonian is a constant of the motion ($dH/dt = 0$) and equal to the total energy (kinetic plus potential) of the system, i.e, $H = T + V$. Further, the system's equations of motion are given by

$$\dot{q}_i = \partial H/\partial p_i, \qquad \dot{p}_i = -\partial H/\partial q_i. \qquad (6.6)$$

These are Hamilton's famous equations. As with the Lagrangian recipes, we shall show examples which mimick hand calculations and examples which make use of the `hamilton_eqs` command found in the `DEtools` library package.

6.3.1 Turning a Bar of Soap into a Conical Pendulum

...conic sections were studied for 1800 years merely as an abstract science...and then...they were found to be the...key with which to attain the knowledge of the most important laws of nature.
Alfred North Whitehead, English philosopher and mathematician (1861–1947)

Jennifer will use the Hamiltonian formulation to derive the equation of motion for a small bar of soap (mass m) which is acted upon by a uniform gravitational field and is free to slide on the inside surface of a slippery conical bowl. This is often referred to as the "conical pendulum" problem. Then she will numerically solve the equation and animate the motion of the soap. Referring to Fig. 6.13, Jennifer enters the Cartesian coordinates x, y, z of the soap at time t. The soap is constrained to move on the conical surface $z = a\,r$, with $a > 0$.

```
> restart: with(plots):
> x:=r(t)*cos(theta(t)); y:=r(t)*sin(theta(t)); z:=a*r(t);
```
$$x := \mathrm{r}(t) \cos(\theta(t))$$
$$y := \mathrm{r}(t) \sin(\theta(t))$$
$$z := a\,\mathrm{r}(t)$$

---
[1] The extension to a system of N particles, with $N > 1$, is straightforward.

6.3. SOME HAMILTONIAN EXAMPLES

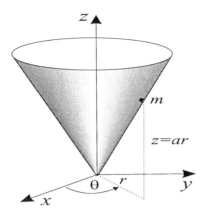

Figure 6.13: Bar of soap constrained to move on a conical surface.

The Cartesian components ($px = m\, dx/dt$, etc.) of the soap's momentum are calculated and the kinetic energy ($KE = (px^2 + py^2 + pz^2)/2m$) evaluated.

> `px:=m*diff(x,t); py:=m*diff(y,t); pz:=m*diff(z,t);`

$$px := m\,((\tfrac{d}{dt}\mathrm{r}(t))\cos(\theta(t)) - \mathrm{r}(t)\sin(\theta(t))\,(\tfrac{d}{dt}\theta(t)))$$

$$py := m\,((\tfrac{d}{dt}\mathrm{r}(t))\sin(\theta(t)) + \mathrm{r}(t)\cos(\theta(t))\,(\tfrac{d}{dt}\theta(t)))$$

$$pz := m\,a\,(\tfrac{d}{dt}\mathrm{r}(t))$$

> `KE:=simplify((px^2+py^2+pz^2)/(2*m));`

$$KE := \tfrac{1}{2} m\,(\mathrm{r}(t)^2\,(\tfrac{d}{dt}\theta(t))^2 + (\tfrac{d}{dt}\mathrm{r}(t))^2 + a^2\,(\tfrac{d}{dt}\mathrm{r}(t))^2)$$

Entering the potential energy (PE) of the bar of soap, Jennifer then calculates the *Lagrangian* and *Hamiltonian* functions.

> `PE:=m*g*z;`

$$PE := m\,g\,a\,\mathrm{r}(t)$$

> `Lagrangian:=KE-PE;`

$$Lagrangian := \tfrac{1}{2} m\,(\mathrm{r}(t)^2\,(\tfrac{d}{dt}\theta(t))^2 + (\tfrac{d}{dt}\mathrm{r}(t))^2 + a^2\,(\tfrac{d}{dt}\mathrm{r}(t))^2) - m\,g\,a\,\mathrm{r}(t)$$

> `Hamiltonian:=KE+PE;`

$$Hamiltonian := \tfrac{1}{2} m\,(\mathrm{r}(t)^2\,(\tfrac{d}{dt}\theta(t))^2 + (\tfrac{d}{dt}\mathrm{r}(t))^2 + a^2\,(\tfrac{d}{dt}\mathrm{r}(t))^2) + m\,g\,a\,\mathrm{r}(t)$$

She introduces the relations $dr(t)/dt = w1$, $d\theta(t)/dt = w2$, $r(t) = R$ in *rel1*. In *rel2*, *w1*, *w2*, and R are replaced with their original time-dependent forms. She then substitutes *rel1* into the Lagrangian and Hamiltonian.

> `rel1:=diff(r(t),t)=w1,diff(theta(t),t)=w2,r(t)=R:`
> `rel2:=w1=diff(r(t),t),w2=diff(theta(t),t),R=r(t):`

> L:=subs(rel1,Lagrangian); H:=subs(rel1,Hamiltonian);

$$L := \frac{m\left(R^2\,w2^2 + w1^2 + a^2\,w1^2\right)}{2} - m\,g\,a\,R$$

$$H := \frac{m\left(R^2\,w2^2 + w1^2 + a^2\,w1^2\right)}{2}) + m\,g\,a\,R$$

The radial generalized momentum is given by $p_r = \partial L/\partial \dot{r}$, or here by $pr = \partial L/\partial w1$. Similarly, the angular generalized momentum is given by $p_\theta = \partial L/\partial \dot{\theta}$ or $ptheta = \partial L/\partial w2$.

> eq1a:=pr=diff(L,w1);

$$eq1a := pr = \frac{m\left(2\,w1 + 2\,a^2\,w1\right)}{2}$$

> eq1b:=ptheta=diff(L,w2);

$$eq1b := ptheta = m\,R^2\,w2$$

Equations $eq1a$ and $eq1b$ are solved for $w1$ and $w2$, the results being labeled $w1b$ and $w2b$. These forms are then substituted into the Hamiltonian H, which is then expressed in terms of the angular and radial generalized momenta.

> w1b:=solve(eq1a,w1); w2b:=solve(eq1b,w2);

$$w1b := \frac{pr}{m\left(1 + a^2\right)}$$

$$w2b := \frac{ptheta}{m\,R^2}$$

> H:=expand(subs({w1=w1b,w2=w2b},H));

$$H := \frac{ptheta^2}{2\,R^2\,m} + \frac{pr^2}{2\,m\left(1 + a^2\right)^2} + \frac{a^2\,pr^2}{2\,m\left(1 + a^2\right)^2} + m\,g\,a\,R$$

Now, Jennifer applies Hamilton's equations of motion. In $eq2a$, $eq2b$, $eq2c$, and $eq2d$, she calculates $\dot{p}_r = -\partial H/\partial r$, $\dot{p}_\theta = -\partial H/\partial \theta$, $\dot{r} = \partial H/\partial p_r$, and $\dot{\theta} = \partial H/\partial p_\theta$.

> eq2a:=prdot=-diff(H,R);

$$eq2a := prdot = \frac{ptheta^2}{R^3\,m} - m\,g\,a$$

> eq2b:=pthetadot=-diff(H,theta);

$$eq2b := pthetadot = 0$$

> eq2c:=rdot=simplify(diff(H,pr));

$$eq2c := rdot = \frac{pr}{m\left(1 + a^2\right)}$$

> eq2d:=thetadot=diff(H,ptheta);

$$eq2d := thetadot = \frac{ptheta}{m\,R^2}$$

From $eq2b$, she notes that the angular momentum p_θ about the z-axis is a constant of the motion. Further, the results in $eq2c$ and $eq2d$ are just duplicates of $w1b$ and $w2b$. Jennifer substitutes $rel2$ into the rhs of $eq1a$ and differentiates

6.3. SOME HAMILTONIAN EXAMPLES

the result with respect to t, thus obtaining and expression for $\dot{p}_r$. She also substitutes *rel2* into the rhs of *eq1b* to obtain the time-dependent form of p_θ.

```
> prdot:=diff(subs(rel2,rhs(eq1a)),t);
```
$$prdot := \frac{1}{2}m\,(2\,(\frac{d^2}{dt^2}\,\mathrm{r}(t)) + 2\,a^2\,(\frac{d^2}{dt^2}\,\mathrm{r}(t)))$$

```
> ptheta:=subs(rel2,rhs(eq1b));
```
$$ptheta := m\,\mathrm{r}(t)^2\,(\frac{d}{dt}\,\theta(t))$$

Substituting *rel2* into *eq2a* divided by m, and expanding, yields *Eq2a*, one of the equations of motion of the soap. The second equation of motion is obtained in *Eq2b* by setting p_θ/m equal to its initial value.

```
> Eq2a:=expand(subs(rel2,eq2a)/m);
```
$$Eq2a := (\frac{d^2}{dt^2}\,\mathrm{r}(t)) + a^2\,(\frac{d^2}{dt^2}\,\mathrm{r}(t)) = \mathrm{r}(t)\,(\frac{d}{dt}\,\theta(t))^2 - g\,a$$

```
> Eq2b:=ptheta/m=r(0)^2*D(theta)(0);
```
$$Eq2b := \mathrm{r}(t)^2\,(\frac{d}{dt}\,\theta(t)) = \mathrm{r}(0)^2\,\mathrm{D}(\theta)(0)$$

To numerically solve the equations of motion, Jennifer takes $a = 1$, $g = 9.8 \text{m/s}^2$, a total time $T = 25$ s, and initial values $r(0) = 1$ m, $\theta(0) = \pi/4$ rad, $\dot{r}(0) = 0$ m/s, and, in *de2*, $\dot{\theta}(0) = 1$. The coupled ODEs *de1* and *de2* are numerically solved with the dsolve command. The option maxfun=0 is included so that the number of function evaluations is not limited to Maple's default number.

```
> a:=1: g:=9.8: T:=25:
> ic:=r(0)=1,theta(0)=Pi/4,D(r)(0)=0;
> de1:=Eq2a;
```
$$de1 := 2\,(\frac{d^2}{dt^2}\,\mathrm{r}(t)) = \mathrm{r}(t)\,(\frac{d}{dt}\,\theta(t))^2 - 9.8$$

```
> de2:=subs(ic,{D(theta)(0)=1},Eq2b);
```
$$de2 := \mathrm{r}(t)^2\,(\frac{d}{dt}\,\theta(t)) = 1$$

```
> sol:=dsolve({de1,de2,ic},{r(t),theta(t)},type=numeric,
> output=listprocedure,maxfun=0):
```
Jennifer creates a graph gr1 of the soap's motion with the odeplot command

```
> gr1:=odeplot(sol,[r(t)*cos(theta(t)),r(t)*sin(theta(t)),
> a*r(t)],t=0..T,style=line,numpoints=1000,axes=boxed,
> orientation=[-40,50],labels=["x","y","z"],tickmarks=[3,3,2]):
```
and then displays it in Figure 6.14. The 3-dimensional plot may be rotated.

```
> display(gr1);
```
Jennifer now animates the motion of the soap, taking $N = 200$ frames and a time step size $T/N = 25/200 = 1/8$ s.

```
> N:=200: step:=T/N;
```

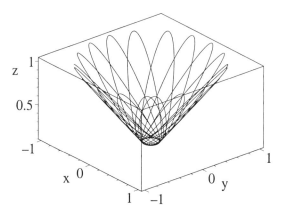

Figure 6.14: Trajectory of the bar of soap in the conical bowl.

She uses the numerical solution to evaluate $r(t)$ and $\theta(t)$ at arbitrary time t and then forms the Cartesian coordinates for the bar of soap.

```
>   R:=eval(r(t),sol): Theta:=eval(theta(t),sol):
>   X:=R(t)*cos(Theta(t)): Y:=R(t)*sin(Theta(t)): Z:=a*R(t):
```

The following do loop animates the motion of the bar of soap, representing it as a black circle, and superimposing its position at each time step on the graph of the complete trajectory. On executing the display command with the insequence=true option, the reader will be able to see the bar of soap move in the conical bowl, following the path shown in Fig. 6.14.

```
>   for i from 0 to N do
>   t:=step*i:
>   gr2||i:=pointplot3d([X(t),Y(t),Z(t)],symbol=circle,
>   symbolsize=16,color=black):
>   pl||i:=display({gr1,gr2||i}):
>   end do:
>   display(seq(pl||i,i=0..N),insequence=true);
```

Jennifer advises the reader to try other parameter values and initial conditions to see how the trajectory can be altered.

6.3.2 Lots of Lissajous Figures

A mathematician, like a painter or poet, is a maker of patterns.
If his patterns are more permanent than theirs,
it is because they are made with ideas.
G. H. Hardy, English mathematician (1877–1947)

In this recipe, we demonstrate how Lissajous figures may be easily created

6.3. SOME HAMILTONIAN EXAMPLES

for the 2-dimensional non-isotropic harmonic oscillator. We load the DEtools package because it will enable us to use the hamilton_eqs command to generate Hamilton's equations.

>   restart: with(plots): with(DEtools):

A functional operator is introduced for creating the non-isotropic potential function V with (in general) different frequencies $\omega 1$ and $\omega 2$.

>   V:=(omega1,omega2)->omega1^2*x^2/2+omega2^2*y^2/2;

$$V := (\omega 1, \omega 2) \rightarrow \frac{1}{2}\omega 1^2\, x^2 + \frac{1}{2}\omega 2^2\, y^2$$

Lissajous figures, which correspond to closed trajectories in x-y space, are produced if $\omega 2 = (n2/n1)\,\omega 1$, where $n1$ and $n2$ are integers. The patterns that are traced out depend on the values of $n1$ and $n2$. If $\omega 2/\omega 1$ is not a ratio of integers, i.e., the frequencies are not commensurate, the path will not be closed. For the sake of definiteness, we shall take $\omega 1 = 6$ and let $\omega 2$ vary. We shall consider one incommensurate case, choosing $\omega 2[0] = \sqrt{15}$ and $N = 10$ other frequencies which produce various Lissajous patterns. We use a do loop to generate the frequencies $\omega 2 = 1, 2, ..., N$.

>   N:=10: omega1:=6: omega2[0]:=sqrt(15):
>   for i from 1 to N do
>   omega2[i]:=i;
>   end do:

A 2-dimensional contour plot of the potential function can be created. For example, let's take $\omega 2=9$ (so $\omega 2/\omega 1 = 3/2$) and produce contours corresponding to $V = 0, 1, 2, ..., 10$. The plotting grid is taken to be 50 by 50 and the scaling is constrained. A plot similar to that shown in Figure 6.15 is produced

>   contourplot(V(omega1,omega2[9]),x=-1..1,y=-1..1,
>   contours=[seq(i,i=0..10)],grid=[50,50],scaling=constrained);

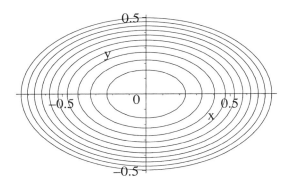

Figure 6.15: 2-dimensional contour plot of $V(x, y)$ for $\omega 1 = 6$, $\omega 2 = 9$.

A 3-dimensional contour plot of $V(x, y)$ may also be produced (not displayed here) by using the `plot3d` command and choosing various stylistic options.

```
> plot3d(V(omega1,omega2[9]),x=-1..1,y=-1..1,contours=
> [seq(i,i=0..10)],view=[-1..1,-1..1,0..10],orientation=
> [32,60],axes=framed,shading=XY,style=PATCHCONTOUR,
> labels=["x","y","V"],tickmarks=[2,2,2]);
```

The force components Fx and Fy are calculated by differentiating V with respect to x and y. Hooke's law prevails here, with unequal spring constants.

```
> Fx:=-diff(V(omega1,omega2[9]),x);
> Fy:=-diff(V(omega1,omega2[9]),y);
```

$$Fx := -36\,x$$
$$Fy := -81\,y$$

Setting the components of the force equal to zero and solving for x and y yields $x=y=0$. This is the minimum of the potential energy function, which is quite obvious from Fig. 6.15, and even more so from the 3-dimensional contour plot.

```
> sol:=solve({Fx=0,Fy=0},{x,y});
```

$$sol := \{x = 0, y = 0\}$$

The kinetic energy of a particle of unit mass moving in the potential V is $T = p1^2/2 + p2^2/2$, where $p1$ and $p2$ are the x- and y-components of momentum. Maple uses $p1$, $p2$, etc., for the momentum components and $q1$, $q2$, etc., for the coordinates in its `hamilton_eqs` command.

```
> T:=p1^2/2+p2^2/2;
```

$$T := \frac{p1^2}{2} + \frac{p2^2}{2}$$

For initial conditions we take $x0 \equiv q1(0) = 0.5$, $y0 \equiv q2(0) = 0$, $py0 \equiv p2(0) = -0.05$, and choose a total energy $E0 = 15$. Since the total energy is conserved, we have $px0^2/2 + py0^2/2 + \omega1^2\,x0^2/2 + \omega2^2\,y0^2/2 = E0$. We could solve for the initial x-component, $px0 \equiv p1(0)$, of the momentum but this is unnecessary as Maple will do the calculation for us with the `generate_ic` command.

```
> x0:=0.5: y0:=0: py0:=-0.05: E0:=15:
```

For each i value from 0 to $N = 10$, the following do loop will calculate the potential energy, produce the Hamiltonian, generate Hamilton's equations, generate the initial conditions (determine $p1(0)$), numerically solve Hamilton's equations, and produce a 3-dimensional plot of the trajectory.

```
> for i from 0 to N do
> V||i:=subs({x=q1,y=q2},V(omega1,omega2[i])):   #potential
> H||i:=T+V||i; #Hamiltonian
> hamilton_eqs(H||i); #Hamilton's equations
```

In the following command line to generate the initial condition(s), the first argument is the ith Hamiltonian, the second is the set of initial $p2$, $q2$, and $q1$ values with the energy fixed at $E0$, and the third argument is the number N of

6.3. SOME HAMILTONIAN EXAMPLES

initial conditions ($N=1$ here) to be generated. More generally, one can provide ranges for the initial ps and qs rather than specific values, and ask Maple to calculate a list of N (e.g., 3) different initial conditions for a given $E0$. See Maple's Help for a more complete discussion of this aspect.

```
> ic||i:=generate_ic(H||i,{t=0,p2=py0,q2=y0,q1=x0,energy=E0},1);
```

In the **poincare** plotting command below, the first argument is the ith Hamiltonian. The second argument is the time interval (say $t = 0...50$). The third is the ith initial condition. The fourth is the time step size (.05) in Maple's numerical solving algorithm. The fifth is the number of iterations (3) in the algorithm. The sixth is the scene command which here will produce a plot of $q2$ vs. $p2$ vs. $q1$ over the ranges indicated. The last number (3) indicates that a 3-dimensional plot is desired.

```
>   display(poincare(H||i,t=0..50,ic||i,stepsize=.05,
>   iterations=3,scene=[q2=-0.1..0.1,p2=-0.1..0.1,q1=-1..1],3),
>   orientation=[-30,75],tickmarks=[3,3,3]);
>   end do;
```

For each i value, an output similar to the following output for $i = 0$ will be generated. The form of the potential is displayed, with $q1 \equiv x$ and $q2 \equiv y$.

$$V0 := 18\, q1^2 + \frac{15\, q2^2}{2}$$

Next, the Hamiltonian is displayed with the frequencies inserted ($\omega1^2/2 = 6^2/2 = 18$ and $\omega2^2/2 = (\sqrt{15})^2/2 = 15/2$ here).

$$H0 := \frac{p1^2}{2} + \frac{p2^2}{2} + 18\, q1^2 + \frac{15\, q2^2}{2}$$

Hamilton's four equations are calculated and displayed. The momentum and coordinate variables are also given as a list.

$$[\tfrac{d}{dt}\, \text{p1}(t) = -36\, \text{q1}(t),\ \tfrac{d}{dt}\, \text{p2}(t) = -15\, \text{q2}(t),\ \tfrac{d}{dt}\, \text{q1}(t) = \text{p1}(t),\ \tfrac{d}{dt}\, \text{q2}(t) = \text{p2}(t)],$$
$$[\text{p1}(t),\ \text{p2}(t),\ \text{q1}(t),\ \text{q2}(t)]$$

The initial conditions are displayed as well as the total energy. We see that here the calculated value of $p1(0)$ (or $px(0)$) is about 4.58. The maximum percentage deviation of the energy from the input value caused by the numerical scheme is given as well as the computer time to produce a plot of the trajectory.

$$ic0 := \{[0.,\ 4.582302914,\ -0.05,\ 0.5,\ 0.]\}$$

$H = 15.000000,\quad \textit{Initial conditions} :,\ t = 0.,$
$p1 = 4.582302914,\ p2 = -0.05,\ q1 = 0.5,\ q2 = 0.$
$\textit{Maximum H deviation} :\ .4156900000e - 2\ \%$

$\textit{Time consumed} :\ 1\ seconds$

Finally, for each i value the do loop produces a 3-dimensional plot of the trajectory such as the one shown on the left of Figure 6.16 for $i = 0$, i.e., for $\omega 1 = 6$ and $\omega 2 = \sqrt{15}$. The 3-dimensional plot[2] may be rotated on the computer screen. The trajectory never closes because the frequencies are incommensurate, but the trajectory remains in a fixed region (the surface of the elliptical cylinder). The plot on the right of Figure 6.16 shows a closed trajectory corresponding to the commensurate frequencies $\omega 1 = 6$ and $\omega 2 = 9$.

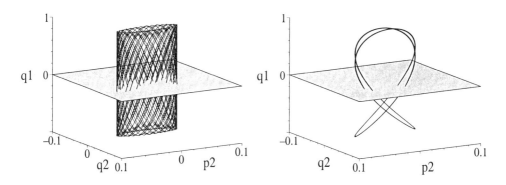

Figure 6.16: Lissajous figures. Left: $\omega 1 = 6$, $\omega 2 = \sqrt{15}$; Right: $\omega 1 = 6$, $\omega 2 = 9$.

The Lissajous figures for $\omega 2 = 1...8$ and $\omega 2 = 10$ may be viewed by executing the recipe. By varying $\omega 1$ and $\omega 2$, other patterns can be created. See if you can find some interesting ones. You may have to change the plot ranges in the scene command, particularly if you wish to view only the *q1-q2* plane.

6.3.3 The KAM Torus and Hamiltonian Chaos

We adore chaos because we love to produce order.
Maurits C. Escher, Dutch artist (1898–1972)

In the recipe on Lissajous figures, we could have obtained the same results without introducing the Hamiltonian formulation. The reason for choosing this approach is that we can now introduce the reader to a concept from the forefront of nonlinear dynamics. Originally motivated to model the motion of a star inside a galaxy, Hénon and Heiles[HH64], introduced a simple Hamiltonian to describe the motion of a particle in an anharmonic potential. The potential V that they chose to study is a generalization of that in the last recipe, cubic terms in the coordinates being added to the harmonic potential. After loading some necessary library packages, we enter the form of $V(x, y)$.

> `restart: with(plots): with(VectorCalculus): with(DEtools):`

---

[2] In standard mechanics texts, one usually sees only a 2-dimensional plot in *q1-q2* space.

6.3. SOME HAMILTONIAN EXAMPLES

```
> V:=omega1^2*x^2/2+omega2^2*y^2/2+a*x^2*y+b*y^3;
```

$$V := \frac{\omega 1^2 \, x^2}{2} + \frac{\omega 2^2 \, y^2}{2} + a\, x^2\, y + b\, y^3$$

For the sake of definiteness, let us take $\omega 1 = 1$, $\omega 2 = 1$, $a = 1$, and $b = -/4$. The potential then is as follows.

```
> omega1:=1: omega2:=1: a:=1: b:=-1/4: V:=V;
```

$$V := \frac{1}{2} x^2 + \frac{1}{2} y^2 + x^2 y - \frac{1}{4} y^3$$

Without the cubic terms present, the potential has the shape of a symmetric bowl and a particle moving in this bowl traces out a closed path (a Lissajous figure). Let's now look at the shape of the potential when the cubic terms are included. We create two- and three-dimensional contour plots with the `contourplot` and `plot3d` commands. In each case, the potential contours are taken in steps of 0.02, from $V = 0$ to $V = 0.36$. The x and y plotting ranges are taken from -2 to $+2$.

```
> contourplot(V,x=-2..2,y=-2..2,contours=
> [seq(0.02*i,i=0..18)],grid=[50,50],scaling=constrained);
> plot3d(V,x=-2..2,y=-2..2,contours=[seq(0.02*i,i=0..18)],
> view=[-2..2,-2..2,0..0.36],orientation=[-32,40],axes=
> framed,shading=XY,style=PATCHCONTOUR,labels=["x","y","V"]);
```

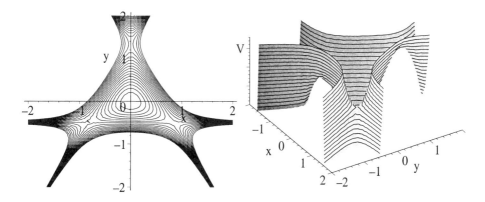

Figure 6.17: Two- and three-dimensional contour plots of $V(x,y)$.

The resulting pictures are shown in Figure 6.17. From Figure 6.17, we see that the potential has a central minimum surrounded by three saddle points. The 3-dimensional plot may be rotated on the computer screen to view each saddle point or to look down at the central minimum. If one does this, one finds that the elevations of the saddle points are not the same, two of the saddle points having a much lower value of V than the third.

The elevations of the saddle points may be found by noting that at these points, as well as at the central minimum, the force on the particle must vanish. We can calculate the force components using the Gradient operator. Because of the cubic terms in the potential, the force F is no longer given by Hooke's force law. Setting the components of F equal to zero, we can solve them for x and y, the coordinates of the minimum and the saddle points.

> F:=-Gradient(V,[x,y]);

$$F := (-x - 2xy)\bar{e}_x + (-y - x^2 + \frac{3}{4}y^2)\bar{e}_y$$

> sol:=solve({F[1]=0,F[2]=0},{x,y});

$$sol := \{x = 0, y = 0\}, \{x = 0, y = \frac{4}{3}\},$$
$$\{x = \frac{1}{4}\text{RootOf}(-11 + \_Z^2, label = \_L3), y = \frac{-1}{2}\}$$

Examining the output of *sol* and referring to the above figure, $x = 0$, $y = 0$ is the location of the central minimum. One of the saddle points must therefore be at $x = 0$, $y = 4/3$. The other pair of saddle points have a y-coordinate, $y = -1/2$, but the x coordinates cannot be analytically determined. We ask Maple to give all values of the third answer in *sol* and numerically evaluate the x and y values.

> sol3:=evalf(allvalues(sol[3]));

The x, y coordinates of the bottom (b) of the central minimum and the three saddle points ($s1$, $s2$, $s3$) are now written out.

> b:=sol[1]; s1:=evalf(sol[2]); s2:=sol3[1]; s3:=sol3[2];

$$b := \{x = 0, y = 0\}$$
$$s1 := \{x = 0., y = 1.333333333\}$$
$$s2 := \{y = -0.5000000000, x = 0.8291561975\}$$
$$s3 := \{y = -0.5000000000, x = -0.8291561975\}$$

The corresponding elevations (V values) at the above points are now evaluated.

> Vb:=subs(b,V); Vs1:=subs(s1,V); Vs2:=subs(s2,V);
> Vs3:=subs(s3,V);

$$Vb := 0$$
$$Vs1 := 0.2962962963$$
$$Vs2 := 0.1562500000$$
$$Vs3 := 0.1562500000$$

For two of the saddle points $V \approx 0.16$, while for the third $V \approx 0.30$. If the particle has a total energy greater than that at the lowest saddle points, clearly it is possible for the particle to escape to infinity and its orbit to become unbounded. In the harmonic case (set $a = b = 0$ in V), the particle could not escape from the potential well, but with the inclusion of the cubic terms this is no longer the case. Further, the presence of the anharmonic terms will increasingly affect

6.3. SOME HAMILTONIAN EXAMPLES

the trajectory of the particle as its total energy is increased from very small V to values approaching those at the lower saddle points. To investigate what happens, we can use the same approach as in the Lissajous figures recipe. We enter the kinetic energy T and re-express the potential energy in terms of the coordinates $q1 \equiv x$ and $q2 \equiv y$. We now have a four-dimensional "phase space", viz., $q1$ vs. $q2$ vs. $p1$ vs. $p2$. In order to make a three-dimensional plot, we will have to hold one of the variables fixed. Traditionally, one plots $q2$ vs. $p2$ vs $q1$.

> `T:=p1^2/2+p2^2/2: V:=subs({x=q1,y=q2},V):`

The Hamiltonian is now entered. Since no frictional effects have been included, the Hamiltonian is a constant of the motion. I.e., once its value is specified, the total energy of the particle is fixed.

> `H:=T+V;`

$$H := \frac{p1^2}{2} + \frac{p2^2}{2} + \frac{q1^2}{2} + \frac{q2^2}{2} + q1^2\, q2 - \frac{q2^3}{4}$$

Next, Hamilton's equations are generated, the dependent variables being $p1(t)$, $p2(t)$, $q1(t)$, and $q2(t)$.

> `hamilton_eqs(H);`

$$[\tfrac{d}{dt} p1(t) = -q1(t) - 2\, q1(t)\, q2(t),\ \tfrac{d}{dt} p2(t) = -q2(t) - q1(t)^2 + \tfrac{3}{4} q2(t)^2,$$
$$\tfrac{d}{dt} q1(t) = p1(t),\ \tfrac{d}{dt} q2(t) = p2(t)],\ [p1(t), p2(t), q1(t), q2(t)]$$

To solve Hamilton's equations, we must specify some initial conditions for the particle. We take $q1(0) = x0 = -0.1$, $q2(0) = y0 = -0.2$, $p2(0) = py0 = -0.05$, and two total energies, $E0 = 0.06$ (well below the lowest saddle point potential energy) and $E0 = 0.14$ (just below the lowest saddle point energy).

> `x0:=-0.1; y0:=-0.2; py0:=-0.05; E0:=0.06; E02:=0.14;`

Specifying the total energy, the following functional operator will solve for the initial momentum $p1(0)$, which was not specified, and express the complete set of initial conditions for the particle. If instead of specifying, say, a fixed value of $q2(0)$, a range of $q2(0)$ could be specified. Then changing the number 1 to, say, 3 in ic, we could request three lists (if they exist) of initial conditions.

> `ic:=e->generate_ic(H,{t=0,p2=py0,q2=y0,q1=x0,energy=e},1):`

Applying the functional operator to $E0$ and $E02$, we find that $p1(0) \approx 0.26$ for the former, and $p1(0) \approx 0.48$ for the latter.

> `ic1:=ic(E0); ic2:=ic(E02);`

$$ic1 := \{[0., 0.2598076211, -0.05, -0.1, -0.2]\}$$
$$ic2 := \{[0., 0.4769696007, -0.05, -0.1, -0.2]\}$$

The following functional operator will numerically solve Hamilton's equations, subject to the specified initial conditions, and generate a plot of the particle's trajectory in $q2$ vs. $p2$ vs. $q1$ space. A time interval $t = 0..300$ is considered, the time step size is taken to be 0.05 in Maple's numerical algorithm, and 3 iterations are performed in the algorithm. The coordinate ranges in the `scene` option are adjusted to include the entire trajectory of the particle. For other initial conditions, the ranges may have to be altered. In the 3-dimensional plot, Maple will draw a shaded plane corresponding to $q1 = 0$.

> P:=IC->poincare(H,t=0..300,IC,stepsize=.05,iterations=3,

> scene=[q2=-0.8..0.8,p2=-0.8..0.8,q1=-0.8..0.8],3):

A second functional operator is used to show the points in the $q1 = 0$ plane where the orbit passes through this plane. This is like the Poincaré sections approach (hence the Maple command name) that we used in some earlier recipes.

> P2:=IC->poincare(H,t=0..300,IC,stepsize=.05,

> iterations=3,scene=[q2,p2]):

To see the resulting pictures, one must use the `display` command. The pictures sometimes are rather small on the computer screen. One could use the magnification command in Maple's tool bar, but an alternate approach is to enter the following interface command, which will produce pictures that fill the entire viewing window. Close this viewing window to continue with the recipe.

> interface(plotdevice=window):

We display the 3-dimensional plot, corresponding to $ic1$.

> display(P(ic1),orientation=[-82,84],tickmarks=[3,3,3]);

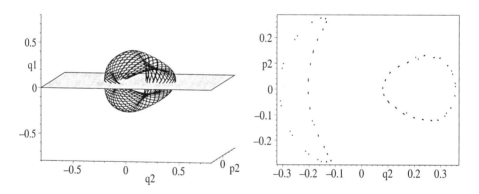

Figure 6.18: Left: Quasiperiodic trajectory on the surface of a KAM torus. Right: Points where the trajectory crosses the $q1 = 0$ plane.

The result is shown on the left-hand-side of Figure 6.18. The particle executes quasi-periodic motion on the surface of a twisted torus. This is referred to as a KAM (Kolmogorov–Arnold–Moser) torus after the Russian mathematicians who studied such tori.

In addition to the picture, the output includes information about the total energy (H), the initial conditions that were used, the maximum percentage deviation ($0.88 \times 10^{-6}\,\%$ here) from the total energy in applying the numerical

6.3. SOME HAMILTONIAN EXAMPLES

scheme, and the cpu time used. The cpu time (55 seconds here) depends on the speed of computer being used. One can decrease the cpu time by taking a shorter time in the poincare command, but the plot may not look quite as nice. You may have to compromise between how "beautiful" you want the plot to be and how long you are willing to wait.

$H = .60000000e - 1,$ Initial conditions :, $t = 0.$,
$p1 = 0.2598076211, p2 = -0.05, q1 = -0.1, q2 = -0.2$

Maximum H deviation : $.8800000000e - 6$ %

Time consumed : 55 seconds

The points in the $q1 = 0$ plane where the trajectory crosses the plane are now displayed for the initial condition $ic1$.

>   display(P2(ic1));

The result is shown on the right-hand side of Figure 6.18. Since the torus is like a donut that has been twisted and had its cross-section distorted, the distribution of points in the $q1 = 0$ "slice" are easy to understand. In the output, in addition to similar information to that given above, the number of points (96) in the plane are given. The cpu time is considerably shorter to draw the 2-dimensional figure than to produce the 3-dimensional plot.

The following two command lines produce pictures similar to those above, but now for the second initial condition $ic2$.

>   display(P(ic2),orientation=[-60,72],tickmarks=[3,3,3]);
>   display(P2(ic2));

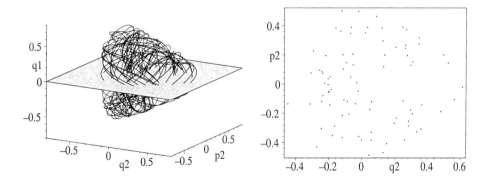

Figure 6.19: Left: Chaotic trajectory. Right: Points in the $q1 = 0$ plane.

The results are shown in Figure 6.19. The trajectory now resembles a tangled ball of yarn. The distribution of points in the $q1 = 0$ plane are now chaotically distributed. This situation is referred to as Hamiltonian chaos. There is a great interest amongst mathematicians in investigating KAM tori behavior and the onset of Hamiltonian chaos as the energy of the system is changed.

The recipe concludes with the following interface command which restores Maple's default settings. Don't forget to execute it.

> `interface(plotdevice=default):`

6.4 Supplementary Recipes

06-S01: Parametric Excitation

Jennifer has given her mechanics class an assignment, the first problem involving the "parametric excitation" of a plane pendulum. The pendulum, consisting of a point mass m on the end of a weightless connecting rod of length r, has its pivot point periodically displaced horizontally by an amount $d = A \sin(\omega t)$. Neglecting all frictional effects and letting $\theta(t)$ be the angle of the connecting rod with the vertical at time t, use the Lagrangian approach to derive the equation of motion for m. Taking $A = 2$ m, $\omega = 1.5$ rad/s, $r = 9.8$ m, $g = 9.8$ m/s^2, $\theta(0) = 30°$, $\dot{\theta}(0) = 0$, numerically solve the equation of motion and plot $\theta(t)$ over the time interval $t = 100$ to $t = T = 200$ seconds. Is the motion periodic or chaotic? Animate the motion of the pendulum over the time interval $t = 0..T$. Represent the oscillating pivot point as a red circle, the swinging connecting rod as a thick green line, and the moving mass m as a blue circle.

06-S02: The Oscillating Spherical Pendulum Ride

Mr. X has an idea for a new ride for his amusement park. A cage of mass m, including the riders, is attached to the lower end of a light connecting arm, which can swing freely in all directions from a fixed pivot point. If the arm had a fixed length, this would be the spherical pendulum problem, as m would then be free to move on the surface of a sphere. In Mr. X's ride, however, the arm length $s(t)$ is a prescribed function of time t.

(a) Using the Lagrangian approach, determine the equations of motion in terms of spherical polar coordinates. Take the z-axis to point downwards and let $\theta(t)$ be the angle that the connecting arm makes with the z-axis and $\phi(t)$ the angle that its projection onto the x-y plane makes with the x-axis at time t.

(b) Show that the oscillating spherical pendulum's motion satisfies

$$\ddot{\theta} + 2\dot{s}\dot{\theta}/s - C^2 \cos\theta/(s^4 \sin^3\theta) + g \sin\theta/s = 0,$$

where C is an angular momentum component.

(c) If $s(t) = R(1 + A \sin(\omega t))$, numerically solve the ODE and animate the motion of the connecting arm and cage for $R = 10$ m, $A = 0.5$, $\omega = 1.5$ rad/s, $g = 10$ m/s^2, $C = 50$, $\theta(0) = 60°$, $\dot{\theta}(0) = 0$, and $\phi(0) = 0$.

6.4 SUPPLEMENTARY RECIPES

06-S03: Coupled Pendulums

The second pendulum problem on Jennifer's mechanics assignment involves two plane pendulums coupled by a linear spring. Pendulum 1 has its pivot point located at $x = -a$, $y = 0$ and has a mass $m1$ attached to a light pivot arm of length $r1$. Pendulum 2 has its pivot point located at $x = a$, $y = 0$ and has a mass $m2$ attached to a light pivot arm of length $r2$. The masses are connected by a linear spring of spring constant k. Frictional effects are neglected.

(a) Using the Lagrangian approach, and letting $\theta_1(t)$ and $\theta_2(t)$ be the angles that the two pivot arms make with the vertical at time t, derive the equations of motion for $m1$ and $m2$.

(b) Taking $a = 1$ m, $r1 = r2 = 2$ m, $m1 = m2 = 1$ kg, $g = 9.8$ m/s², $k = 0.5$ N/m, $\theta_1(0) = -0.2$ rad, $\theta_2(0) = 0.2$ rad, $\dot{\theta}_1(0) = \dot{\theta}_2(0) = 0$ rad/s, numerically solve the coupled ODEs and use the odeplot command to create a 3-dimensional plot of θ_1 vs. θ_2 vs. t. Take a total time of 20 s.

(c) Animate the motion of the coupled pendulums for the above time interval. Take 200 frames, and represent the masses by blue circles, the pivot arms by thick green lines, and the spring by a thick red line.

06-S04: Eulerian Angles

The fixed (space) axes x', y', z' are rotated into the body axes x, y, z with three consecutive rotations. In the first stage, the space axes are rotated counterclockwise through an angle ϕ about the z' axis. The new axes are labeled x'', y'', $z'' = z'$. In the second stage, the axes x'', y'', z'' are rotated counterclockwise through an angle θ about the x'' axis. The new axes are labeled $x''' = x''$, y''', z'''. The third rotation is counterclockwise through an angle ψ about the z''' axis. The resulting axes are the body axes, x, y, $z = z'''$.

(a) Determine the total rotation matrix M for rotating x', y', z' into x, y, z.

(b) Determine the inverse matrix M^{-1} for going from the body to the space axes. Confirm that the product of M and M^{-1} yields the identity matrix.

(c) Calculate M for $\phi = \pi/6$, $\theta = \pi/6$, $\psi = \psi/6$. Label this matrix $M1$.

(d) Apply $M1$ to the unit vector along each space axis. Confirm that the resulting three vectors are also unit vectors.

(e) Making use of the above unit vectors, create a 3-d figure showing the space and body axes for the rotation $M1$.

06-S05: George's Nonlinear Inchworm

In supplementary recipe 04-S04, George introduced the "linear inchworm", consisting of three airtrack gliders on a linear airtrack, the gliders connected by two linear springs. The motion of the linear inchworm was analytically determined using a Newtonian approach and then animated for some specified initial conditions and parameter values. In the following problem, George is asking you to solve the "nonlinear inchworm" problem, the gliders now being connected by two nonlinear springs, each governed by a potential energy of the form $V = (1/2)\,k\,x^2 + (1/2)\,a\,x^4$, with $k > 0$, $a > 0$.

Three airtrack gliders (masses m_1, m_2, m_3) are connected by two nonlinear springs. The first spring, connecting m_1 and m_2, has $k = k_1$, $a = a_1$, while the second spring, connecting m_2 and m_3, has $k = k_2$, $a = a_2$. Each spring has an unextended length s. The gliders are placed on an airtrack of length L. Initially m_1 is at $x10 = 0$ with velocity $v10 = 0$, mass m_2 at $x20 = s$ with velocity $v20 = 0$, and m_3 at $x30 = 2\,s$ with non-zero initial velocity $v30$.

(a) Determine the Lagrangian for the nonlinear inchworm and derive the equation of motion for each of the three gliders.

(b) Given $m_1 = 3$ kg, $m_2 = 2$ kg, $m_3 = 1$ kg, $s = 0.4$ m, $L = 3$ m, $k_1 = 4$ N/m, $a_1 = 4$ N/m^3, $k_2 = 8$ N/m, $a_2 = 8$ N/m^3, and $v_30 = 0.5$ m/s, numerically determine the position of each glider at time $t > 0$.

(c) Calculate the time that it takes m_3 to reach the other end of the airtrack.

(d) Animate the motion of the gliders over this time interval, representing each glider by a differently colored box and the two connecting springs by heavy colored lines.

06-S06: The Fermi–Pasta–Ulam Problem

Research on nonlinear lattice dynamics began seriously in the 1950s when Enrico Fermi, J. Pasta, and Stan Ulam [FPU55] used the MANIAC I computer at Los Alamos to numerically investigate the longitudinal vibrations of a linear chain of identical atoms connected by identical nonlinear springs. One can create a simplified version of the so-called Fermi–Pasta–Ulam problem by using airtrack gliders, connected by nonlinear springs, free to move on a linear airtrack. Three airtrack gliders are connected to each other and to fixed end points by four nonlinear springs governed by the anharmonic potential $V = (1/2)\,k\,x^2 + (1/2)\,a\,x^4$. The equilibrium positions of the gliders are as follows: mass m_1 is at $x = s$, m_2 at $x = 2\,s$, and m_3 at $x = 3\,s$, where s is the unstretched spring length. Initially m_1 and m_2 have zero velocity, but m_3 has an initial velocity of 0.1 m/s. Determine the Lagrangian and derive the equations of motion. If $m_1 = m_2 = m_3 = 1$ kg, $k = 8$ N/m, $a = 0.1$ N/m^3, $s = 0.3$ m, numerically solve the equations of motion and animate the gliders' motions.

06-S07: The Rotating Pendulum

The third pendulum problem on Jennifer's assignment involves a "rotating pendulum". A bead of mass m is allowed to slide along a thin frictionless circular ring of radius r which is rotating with constant angular velocity ω as shown in Fig. 6.20. At time t, the bead makes an angle θ with the downward vertical.

(a) If the plane of the circular ring is oriented along the y-axis at $t = 0$, use the Lagrangian approach to show that the bead's equation of motion is

$$\ddot{\theta} - (\omega^2/2)\sin(2\theta) + \omega_0^2 \sin\theta = 0, \quad \text{with } \omega_0 = \sqrt{g/r}.$$

6.4 SUPPLEMENTARY RECIPES

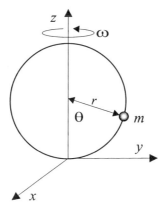

Figure 6.20: The rotating pendulum.

(b) Taking $\omega_0 = 2$ rad/s, $\omega = 1$ rad/s, $g = 9.8$ m/s^2, $\theta(0) = 2\pi/3$ rad, and $\dot{\theta}(0) = 0$, numerically solve the equation of motion and plot t vs. $\theta(t)$ vs. $\dot{\theta}(t)$ over the time interval $t = 0$ to 50 s. Plot the trajectory of the bead in x-y-z space over the same interval.

(c) Create a 3-dimensional animation of the bead sliding on the rotating ring.

06-S08: The Toda Potential
The Hamiltonian for a three-particle "molecule", with the forces between particles governed by the Toda potential, can be written as

$$H = p1^2/2 + p2^2/2 + (1/3)[e^{(q2+\sqrt{3}q1)} + e^{(q2-\sqrt{3}q1)} + e^{(-2q2)}] - 1,$$

where $p1$ and $p2$ are momenta and $q1$ and $q2$ are coordinates. The interested reader may see *Perspectives of Nonlinear Dynamics* [Jac90] for the derivation.

(a) Create 2- and 3-dimensional contourplots of the potential energy.

(b) Generate Hamilton's equations as well as the initial condition for an energy $E = 1.0$, and initial values $p2(0) = -0.05$, $q2(0) = -0.2$, $q1(0) = -0.2$.

(c) Make a plot of the system's trajectory in the $q2$ vs. $p2$ vs. $q1$ space.

(d) Form a "Poincaré section" by viewing the $q2$-$p2$ plane sliced at $q1 = 0$. Interpret the picture in terms of the 3-dimensional plot.

(e) Explore the Toda potential problem for other energy values. Is there any evidence of Hamiltonian chaos?

06-S09: Abby Moves in Great Circles
On the surface of the spherical beach ball where Abigail Ant is still roaming,

an element of arclength is given by $ds = \sqrt{a^2 (d\theta)^2 + a^2 \sin^2\theta (d\phi)^2}$ where a is the radius, θ is the angle with the z-axis, and ϕ the angle from the x-axis to the projection of the radius vector onto the x-y plane. Show that the path that Abby should walk to minimize the distance between two distinct points on the surface is the arc of a great circle (circle of radius a on the sphere). Plot a representative great circle on the beach ball's surface.

06-S10: Suzy Spider's Alternate Route

Suppose that the fence to which Suzy Spider (the same Suzy as in text recipe **06-2-2**) wants to attach a strand as an escape route is leaning because the support posts have rotted. The top of the strand is taken to be the origin with the y-coordinate measured downwards. The top of the fence is located at $x = 1$ m, $y = 0$ m, while the bottom of the straight fence is located at $x = 0$ m, $y = 2$ m. Suzy wishes to slide down the strand, starting from rest, along a 2-dimensional path that will minimize the time of descent to the fence. It can be shown that in this case the path must intercept the fence with a slope $y' = (\partial G/\partial y)/(\partial G/\partial x)$ where $G(x, y) = 0$ is the (straightline) equation of the fence. Neglecting friction and making use of the parametric forms of x and y derived in **06-2-2**, through what vertical distance will Suzy drop in reaching the fence? What is her horizontal displacement from her starting point? How long will it take her to reach the fence? Plot Suzy's path, the fence, and the end point in the same figure.

06-S11: Fermat's Principle and the Bending of Light

If c is the vacuum speed of light, the speed of light in a medium of refractive index n is $v = c/n$. Fermat's principle states that a ray of light in a medium with a variable refractive index will follow the path which requires the shortest traveling time. Consider a refractive index which varies linearly with height y in the following way: $n = n_0(1 + \alpha y)$ with $n_0 > 0$ and α non-zero. Use Fermat's principle and the Euler–Lagrange equation to determine the path followed by a light ray which is emitted with slope of magnitude S by a small beacon located on top of a cliff at $x = 0$, $y = 0$. If the cliff edge is a distance $H = 500$ m above the valley floor and $\alpha = 10^{-6}$ m^{-1}, plot the paths traveled by two light rays emanating from the source, with $S1 = 0.006$ and $S2 = 0.0006$, over a horizontal distance $d = 20$ km. Use unconstrained scaling. Determine the tangent lines to the light ray curves at $x = d$ and plot them in the same figure as the light rays. Interpret the tangent curves and determine the angle in degrees that they make with the horizontal.

Bibliography

[AS72] M. Abramowitz and I. A. Stegun. *Handbook of Mathematical Functions with Formulas, Graphs, and Mathematical Tables.* National Bureau of Standards, Washington, DC, 1972.

[Cha02] B. W. Char. *Maple 8 Learning Guide.* Waterloo Maple, Waterloo, ON, Canada, 2002.

[EM00] R. H. Enns and G. C. McGuire. *Nonlinear Physics with Maple for Scientists and Engineers,* 2nd ed. Birkhäuser, Boston, MA, 2000.

[EM01] R. H. Enns and G. C. McGuire. *Computer Algebra Recipes, A Gourmet's Guide to the Mathematical Models of Science.* Springer, New York, 2001.

[Erl83] H. Erlichson. Maximum projectile range with drag and lift, with particular application to golf. *Amer. J. Phys.*, 51:357, 1983.

[FC99] G. R. Fowles and G. L Cassiday. *Analytical Mechanics,* 6th ed. Saunders College, Orlando, FL, 1999.

[FPU55] E. Fermi, J. R. Pasta, and S. M. Ulam. Studies of nonlinear problems. *Los Alamos Science Laboratory Report*, LA-1940, 1955.

[GPS02] H. Goldstein, C. Poole, and J. Safco. *Classical Mechanics,* 3rd ed. Addison Wesley, New York, 2002.

[Gri99] D. J. Griffiths. *Introduction to Electrodynamics,* 3rd ed. Prentice-Hall, Upper Saddle River, N.J., 1999.

[HH64] M. Hénon and C. Heiles. The applicability of the third integral of motion: some numerical experiments. *Astrophys. J.*, 69:73, 1964.

[Jac90] E. A. Jackson. *Perspectives of Nonlinear Dynamics.* Cambridge University Press, Cambridge, 1990.

[MA88] J. J. McPhee and G. C. Andrews. Effect of sidespin and wind on projectile trajectory, with particular application to golf. *Amer. J. Phys.*, 56:933, 1988.

[MBM02a] M. B. Monagan, K. O. Geddes, K. M. Heal, G. Labahn, S. M. Vorkoetter, J. McCarron, and P. DeMarco. *Maple 8 Introductory Programming Guide.* Waterloo Maple, Waterloo, ON, Canada, 2002.

[MBM02b] M. B. Monagan, K. O. Geddes, K. M. Heal, G. Labahn, S. M. Vorkoetter, J. McCarron, and P. DeMarco. *Maple 8 Advanced Programming Guide.* Waterloo Maple, Waterloo, ON, Canada, 2002.

[MH91] W. M. MacDonald and S. Hanzely. The physics of the drive in golf. *Amer. J. Phys.*, 59:213, 1991.

[MT95] J. B. Marion and S. T. Thornton. *Classical Dynamics of Particles and Systems,* 4th ed. Saunders College, Orlando, FL, 1995.

[MW70] J. Mathews and R. L. Walker. *Mathematical Methods of Physics,* 2nd ed. W. A. Benjamin, New York, 1970.

[PFTV90] W. H. Press, B. P. Flannery, S. A. Teukolsky, and W. T. Vetterling. *Numerical Recipes.* Cambridge University Press, Cambridge, 1990.

[Ste87] J. Stewart. *Calculus.* Brooks/Cole, Pacific Grove, CA, 1987.

[Tip91] P. A. Tipler. *Physics for Scientists and Engineers.* Worth, New York, 1991.

Index

acceleration
    angular, 89
    gravitational, 34
    linear, 89
acceleration magnitude, 38
acceleration vector, 42
air density, 149
air resistance
    Newton's law, 148
    Stokes' law, 143
airtrack, 159, 216, 252
amplitude, 142, 173
angle of elevation, 35
angular acceleration, 83
angular heading, 25
angular momentum, 186, 220
angular velocity, 84, 173
anharmonic oscillator, 178
anharmonic potential, 178, 189, 244
Archimedes' principle, 136
arclength, 76
asteroid, 169
astronomical unit, 172
Atwood's machine, 87
Atwood, George, 87
average speed, 24
axis of rotation, 83

back quotes, 71
backspin, 148
Bessel function, 186
blackout, 76
body axes, 219
bogey, 11
booster rocket, 39
buoyant force, 135

calculus of variations, 227
center of mass, 83, 212
central force, 166
central force law, 178
centrifugal force, 198, 202
centripetal force, 62, 76
chaotic solution, 194
coefficient of static friction, 87
column vector, 154
Compton effect, 70
Compton formula, 71
Compton, Arthur, 70
constant of motion, 220, 247
constraint equation, 58, 89, 216, 236
contour plot, 124
coordinates
    curvilinear, 103
    cylindrical polar, 132
    orthogonal, 115
    plane polar, 104, 166
    spherical polar, 109
    toroidal, 115
Coriolis force, 198, 202
Coulomb's law, 48
cpu time, 248
cross product, 29, 39
    double, 42
cue angle, 96
cue ball, 92
curl, 115, 126
curvature, 76

degrees of freedom, 236
diagonal matrix, 155
discrete Fourier transform, 208
displacement, 140

displacement vector, 13
divergence, 115, 132
divergence theorem, 133
dot product, 29
double well potential, 78
doughnut, 115
drag
    linear, 135
    nonlinear, 176
drag coefficient, 135, 148
drag force, 135
driving frequency, 206
Duffing oscillator, 206
dust ring, 179

eccentricity, 168
effective weight, 80
eigenfrequency, 156
eigenvalues, 155
ejection velocity, 39
elastic collision, 65
elastic fatigue, 205
electric field, 182
electric force, 48
electron rest mass, 71
elliptic integral
    complete, 164
    first kind, 164, 174
    incomplete, 164
elliptical orbit, 171
energy conservation, 64, 74
energy transfer, 65
envelope of safety, 48, 143
equation
    Euler–Lagrange, 227, 228
    forced Duffing, 206
    Hamilton's, 236
    hard spring, 162
    Lagrange, 211
    nonlinear, 173
    recurrence, 157
equipotential, 128
Erehwon, 4, 30
Eros, 165
Euler's constant, 139

Euler–Lagrange equation, 227, 228
Eulerian angles, 251
exhaust speed, 100

Falkland islands, 202
falling pencil, 83
fast Fourier transform (FFT), 208
Fermat's principle, 254
Fermi–Pasta–Ulam, 252
flux, 133
force
    centrifugal, 198
    components, 49
    constant, 49
    Coriolis, 198
    drag, 148
    frictional, 50, 85, 93
    gravitational, 148
    lift, 148
    minimum, 50
    normal, 50, 58, 85
    time-dependent, 208
    velocity-dependent, 135
forces
    resolving, 50
free body diagram, 50, 57, 62, 83
free fall, 30
frictional force, 74

G force, 76
Gamma function, 178
Gauss's theorem, 133
Gauss, Karl Friedrich, 133
generalized coordinates, 211
generalized momentum, 236, 238
generalized velocities, 211
golf ball dynamics, 148
golf ball trajectory, 150, 176
governor, 62
gradient, 115, 122
graphical vector addition, 13
gravitational acceleration, 74
gravitational force, 202
great circle, 254

Hénon and Heiles, 244

INDEX

Halley's comet, 165
Hamilton's equations, 236
Hamiltonian
    function, 236
    Toda, 253
Hamiltonian chaos, 249
hang glider, 30
hard spring, 159
hole in one, 150
Hooke's force law, 62, 153
horizontal range, 100

impulse, 96
inchworm
    linear, 177
    nonlinear, 252
inclined plane, 57
incommensurate, 241
inelastic collision, 99
inertial frame, 57, 198
inflection point, 78
inverse square law, 166
isotherm, 134

Jacobian elliptic sine function, 164
Justine's dive, 135

KAM torus, 248
kinematic formula, 34
kinematic relation, 34, 53
kinetic energy, 65, 213
kinetic friction, 50, 53
knot, 19
Kolmogorov–Arnold–Moser, 248

ladder problem, 98
Lagrangian, 211, 236
    air track, 216
    double pendulum, 224
    rotating pendulum, 252
    spinning top, 220
Laplacian, 115
latitude, 202
Learnu molecule, 48
Legendre transformation, 236
lift coefficient, 148

lift force, 148
light bending, 254
line of nodes, 219
linear atomic chain, 153
linear density, 83
linear momentum, 65
linear superposition, 153
Lissajous figures, 241
local maximum, 134
longitudinal vibrations, 153
Lorentz force, 182

Machu Pichu, 42
magnetic field, 182
Maple
    activating animation, 29
    adding coordinates, 115
    angular coordinate window, 18
    arrow operator, 16
    assignment operator, 5
    case sensitivity, 12
    clearing internal memory, 5
    comment, 16
    concatenation operator, 44
    constrained plots, 14
    context bar, 6
    controlling accuracy, 25
    copying examples, 3
    default accuracy, 7
    differential operator, 136
    do loop, 44
    dot product, 31
    double quotes, 6
    functional operator, 16
    Help
       Full Text Search, 3
       Topic Search, 3
    library package, 12
    LinearAlgebra package, 12
    list, 6, 14
    list of lists, 21
    parametric plot, 21
    producing random colors, 28
    recipes on CD, 2
    removing warnings, 8

rotating 3-d plot, 33
selection operation, 17, 50
set, 14
string, 6, 14, 25
syntax, 4
tool bar, 8
using Help, 3
using the hyperlink, 2
VariationalCalculus package, 216
VectorCalculus package, 12
Maple Command
   ||, 44
   ^, 5
   ' ', 96
   *, 5
   +, 5
   ->, 16, 28, 40, 53, 79
   -, 5
   ., 31, 43
   :=, 5
   :, 5
   ;, 5
   < , >, 13
   <|>, 104
   ?, 4
   Arc, 130
   BandMatrix, 155
   COLOR(RGB,rand()), 28
   CharacteristicMatrix, 155
   Circle, 130
   CrossProduct, 39
   Curl, 126
   Delta, 139
   Del, 128
   Determinant, 156
   Digits, 156, 175
   Divergence, 128
   DotProduct, 31, 107
   D, 136
   Eigenvalues, 156
   EulerLagrange, 217, 232
   Gamma, 186
   Gradient, 122
   Int, 118
   LineInt, 126, 130
   MapToBasis, 128
   Multiply, 155
   Nabla, 128
   Norm(,2), 105
   Pi, 12
   RootOf, 70
   SetCoordinates, 128
   SurfaceOfRevolution, 231
   Theta, 84
   Vector([,]), 13
   VectorField, 110, 126
   [], 17, 50
   #, 16
   $, 149
   %, 24
   &x, 39, 43
   { }, 14
   abs, 13, 79, 142
   adaptive=false, 174
   addcoords, 116
   additionally, 173
   alias, 199, 228
   align=RIGHT, 41
   allvalues, 71, 246
   alpha, 84
   animate, 8, 29, 138
   arccos, 37, 43, 107
   arcsin, 13, 96, 187
   arctan, 37
   arrows=MEDIUM, 190
   arrows=THICK, 124, 128
   arrow, 14, 44
   assign, 17, 51, 105
   assume, 39, 104, 163
   assuming, 39
   axes=FRAME, 110
   axes=boxed, 21
   axes=framed, 18, 32
   axes=normal, 60
   break, 54
   circle, 46, 95
   coeff, 106
   collect, 71, 106, 217
   color, 6, 14
   combine(trig), 75, 140, 214

INDEX

contourplot, 124, 241
contours, 121
convert(list), 32
convert(polynom), 160, 203
convert(radian,degree), 51
convert(sincos), 168
convert(units), 20, 34, 35
convert(units,m/s,km/h), 53
coordplot3d, 110, 115
coords=spherical, 113
cos, 5, 126
cylindrical_arrow, 18
delta, 96
denom, 52
diff, 6, 20, 42, 51
dirgrid, 161, 190
display, 14, 18, 21, 32
dsolve(method=laplace), 140
dsolve(numeric), 151, 170
dsolve, 136, 162
end do, 44
evalf, 12, 20
eval, 7, 20, 21, 32, 151
expand, 16, 71, 140, 187
exp, 5, 126
factor, 75
fieldplot3d, 128
fieldplot, 124
filled=true, 60, 151
font=[SYMBOL,12], 83
for...from...do, 44, 54
frames, 8, 29
fsolve, 7, 17, 35
gamma, 139
generate_ic, 243, 247
grid, 124, 128
hamilton_eqs, 243, 247
head_length, 14
head_width, 14
if...elif...else, 150
if...then...else, 95
if...then...end if, 54
implicitplot3d, 128, 230
insequence=true, 35, 60
interface

imaginaryunit=j, 83
plotdevice=default, 61
plotdevice=window, 60
showassumed=0, 104
warnlevel=0, 8
int, 25
isolate, 68, 77, 168
iterations, 243
kappa, 154
labelfont, 175
labels, 6, 18, 21, 142
lambda, 140
lhs, 22
linecolor, 161, 190
linestyle, 6, 27
map, 104
maxfun=0, 151, 191
max, 157
min, 22
mu, 50, 94
numer, 228
numpoints, 26, 32
odeplot, 191, 214
omega, 84, 112
op, 157
orientation, 18, 41
output=listprocedure, 151
phaseportrait, 161
phi, 50
piecewise, 164
plot3d, 113, 121, 242
plot, 6, 21, 52
poincare, 243, 247
pointplot3d, 41, 113, 121
pointplot, 14, 21, 44, 54
polarplot, 45, 108, 169
polygonplot, 195
polygon, 48
radnormal, 157
radsimp, 26, 32, 59, 68
rand(), 196
randomize(), 196
remove, 217
restart, 5
rho, 149

rhs, 72
round, 112, 127, 157
rsolve, 156
scaling=constrained, 14, 21
scatterplot, 157
scene, 161, 190
select, 66, 232
seq, 17, 28, 35, 154
shading=zhue, 128
shape, 14, 18
simplify(radical), 96
simplify(symbolic), 43, 58
simplify(trig), 96
simplify, 20
sin, 53, 126
solve, 20, 34
sort, 156
spacecurve, 32, 40, 112, 129
sqrt, 5, 20
stepsize, 161
style=line, 27
style=patchcontour, 121, 128
style=point, 8, 14, 21, 29
style=wireframe, 113
subs, 20, 22, 34
symbol=circle, 8, 14, 29
symbol=cross, 158
symbolsize, 14, 21, 29
tan, 58
tau, 84
taylor, 160, 203
textplot3d, 41, 129
textplot, 14, 45, 83
theta, 26, 50
thickness, 6, 26
tickmarks, 14, 18
title, 25
type, 104
unapply, 77, 174
unassign, 52, 67, 128
unprotect, 139
value, 118
view, 14, 21, 27
while...do, 196
width, 14

with(DEtools), 160, 189
with(LREtools), 154
with(LinearAlgebra, 104
with(VariationalCalculus), 216
with(VectorCalculus), 12, 30
with(plots), 8, 12, 30
with(plottools), 46, 92
with(stats[statplots]), 154
matrix
    characteristic, 155
matrix approach, 154
maximum speed, 24
maximum volume, 134
mechanical energy, 79
menu file, 2
merry-go-round, 199
minimum area, 229
minimum distance, 254
    between lines, 48
minimum time, 232
moment arm, 84
moment of inertia, 83, 212
    disc, 89
    serving platter, 99
    sphere, 93
momentum, 70
momentum conservation, 64, 74
Monte Carlo approach, 195
motion
    circular, 49
    translational, 49
Mr. X's ride, 57
multi-car crash, 73
muzzle speed, 34

nautical mile, 19
net force, 57
Newton's first law, 50
Newton's laws, 49
Newton's second law, 53, 57
Newton's third law, 80
non-inertial frame, 198, 202
non-isotropic oscillator, 241
nonlinear ODE, 149, 229
nonlinear spring, 252

INDEX

normal force, 80
normal mode, 153
normal weight, 76
nutation, 219

Olympic war games, 34
orbital precession, 178
ornithopter, 30

parallel axis theorem, 99, 212
parametric excitation, 250
Parseval's theorem , 208
particle model, 70
Pegasus, 38
pendulum
    conical, 236
    coupled, 251
    double, 224
    oscillating spherical, 250
    plane, 172
    rotating, 252, 253
period, 159, 173
period 1, 192
period 2, 192, 206
period 4, 192
perturbation method, 179
phase space, 246
phaseplane portrait, 161, 190
photon, 70
piecewise function, 164
Planck's constant, 71
planetary motion, 165
Poincaré section, 205, 206
position vector, 30
potential energy, 213
power spectrum, 208
pulley, 89
pure rotation, 82
pursuit problem, 179

quasi-periodic, 248

radius of curvature, 76
raindrop, 180
    falling, 195
random number generator, 195

random number seed, 196
recoil electron, 70
reference frame
    accelerated, 198
relative velocity, 24, 26, 46
relativistic energy, 70
relativity
    general theory of, 178
resultant force, 16
rigid body, 49
rocket trajectory, 100
roller coaster, 76
rolling wheel, 46
rolling without slipping, 93
rotational dynamics, 82
route to chaos, 189
Runge–Kutta–Fehlberg, 151

saddle point, 245
sampling theorem, 208
scale factor, 115
second derivatives test, 134
semimajor axis, 168
simple harmonic oscillator, 173
skid marks, 74
slipping, 94
slope, 77, 122
smoke jumpers, 32
snooker, 92
space axes, 219
specific gravity, 136
speed, 43
    average, 46
speed of light, 71
spin axis, 223
spinning top, 219
spring constant, 62, 139
static friction, 98
stationary observer, 43
statistical approach, 206
steady-state solution, 139
Stokes' law, 143
Stokes' theorem, 126
strange attractor, 206
surface of revolution, 228

suspension bridge, 98

tangential velocity, 166
Tarzan, 186
Taylor series, 160, 203
tension, 90
theorem of Pythagorus, 15, 216
thrust force, 100
toboggan, 53
Toda Hamiltonian, 253
Toda potential, 253
Tokamak, 115
torque, 84, 89, 93, 96, 186
torus, 115
toy car, 97
transcendental equation, 35, 100
transient solution, 139
trick shot, 92
tridiagonal matrix, 155
tug of war, 206
turning points, 166

unit vector, 16, 36, 48, 104, 148

variational calculus, 227
vector
    column, 104
    row, 104
vector addition
    analytical, 11
    graphical, 11
vector difference, 31
vector field
    conservative, 126, 132
vector identity, 47, 133
vector multiplication
    cross product, 29
    dot product, 29
vector operator, 120
velocity vector, 25, 31, 42

wave model, 70
wavelength, 70
wavelength shift, 70
wheel
    rolling lopsided, 211

work, 51

X-ray, 70

© 2003 Birkhäuser Boston

This electronic component package is protected by federal copyright law and international treaty. If you wish to return this book and the CD-ROM to Birkhäuser, do not open the disc envelope or remove it from the book. Birkhäuser will not accept any returns if the package has been opened and/or separated from the book. The copyright holder retains title to and ownership of the package. U.S. copyright law prohibits you from making any copy of the entire CD-ROM for any reason, without the written permission of Birkhäuser, except that you may download and copy the files from the CD-ROM for your own research, teaching, and personal communications use. Commercial use without the written consent of Birkhäuser is strictly prohibited. Birkhäuser or its designee has the right to audit your computer and electronic components usage to determine whether any unauthorized copies of this package have been made.

Birkhäuser or the authors make no warranty or representation, either express or implied, with respect to this CD-ROM or book, including their quality, merchantability, or fitness for a particular purpose. In no event will Birkhäuser or the authors be liable for direct, indirect, special, incidental, or consequential damages arising out of the use or inability to use the CD-ROM or book, even if Birkhäuser or the authors have been advised of the possibility of such damages.